Lecture Notes
in Control and Information Sciences 249

Editor: M. Thoma

Springer-Verlag London Ltd.

G. Cooperman, E. Jessen and G. Michler (Eds)

Workshop on Wide Area Networks and High Performance Computing

With 136 Figures

 Springer

British Library Cataloguing in Publication Data
Workshop on wide area networks and high performance
   computing. - (Lecture notes in control and information
   sciences)
   1.Wide area networks (Computer networks) - Congresses
   2.High performance computing - Congresses
   I.Cooperman, G. (Gene), 1952- II.Jessen, E. III.Michler, G.
   O. (Gerhard O), 1938-
   004.6'7
   ISBN 978-1-85233-642-4   ISBN 978-1-84628-578-3 (eBook)
   DOI 10.1007/978-1-84628-578-3

Library of Congress Cataloging-in-Publication Data
A catalog record for this book is available from the Library of Congress

Typesetting: Camera ready by contributors

69/3830-543210 Printed on acid-free paper SPIN 10715908

Preface

A workshop on "High Performance Computing and Gigabit Wide Area Networks" was held from 1 until 5 September 1998 at the Institute for Experimental Mathematics of Essen University, Gemany. It was organized by Professor Gene Cooperman (Northeastern University, Boston), Professor E. Jessen (Technical University, Munich) and Professor G. Michler (Institute for Experimental Mathematics, Essen). The 36 participants came from Australia, Germany, the Netherlands, Norway, Spain, Switzerland and the United States.

Distributed high performance computation in wide area high speed networks is a new interdisciplinary area of research. It requires the co-operation of mathematicians with computer scientists and communication engineers, because the technical problems caused by the latency cannot be solved by the mathematicians. The mathematical problems produce large amounts of data which have to be communicated quickly and correctly between the different computers at the various locations. Therefore new protocols and management tools have to be developed. Also it turns out that the network hardware has to be chosen carefully, and in particular routers should be avoided wherever possible. The main aim of the meeting was a careful analysis of the interplay between computer network technology, communication software and the performance of mathematical algorithms used for large scale computations in wide area high speed computer networks.

At least in the United States the number of supercomputer centers is shrinking, and also in Germany it will be difficult to provide enough resources for high performance computations in basic University research areas. It has been known for a long time that distributed high performance computing offers an alternative to demanding parallel computations on supercomputers. There are many mathematical problems for which it is possible to develop efficient algorithms which are suitable for distributed computations on middle sized parallel computers of different institutions connected by means of a high speed network. However, so far only a few scientists have access to such wide area high speed networks.

Another goal of the meeting was to give a survey about the existing and planned high speed networks in the United States, Europe, and in particular in Germany.

By bringing together scientists working in computer networks, network protocols and numerical and linear algebra, the meeting has given new insights that would not have been attainable otherwise. Only by interdisciplinary research through the co-operation of electrical engineers, mathematicians and computer scientists will it be possible to determine the most efficient combinations of parallel algorithms, protocols and network hardware. This is a new and fruitful research area to which the 22 refereed research articles of these proceedings contribute.

At the end of these proceedings we list the titles of the 29 original invited lectures of the workshop, along with the addresses of all the participants.

The conference was supported in part by the Volkswagen Foundation, Deutsche Forschungsgemeinschaft, National Science Foundation (USA) and the University of Essen. The editors of these proceedings thank these institutions and corporations for their generous help.

Boston, München, Essen, June 30, 1999
G. Cooperman
E. Jessen
G. Michler

Contents

Introduction: Distributed High Performance Computing and Gigabit Wide Area Networks

G. Cooperman, E. Jessen and G. Michler

College of Computer Science
Northeastern University
Boston, MA 02115, USA

Technische Universität München
Institut für Informatik
Lehrstuhl VIII – Rechnerstruktur/architektur
D-80290 München

Institut für Experimentelle Mathematik
Universität Essen
Ellernstraße 29, D-45326 Essen

In many countries there are nation wide high performance wide area computer networks, e.g. the German broad band science network B-Win or the very high speed backbone network services vBNS of the American National Science Foundation (NSF). They form a part of the Internet, the widest wide area network of them all. Experience with these networks provides a wealth of information, in particular on hardware and software for wide area networking and high performance computing.

The lectures and discussions at the meeting have inspired several fruitful international research collaborations on distributed information technologies and applications. They emphasize the interdisciplinary cooperation of electrical engineers, mathematicians and computer scientists in this very active area of research. In particular, this meeting surveys recent research directions in large scale networking:

a) Technologies enabling optical, electrical and wireless communications;
b) Network engineering and network management systems;
c) System software and program development environments for distributed high
    performance computing;
d) Distributed applications such as electronic commerce, distance learning and digital
    libraries;
e) High confidence systems with secure access, high availability, and strong privacy
    guarantees.

Experimental results for such systems of the future can only be gained through access to today's gigabit wide area networks. Therefore another important section of this workshop was devoted to reports about existing gigabit testbeds and planned high speed networks in the United States, Europe, and in particular in Germany. These testbeds include: the American initiatives for the next generation internet (NGI) and the Internet 2 of about 120 Universities in the United States (cf. K. Abdali); the Norwegian National Research Network (cf. T. Plagemann); and the planned German Gigabitwissenschaftsnetz (GWin) (cf. E. Jessen).

The NGI initiative is financially supported by the American federal agencies DARPA, DOE, NASA, NIH, NIST and NSF. It aims:

a) to promote research, development and experimentation in advanced networking
    technologies;
b) to deploy a NGI testbed emphasizing end-to-end performance, end-to-end quality of
    service and security;
c) to develop ultra high speed switching and transmission technologies;
d) to develop demanding applications that make use of the advances in network
    technologies. Among the proposed application areas are: health care, crisis management
    and response, distance learning, and distributed high performance computations for
    biomedicine, climate modeling, and basic science.

The Internet 2 project is funded by 77 American universities and some industrial partners. It is driven by education and research. Internet 2 will include a gigabit network (Project Abilene), which will operate in 1999. Of course it will benefit from the experiments and results of the NGI initiative. In Germany, in spring 2000, GWin, the gigabit network of DFN (Deutsches Forschungsnetz; German national scientific networking association) will start its operation.

As a forerunner for its gigabit network, DFN is supporting two gigabit testbeds in West and South Germany (with a link to Berlin) where experiments are performed. Several articles of this volume report both on planned and on completed experiments. The gigabit testbed West connects the research centers GMD St. Augustin and FZ Jülich in North Rhine Westphalia with a bandwidth of 2.5 Gbps. It has broadband connections to the computer centers of the DLR (Deutsches Zentrum für Luft- und Raumfahrt) in Cologne-Porz and the Universities of Cologne and Essen. The gigabit testbed Süd connects the University of Erlangen and the Technical University of Munich. It currently consists of a dark fiber connection between the computer centers of these universities. The bandwidth of this switched ATM network is initially 3 times 2.5 Gbps, and has a capacity many times larger through the use of wave length division multiplexing. The gigabit testbed Süd will be extended to Berlin and Stuttgart, in order to connect the supercomputers at the Konrad Zuse Institute in Berlin, Leibniz Computing Center in Munich and the Computing Center of Stuttgart University. This wide area network of supercomputers will be used for demanding distributed high performance computations in applied mathematics, physics, chemistry, engineering, economics and medicine.

In 1997 the United States has established the National Computational Science Alliance. It is led by the Supercomputer Centers at the University of Illinois at Urbana and the University of California at San Diego. Each alliance consists of more than 60 partner institutions, including academic and government research labs and industrial organizations. These cooperating institutions all benefit from a metacomputing environment based on high speed networks. K. Abdali of the NSF provides further details in the article, "Advanced computing and communications research under NSF support".

The strictest requirements for high bandwidth applications can currently be found in the areas of metacomputing and distributed high performance computation. These applications serve a secondary purpose in stress testing the network, to help the engineers and computer scientists design better ones. However, the number of such large scale experiments is currently rather small. As long as the cooperating institutions interconnected by a wide area high speed network are not given extra resources for distributed computer applications this situation is likely to continue.

Another set of lectures at the meeting was devoted to the interplay between communication hardware and software for high speed computer networks and mathematical algorithm development for distributed high performance computations. In particular, the implementations of the parallel linear algebra algorithms help to create experiments checking the technical communication properties of a broadband computer network with 155 Mbit/s bandwidth and higher. On the other side such benchmarks also help to analyze the effficiency of a mathematical algorithm.

This volume also contains several contributions concerning the very organization of scientific knowledge, itself. Many scientific publications quoting high performance computer applications lack proper documentation of the original computer programs and of the memory intensive output data. Recently many mathematical and other scientific journals have begun offering both paper and digital formats. The digital versions offer many advantages for the future. They have the potential for being searched, and they can be incorporated into a distributed library system. Over scientific wide area networks such as the planned German Gigabitwissenschaftsnetz, libraries of universities and research institutes in digital form can be combined into a national distributed research library. The members of these institutions can be allowed to search, read, print and even annotate the digital texts (and their computational appendices containing the original programs and the output data) at their personal computers. The wide area networks offer not only distributed library applications but they also offer completely new applications making use of multimedia, distance learning, and computer-aided teaching. Therefore this volume also contains several contributions describing such applications.

Advanced Computing and Communications Research
under NSF Support

S. Kamal Abdali

National Science Foundation, Arlington, VA 22230, USA

Abstract. This paper discusses the research initiatives and programs supported by the National Science Foundation to promote high-end computing and large-scale networking. This work mainly falls under the US interagency actvity called High Performance Computing and Communications (HPCC). The paper describes the Federal geovernment context of HPCC, and the HPCC programs and their main accomplishemnts. Finally, it decribes the recommendations of a recent high-level advisory committee on information technology, as these are likely to have a major impact on the future of government initiatives in high-end computing and networking.

1 Introduction

A previous paper [1] described the activities of the National Science Foundation (NSF) in the U.S. High Performance Computing and Communications (HPCC) program until 1996. The purpose of the present paper is to update that description to cover the developments since then. While some management changes have taken place during this period, and there is some redirection of its thrusts, the HPCC program continues to flourish, to say the least. The main new activities at the NSF are Partnerships for Advanced Computing Infrastructures (PACIs), the Next Generation Internet (NGI), and the Knowledge and Distributed Intelligence (KDI) initiative, and there are renewed programs for Science and Technology Centers and Digital Libraries. New initiatives that may replace the program or change its direction substantially are also expected to result from the recommendations of the Presidential Information Technology Advisory Committee (PITAC). The paper is mainly concerned with these new issues. But to make it self-contained, the entire HPCC context is briefly described also.

2 The HPCC program

The US High Performance Computing and Communication (HPCC) program was launched in 1991. It operated as a congressionally mandated initiative from October 1991 through September 1996, following the enactment of the

High Performance Computing Act of 1991. Since October 1996, it has continued as a program under the leadership of the Computing, Information, and Communications (CIC) Subcommittee of the Committee on Technology (CT) which is itself overseen by the National Science and Technology Council, a US Cabinet-level organization. Instrumental in the establishment of the program was a series of national-level studies of scientific and technological trends in computing and networking [2–5]. These studies concluded and persuasively argued that a federal-level initiative in high-performance computing was needed to ensure the preeminence of American science and technology. Solving the challenging scientific and engineering problems that were already on the horizon required significantly more computational power than was available. Another factor was the progress made abroad, especially the Japanese advances in semiconductor chip manufacture and supercomputer design, and the Western European advances in supercomputing applications in science and engineering. It was also clear that the advances in information technology would have a far reaching impact beyond science and technology, and would affect society in general in profound, unprecedented ways. The HPCC program was thus established to stimulate, accelerate, and harness these advances for coping with scientific and engineering challenges, solving societal and environmental problems, meeting national security needs, and in improving the nations economic productivity and competitiveness.

As late as 1996, the goals of the HPCC initiative were stated separately (e.g., in [10]) from the CIC mission descriptions. Now that HPCC has become a CIC research and development (R&D) program, its goals are subsumed in the CIC goals, which are formally stated as follows ([13]):

- Assure continued US leadership in computing, information, and communications technologies to meet Federal goals and to support U.S. 21st century academic, defense, and industrial interests
- Accelerate deployment of advanced and experimental information technologies to maintain world leadership in science, engineering, and mathematics; improve the quality of life; promote long term economic growth; increase lifelong learning; protect the environment; harness information technology; and enhance national security
- Advance U.S. productivity and industrial competitiveness through long-term scientific and engineering research in computing, information, and communications technologies

3 HPCC Participants and Components

The HPCC program at present involves 12 Federal agencies, each with its specific responsibilities. In alphabetical order, the participating agencies are: Agency for Health Care Policy and Research (AHCPR), Defense Advanced Research Projects Agency (DARPA), Department of Energy (DOE), Department of education (ED), Environmental Protection Agency (EPA), National Aeronautics and Space Administration (NASA), National Institute of Health (NIH), National Institute of Standards and Technology (NIST), National Oceanic and Atmospheric Administration (NOAA), National Security Agency (NSA), National Science Foundation (NSF), and Department of Veteran Affairs (VA). The activities sponsored by these agencies have broad participation by universities as well as the industry. The program activities of the participating organizations are coordinated by the National Coordination Office for Computing, Information, and Communications (NCO), which also serves as the liaison to the US Congress, state and local governments, foreign governments, universities, industry, and the public. The NCO disseminates information about HPCC program activities and accomplishments in the form of announcements, technical reports, and the annual reports that are popularly known as "blue books" [6–13]. The NCO also maintains the web site *http://www.ccic.gov* to provide up-to-date, online documentation about the HPCC program, as well as links to the HPCC-related web pages of all participating organizations.

The program currently has five components: 1) High End Computing and Computation, 2) Large Scale Networking, 3) High Confidence Systems, 4) Human Centered Systems, and 5) Education, Training, and Human Resources. Together, these components are meant to foster, among other things, scientific research, technological development, industrial and commercial applications, growth in education and human resources, and enhanced public access to information. In addition to these components, there is a Federal Information Services and Applications Council to oversee the application of CIC-developed technologies for federal information systems, and to disseminate information about HPCC research to other Federal agencies not formally participating in the program.

The goals of the HPCC components are as follows (see the "Blue Book" 99 [13] for an official description):

1. **High End Computing and Computation:** To assure US leadership in computing through investment in leading-edge hardware, software, and algorithmic innovations. Some representative research directions are: computing devices and storage technologies for high-end computing systems; advanced computing architectures; advanced software systems, algorithms, and software for modeling and simulation. This component also supports investigation of ideas such as optical, quantum, and biomolecular computing that are quite speculative at present, but may lead to feasible computing technologies in the future, and may radically change the nature of computing.

2. **Large Scale Networking:** To assure US leadership in high-performance communications. This component seeks to improve the state-of-the-art in communications by investing in research on networking components, systems, services, and management. The supported research directions include: advanced technologies that enable wireless, optical, mobile, and

wireline communications; large-scale network engineering; system software and program development environments for network-centric computing; and software technology for distributed applications, such as electronic commerce, digital libraries, and health care delivery.

3. **High Confidence Systems:** To develop technologies that provide users with high levels of security, protection of privacy and data, reliability, and restorability of information services. The supported research directions include: system reliability issues, such as network management under overload, component failure, and intrusion; survival of threatened systems by adaptation and reconfiguration; technologies for security and privacy assurance, such as access control, authentication, and encryption.

4. **Human Centered Systems:** To make computing and networking more accessible and useful in the workplace, school, and home. The technologies enabling this include: knowledge repositories and servers; collaboratories that provide access to information repositories and that facilitate sharing knowledge and control of instruments at remote labs; systems that allow multi-modal human- system interactions; and virtual reality environments and their applications in science, industry, health care, and education.

5. **Education, Training, and Human Resources:** To support HPCC research that enables modern education and training technologies. All levels and modes of education are targeted, including elementary, secondary, vocational, technical, undergraduate, graduate, and career-enhancing education. The education and training also includes the production of researchers in HPCC technologies and applications, and a skilled workforce able to cope with the demands of the information age. The supported research directions include information-based learning tools, technologies that support lifelong and distance learning for people in remote locations, and curriculum development.

4 HPCC at NSF

As mentioned above, NSF is one of the 12 Federal agencies participating in the HPCC program. The total HPCC budget and the NSF share in it since the inception of the program are shown in Table 1. Thus, during this period, NSF's share has ranged approximately between one-fourth and one-third of the total Federal HPCC spending. The HPCC amount has remained approximately 10% of the NSF's own total budget during the same period.

Table 1. HPCC Investment: Total budget and NSF's share (in $M)

Fiscal Year	1992	1993	1994	1995	1996	1997	1998	1999
Total HPCC budget	655	803	938	1039	1043	1009	1070	830
NSF's HPCC share	201	262	267	297	291	280	284	297

The NSF objectives for its HPCC effort are:

- Enable U.S. to uphold a position of world leadership in the science and engineering of computing, information and communications.
- Promote understanding of the principles and uses of advanced computing, communications, and information systems in service to science and engineering, to education, and to society.
- Contribute to universal, transparent, and affordable participation in an information-based society.

Thus NSF's HPCC-related work spans across all of the five HPCC program components.

HPCC research penetrates to varying depth nearly all the scientific and engineering disciplines at NSF. But most of this research is concentrated in the NSF's Directorate of Computer and Information Science and Engineering (CISE). This directorate is organized into 5 *divisions* each of which is, in turn, divided into 2–8 *programs*. The work of the CISE divisions can be, respectively, characterized as: fundamental computation and communications research; information, knowledge, intelligent systems, and robotics research; experimental systems research and integrative activities; advanced computational infrastructure research; and advanced networking infrastructure research. While the phrase "high performance" may not be explicitly present in the description of many programs, the actual research they undertake is very much focused on HPCC. Indeed, the CISE budget is almost entirely attributed to HPCC. Representative ongoing research topics include: scalable parallel architectures; component technologies for HPCC; simulation, analysis, design and test tools needed for HPCC circuit and system design; parallel software systems and tools, such as compilers, debuggers, performance monitors, program development environments; heterogeneous computing environments; distributed operating systems, tools for building distributed applications; network management, authentication, security, and reliability; intelligent manufacturing; intelligent learning systems; problem solving environments; algorithms and software for computational science and engineering; integration of research and learning technologies; very large data and knowledge bases; visualization of very large data sets.

5 Large HPCC Projects

The HPCC program has led to several innovations in NSF's mechanisms for supporting research and human resources development. The traditional manner of funding individual researchers or small research teams continues to be applied for HPCC work too. But to meet special HPCC needs, NSF has initiated a number of totally new programs, such as supercomputing centers, partnerships for advanced computational infrastructures, science and technology centers, and various "challenges". Also launched were special initiatives such as digital libraries, knowledge and distributed intelligence, and the next generation internet. These projects are much larger than the traditional ones in the scope of research, number of participating investigators, research duration, and award size.

5.1   Science and Technology Centers (STCs)

The purpose, structure, and HPCC contributions of STCs were described in [1]. So here we mainly state the developments that have taken place since.

STCs are intended to stimulate "integrative conduct of research, education, and knowledge transfer." They provide an environment for interaction among researchers in various disciplines and across institutional boundaries. They also provide the structure to identify important complex scientific problems beyond disciplinary and institutional limits and scales, and the critical mass and funding stability and duration needed for their successful solution. They carry out fundamental research, facilitate research applications, promote technology transfer through industrial affiliations, disseminate knowledge via visitorships, conferences and workshops, educate and train people for scientific professions, and introduce minorities and underrepresented groups to science and technology through outreach activities.

STCs are large research projects each of which involves typically 50+ principal investigators from 10+ academic institutions, and also has links to the industry. The participants work together on interdisciplinary research unified by a single theme, such as parallel computing or computer graphics. The projects are awarded initially for 5 years, are renewable for another 5 years, and are finally given an extra year for orderly phaseout. There is no further renewal, so a center has to shutdown definitely in at most 11 years. Of course, the investigators are free to regroup and compete again in the program in the future if it continues.

As a result of the competitions that took place in 1989 and 1991, 25 STCs were established by NSF. All of them have entered their final year now. The following four of those STCs were supported by the HPCC program: The Center for Research in Parallel Computation (CRPC) at Rice University; The Center for Computer Graphics and Scientific Visualization at the University of Utah; The Center for Discrete Mathematics and Theoretical Computer Science (DIMACS) at Rutgers University; and The Center for Cognitive Science at the University of Pennsylvania. These STCs have contributed numerous theoretical results, algorithms, mathematical and computer science techniques, libraries, software tools, languages, and environments. They have also made significant advances in various scientific and engineering application areas. Their output has been impressive in quality, quantity, and impact.

In 1995, NSF undertook a thorough evaluation of the STC program. For one study [14], Abt Associates, a private business and policy consulting firm was commissioned to collect various kind of information about the STCs, and the National Academy of Science was asked to examine that data and evaluate the program. Another study [15] was conducted by the National Academy of Public Administration. Both studies concluded that the STC program represented excellent return on federal research dollar investment, and recommended that the program be continued further. The studies also endorsed most of the past guidelines regarding the funding level, award duration, emphasis on education and knowledge transfer (additionally to research), review and evaluation criteria, and management structure.

Based on these findings, NSF has decided to continue the STC program. A new round of proposal solicitations took place in 1998. The submitted proposal have been evaluated, and the awards are expected to be announced soon (as of March 1999).

5.2 Partnerships for Advanced Computational Infrastructures (PACIs)

The precursor to PACIs was a program called *Supercomputing Centers (SCs)* that was established by NSF in 1985 even before the start of the HPCC initiative. But the SC program greatly contributed to the momentum behind HPCC, and, since its launch, became a significant part of the initiative. For a 10-year duration, the program funded four SCs: Cornell Theory Center, Cornell University; National Center for Supercomputing Applications, University of Illinois at Urbana-Champaign; Pittsburgh Supercomputer Center, University of Pittsburgh; and San Diego Supercomputer Center, University of California–San Diego. Several of their accomplishments and HPCC contributions have been reported in [1].

A Task Force to evaluate the effectiveness of the SC program was commissioned by NSF in 1995. This resulted in a document which is popularly known as the "Hayes Report [16]. The study considered the alternatives of renewing the SCs or having a new competition, and recommended the latter. For a more effective national computing infrastructure development, it also recommended funding fewer but larger alliances of research and experimental facilities and national and regional high- performance computing centers. Based on these findings, NSF instituted the PACI program in 1996, as the successor to the SC program. The aim of the PACIs is to help maintain US world leadership in computational science and engineering by providing access nationwide to advanced computational resources, promoting early use of experimental and emerging HPCC technologies, creating HPCC software systems and tools, and training a high quality, HPCC-capable workforce.

After holding a competition, NSF made two PACI awards in 1997. These are the National Computational Science Alliance (Alliance) led by the National Center for Supercomputing Applications (NCSA) at the University of Illinois at Urbana-Champaign, and the National Partnership for Advanced Computational Infrastructure (NPACI) led by the San Diego Supercomputer Center at the University of California at San Diego. Each consists of more than 60 partner institutions, including academic and government research labs, national, state-level and local computing centers, and business and industrial organizations. The leading sites, which maintain a variety of high-performance computer systems, and the partners which maintain smaller configurations of similar systems, jointly constitute a metacomputing environment connected via high-speed networks. The partners contribute to the infrastructure by developing in-house, using, and testing the necessary software, tools, environments, applications, algorithms, and libraries, thereby contributing to the further growth of a "national grid" of networked high-performance computers.

The initial mission of the SCs was to satisfy the supercomputing needs of US computational scientists and engineers. The major role of the PACIs continues to be to provide supercomputing access to the research community in all branches of science and engineering. But their expanded mission puts a heavy emphasis on education and training at all levels.

5.3 Next Generation Internet (NGI)

The NGI initiative, a multi-agency Federal R&D program that began in October 1997, is the main focus of LSN. It represents consolidation and refinement of ideas behind the vision of a National Information Infrastructure. This infrastructure is a subject of various studies, most importantly [17,18]. The NGI initiative supports foundational work to lead to much more powerful and versatile networks than the present-day Internet. To advance this work, the initiative fosters partnerships among universities, industry and the government. The participating federal government agencies include: DARPA, DOE, NASA, NIH, NIST and NSF. The NGI goals are:

1. Promote research, development, and experimentation in networking technologies.
2. Deploy testbeds for systems scale testing of technologies and services.
3. Develop "revolutionary" applications that utilize the advancements in network technologies and exercise the testbeds.

The aim of the advancement stipulated in Goal 1 is to dramatically improve the performance of networks in reliability, security, quality of service/differentiation of service, and network management. Two testbeds are planned for Goal 2. The first testbed is required to connect at least 100 sites and deliver speeds that are at least 100 times faster end-to-end than the present-day Internet. The second testbed is required to connect about 10 sites with end-to-end performance speed faster than the present Internet by at least a factor of 1000. The "revolutionary" applications called for in Goal 3 are to range over enabling applications technologies as well as disciplinary applications. Suggested examples of the former include collaboration technologies, digital libraries, distributed computing, virtual reality, and remote operation and simulation. Suggested application areas for the latter include basic science, education, health care, manufacturing, electronic commerce, and government information services.

The NGI work in progress was showcased in the *Supercomputing 98* conference in a special session called *Netamorphosis*. The "Netamorphosis" demonstrations consisted of 17 significant NGI applications, ranging over visualization, scene analysis, simulation, manufacturing, remote operation, etc. For example, a demonstration entitled "Real-Time Functional MRI: Watching the Brain in Action" showed how one could remotely view brain activity while a patient was performing cognitive or sensory- motor tasks. The system could process functional MRI data in real-time, though the data acquisition, main computations, and visualization all took place at different sites connected by advanced networks. Another demonstration entitled "Distributed Image Spreadsheet: Earth Data from Satellite to Desktop" showed how scientists could analyze, process, and visualize massive amounts of geologic, atmospheric, or oceanographic data transmitted to their workstations from Earth Observing System satellites.

5.4 Digital Libraries Initiative (DLI)

The original DLI, now referred to as DLI Phase 1, started as a joint venture of NSF, DARPA, and NASA. Now the initiative is in Phase 2, and includes as sponsors those agencies as well as the National Library of Medicine, the Library of Congress, and the National Endowment for the Humanities.

The initiative seeks to advance the technologies needed to offer information essentially about anything, to anyone, located anywhere around the nation and the world. A digital library is intended to be a very large-scale storehouse of knowledge in multimedia form that is accessible over the net. The construction and operation of digital libraries requires developing technologies for acquiring information, organizing this information in distributed multimedia knowledge bases, extracting information based on requested criteria, and delivering it in the form appropriate for the user. Thus, the DLI promotes research on information collection, analysis, archiving, search, filtering, retrieval, semantic conversion, and communication.

The Phase 1 is supporting 6 large consortia consisting of academic and industrial partners. Their main project themes and their lead institutions are: geographic information systems, maps and pictures, content-base retrieval (University of California–Santa Barbara); intelligent internet search, semantic retrieval, scientific journal publishing alternatives (University of Illinois); media integration and access, new models of "documents," natural language processing (University of California–Berkeley); digital video libraries, speech, image and natural language technology integration (Carnegie Mellon University); intelligent agent architecture, resource federation, AI service market economies, educational impact (University of Michigan); uniform access, distributed object architectures, interface for distributed information retrieval (Stanford University).

The Phase 1 of the initiative was mainly concerned with learning, prototyping, and experimenting in the small. The Phase 2 expects to put this experience into actually building larger, operational, and usable systems and testbeds. There is emphasis on larger contents and collections, interoperability and technology integration, and expansion of domains and user communities for digital libraries. The supported activities are expected to range through the full spectrum of fundamental research, content and collections development, domain applications, testbeds, operational environments, and applications for developing educational resources and preserving the national cultural heritage.

5.5 Knowledge and Distributed Intelligence (KDI)

KDI is a new initiative that NSF established in 1998. The HPCC research has traditionally been concentrated in the NSF's Computer and Information Science and Engineering directorate. The KDI initiative stems from the realization that the advances in computing, communications, and information technologies provide unprecedented possibilities for accelerating progress in all spheres of human thought and action. KDI stresses knowledge as opposed to information, but realizes, of course, that intelligent gathering of information is a prerequisite to creating knowledge. Thus, a goal of KDI is to improve the human ability to discover, collect, represent, store, apply, and transmit information. This is to lead to improvements in the ways to create knowledge and in the actual acquisition of new knowledge. The KDI research is classified into three components:

1. Knowledge Networking (KN)
2. Learning and Intelligent Systems (LIS)
3. New Computational Challenges (NCC)

The KN component aims at building an open and context-rich environment for online interactions among individuals as well as groups. For such an environment to arise, advances have to be made in the techniques for collecting and organizing information and discovering knowledge from it. The KN-enabled vast scale of information acquisition and the power to uncover knowledge buried in collected data has grave implications for privacy and other human interest matters. Hence, KN is also concerned with research on social, societal, ethical, and other aspects of networked information.

The focus of the LIS component of KDI is to better understand the process of learning itself, as it occurs in humans, animals, and artificial systems. This understanding is to be used for improving our own learning skills, developing better teaching methods, and creating intelligent artifacts.

The NCC component is in the spirit of NSFs "Challenges programs, such as Grand Challenges, National Challenges, and Multidisciplinary Challenges.

In [1], these programs were described, and their impact and some of their accomplishments were stated. The NCC component continues to seek solutions of very complex scientific and engineering problems, ones that are computationally expensive, data intensive, and require multidisciplinary team approaches. The Challenges research and the advance in high-performance computing and communications system have a mutually benefiting push-pull relationship; the former stress tests the latter, and the latter helps the former grow in scale and scope. NCC research aims to improve our ability to model and simulate complex systems such as the oceans or the brain. In adopting the Challenges research, the KDI initiative sees it as another knowledge creation activity.

In 1998, NSF made 40 awards for KDI research for a total funding of $51.5M. The awards span a broad range of topics, vast scopes of research, and investigators representing diverse disciplines and institutions. The 1999 KDI competition is in process.

6   HPCC Evaluation

General directions as well as clear objectives were defined for the HPCC program from the very beginning. Thus, some evaluation is built into the program. Some objectives naturally lead to quantifiable measures of progress, such as computation speeds in teraflops, communication bandwidth in gigabits, network extent in number of connected nodes, etc. On the other hand, there are qualitative aspects of progress, such as scientific breakthroughs, innovative industrial practices, societal penetration of knowledge and technology, quality of work force trained, etc.

The evaluation of the STC and SC programs has already been mentioned. Other parts of the NSF HPCC program have also produced impressive results. For the effectiveness of the HPCC program as whole, a number of evaluation studies have been done. The "Branscomb Report [19], is devoted to studying the means for making the program more productive. A thorough assessment of the effectiveness of the program is undertaken in the "Brooks-Sutherland Report" [20]. The purpose of a more recent recent study [21] is to suggest the most important future HPCC applications, specially the ones with highest national, societal, and economic impact.

There is consensus that the HPCC program has been successful on most fronts. Not only the year by year milestones for quantifiable progress have been met, but the activities undertaken by the program have led to several significant, unanticipated beneficial developments. The launch of new important HPCC-inspired initiatives witnesses the programs strong momentum. But as the next section shows, there is a perception that the HPCC program is underfunded and the progress resulting from it is going to decelerate unless newer and larger investments are added to it.

7 Presidents Information Technology Advisory Committee (PITAC)

PITAC was established in February 1997 to provide advice to the Administration on all areas of computing, communications, and information technology. This committee at present consists of 26 research leaders representing academia and the industry. It issued an interim report in August 1998 and a final one in February 1999 [22], after a series of meetings and broad consultations with the research community. This report examines the impact of R&D in Information Technology (IT) on US business and science, and makes a number of recommendations for further work.

The PITAC report observes that the past IT R&D through HPCC and other programs is a significant factor in the nations world leadership position in science, industry, business, and the general well-being of the citizenry. IT advances are responsible for a third of the US economic growth since 1992, and have created millions of high-paying new jobs. The computational approach to science in conjunction with the HPCC algorithms, software, and infrastructure have helped the US scientists make new discoveries. The competitiveness of US economy is owed much to the efficiencies resulting from IT in engineering design, manufacturing, business, and commerce.

If IT is the engine that is driving the economy, then obviously it needs to be kept running by further investment. The PITAC report argues that the IT industry is spending the bulk of its own resources, financial and human, on near-term development of new products for an exploding market. The IT industry can contribute only a small fraction of the long-term R&D investment needed. Moreover, the industry does not see any immediate benefits of the scientific and social components of IT, and therefore has no interest in pursuing them. After estimating the total US R&D expenditure on IT, and the Federal and industrial shares of it, the PITAC conclusion is that the Federal support of the Information Technology (IT) R&D is grossly inadequate. Moreover, it is focused too much on near-term and applied research.

PITAC has recommended increments of about $1.3 billion per year for the next 5 years. PITAC has also identified the following four high priority areas as main targets of increased investment.

Software: Software production methodologies have to be dramatically improved, by fundamental research, to deliver robust, usable, manageable, cost-effective software.

Scalable Information Infrastructure: With the ever increasing size, complexity, and sheer use of networks, research is needed on how to build networks that can be easily extended yet remain reliable, secure, and easy to use.

High-End Computing: Scientific research and engineering design are becoming more and more computational. The increasing complexity of problems demand ever faster computing and communications. Thus, sustained research is needed on high performance architectures, networks, devices, and systems.

Socioeconomic Impact: Research is needed to exploit the IT advances to serve the society and to spread its benefits to all citizens. The accompanying social, societal, ethical, and legal issues have to be studied, and ways have to be sought for mitigating any potential negative impact.

Based on the PITAC recommendations, a new Federal interagency initiative called Information Technology for the Twenty-first Century (IT^2) is being developed, as a possible successor to the HPCC program.

8 Conclusion

Scientific and engineering work is becoming more computational, because, increasingly, computation is replacing physical experimentation and the construction and testing of prototypes. (Indeed, the US Accelerated Strategic Computing Initiative plans to depend totally on computational simulation in its weapons research program for those weapons whose physical testing is banned by international treaties.) Several recent scientific discoveries have been possible because of computation. The HPCC program has played a key role in the rise of computational science and engineering.

In [1], it was observed that collaboration and team work emerged as an important modality of HPCC research. In particular, the HPCC programs have emphasized 1) multi-disciplinary, multi-investigator, multi-institution teams, 2) partnerships among academia, business, and industry, and 3) cooperative, interagency sponsorship of research. In recent years, the collaboration has increased in intensity and scale. The transition from SCs to PACIs is a good example.

The previous Challenge projects tended to be computation-intensive. In a number of NCC projects, the data-intensive aspect dominates the computation-intensive one. Because of this situation, data mining has emerged as a key solution strategy for many Challenge-scale problems.

In practice, the HPCC program has so far been focused on applications and infrastructure development. Partly this is because most of the participating agencies in the HPCC program have special missions, and have rightly emphasized the fulfillment of their missions rather than basic research. The development of high performance computing infrastructure has also served some critical research needs. But there is need now to bolster fundamental research in order to stimulate further progress towards the original HPCC goals. The PITAC report urges this.

References

1. Abdali S.K.: High Performance Computing Research at NSF, In G. Cooperman, G. Michler and H. Vinck (Eds.), *Proc. Workshop on High Performance*

Computation and Gigabit Local Area Networks, Lect. Notes in Control and Information Sci. # 226, Springer-Verlag Berlin, 1997.

2. *A National Computing Initiative: The Agenda for Leadership*, Society for Industrial and Applied Mathematics, Philadelphia, PA, 1987.

3. *Toward a National Research Network*, National Academy Press, Washington, D.C., 1988.

4. *Supercomputers: Directions in Technology and Applications*, National Academy Press, Washington, D.C., 1989.

5. *Keeping the U.S. Computer Industry Competitive: Defining the Agenda*, National Academy Press, Washington, D.C., 1990.

6. *Grand Challenges: High Performance Computing and Communications* ("FY 1992 Blue Book"), Federal Coordinating Council for Science, Engineering, and Technology, c/o National Science Foundation, Washington, D.C., 1991.

7. *Grand Challenges 1993: High Performance Computing and Communications* ("FY 1993 Blue Book"), Federal Coordinating Council for Science, Engineering, and Technology, c/o National Science Foundation, Washington, D.C., 1992.

8. *High Performance Computing and Communications: Toward a National Information Infrastructure* ("FY 1994 Blue Book"), Office of Science and Technology Policy, Washington, D.C., 1993.

9. *High Performance Computing and Communications: Technology for a National Information Infrastructure* ("FY 1995 Blue Book"), National Science and Technology Council, Washington, D.C., 1994.

10. *High Performance Computing and Communications: Foundation for America's Information Future* ("FY 1996 Blue Book"), National Science and Technology Council, Washington, D.C., 1995.

11. *High Performance Computing and Communications: Advancing the Frontiers of Information Technology* ("FY 1997 Blue Book"), National Science and Technology Council, Washington, D.C., 1996.

12. *Technologies for the 21st Century* ("FY 1998 Blue Book"), National Science and Technology Council, Washington, D.C., 1997.

13. *Networked Computing for the 21st Century* ("FY 1999 Blue Book"), National Science and Technology Council, Arlington, VA, 1998.

14. *National Science Foundation's Science and Technology Centers: Building an Interdisciplinary Research Program*, National Academy of Public Administration, Washington, D.C., 1995.

15. *An Assessment of the National Science Foundation's Science and Technology Centers Program*, National Research Council, National Academy Press, Washington, D.C., 1996.

16. *Report of the Task Force on the Future of the NSF Supercomputing Centers* ("Hayes report"), Pub. NSF 96-46, National Science Foundation, Arlington, VA.

17. *The Unpredictable Certainty: Information Infrastructure through 2000*, National Research Council, National Academy Press, Washington, D.C., 1996.

18. *More Than Screen Deep: Toward Every-Citizen Interfaces to the Nation's Information Infrastructure* ("Biermann Report"), National Research Council, National Academy Press, Washington, D.C., 1997.

19. *From Desktop to Teraflop: Exploiting the U.S. Lead in High Performance Computing* ("Branscomb Report"), Pub. NSB 93-205, National Science Foundation, Washington, D.C., August 1993.

20. *Evolving the High Performance Computing and Communications Initiative to Support the Nation's Information Infrastructure* ("Brooks-Sutherland Report"), National Research Council, National Academy Press, Washington, D.C., 1995.
21. *Computing and Communications in the Extreme: Research for Crisis Management and Other Applications*, National Research Council, National Academy Press, Washington, D.C., 1996.
22. *Information Technology Research: Investing in Our Future*, President's Information Technology Advisory Committee Report to the President, National Coordination Office, Arlington, VA, 1999.

SRP: a Scalable Resource Reservation Protocol for the Internet

Werner Almesberger[1], Tiziana Ferrari[2], and Jean-Yves Le Boudec[1]

[1] EPFL ICA, INN (Ecublens), CH-1015 Lausanne, Switzerland
[2] DEIS, University of Bologna, viale Risorgimento, 2, I-40136 Bologna, Italy; and
Italian National Inst. for Nuclear Physics/CNAF, viale Berti Pichat, 6/2,
I-40127 Bologna, Italy

Abstract. The Scalable Reservation Protocol (SRP) provides a light-weight reservation mechanism for adaptive multimedia applications. Our main focus is on good scalability to very large numbers of individual flows. End systems (i.e. senders and destinations) actively participate in maintaining reservations, but routers can still control their conformance. Routers aggregate flows and monitor the aggregate to estimate the local resources needed to support present and new reservations. There is neither explicit signaling of flow parameters, nor do routers maintain per-flow state.

1 Introduction

Many adaptive multimedia applications [1] require a well-defined fraction of their traffic to reach the destination and to do so in a timely way. We call this fraction the *minimum rate* these applications need in order to operate properly. SRP aims to allow such applications to make a dependable reservation of their minimum rate.

The sender can expect that, as long as it adheres to the agreed-upon profile, no *reserved* packets will be lost due to congestion. Furthermore, forwarding of *reserved* packets will have priority over best-effort traffic.

Traditional resource reservation architectures that have been proposed for integrated service networks (RSVP [2], ST-2 [3], Tenet [4], ATM [5,6], etc.) all have in common that intermediate systems (routers or switches) need to store per-flow state information. The more recently designed Differentiated Services architecture [7] offers improved scalability by aggregating flows and by maintaining state information only for such aggregates. SRP extends upon simple aggregation by providing a means for reserving network resources in routers along the paths flows take.

Recently, hybrid approaches combining RSVP and Differentiated Services have been proposed (e.g. [8]) to overcome the scalability problems of RSVP. Unlike SRP, which runs end-to-end, they require a mapping of the INTSERV services onto the underlying Differentiated Services network, and a means to tunnel RSVP signaling information through network regions where QoS is provided using Differentiated Services.

Reservation mechanism In short, our reservation model works as follows. A source that wishes to make a reservation starts by sending data packets marked as *request* packets to the destination. Packets marked as *request* are subject to packet admission control by routers, based on the following principle. Routers monitor the aggregate flows of *reserved* packets and maintain a running estimate of what level of resources is required to serve them with a good quality of service. The resources are bandwidth and buffer on outgoing links, plus any internal resources as required by the router architecture. Quality of service is loss ratio and delay, and is defined statically. When receiving a *request* packet, a router determines whether hypothetically adding this packet to the flow of *reserved* packets would yield an acceptable value of the estimator. If so, the *request* packet is accepted and forwarded towards the destination, while still keeping the status of a *request* packet; the router must also update the estimator as if the packet had been received as *reserved*. In the opposite case, the *request* packet is degraded and forwarded towards the destination, and the estimator is not updated. Degrading a *request* packet means assigning it a lower traffic class, such as best-effort. A packet sent as *request* will reach the destination as *request* only if all routers along the path have accepted the packet as *request*. Note that the choice of an estimation method is local to a router and actual estimators may differ in their principle of operation.

The destination periodically sends feedback to the source indicating the rate at which *request* and *reserved* packets have been received. This feedback does not receive any special treatment in the network (except possibly for policing, see below. Upon reception of the feedback, the source can send packets marked as *reserved* according to a profile derived from the rate indicated in the feedback. If necessary, the source may continue to send more *request* packets in an attempt to increase the rate that will be indicated in subsequent feedback messages.

Thus, in essence, a router accepting to forward a *request* packet as *request* allows the source to send more *reserved* packets in the future; it is thus a form of implicit reservation.

Aggregation Routers aggregate flows on output ports, and possibly on any contention point as required by their internal architecture. They use estimator algorithms for each aggregated flow to determine their current reservation levels and to predict the impact of accepting *request* packets. The exact definition of what constitutes an aggregated flow is local to a router.

Likewise, senders and sources treat all flows between each pair of them as a single aggregate and use estimator algorithms for characterizing them. The estimator algorithms in routers and hosts do not need to be the same. In fact, we expect hosts to implement a fairly simple algorithm, while estimator algorithms in routers may evolve independently over time.

Fairness and security Denial-of-service conditions may arise if flows can reserve disproportional amounts of resources or if flows can exceed their reservations. We presently consider fairness in accepting reservations a local policy issue (much like billing) which may be addressed at a future time.

Sources violating the agreed upon reservations are a real threat and need to be policed. A scalable policing mechanism to allow routers to identify non-conformant flows based on certain heuristics is the subject of ongoing research. Such a mechanism can be combined with more traditional approaches, e.g. policing of individual flows at locations where scalability is less important, e.g. at network edges.

The rest of this paper is organized as follows. Section 2 provides a more detailed protocol overview. Section 3 describes a simple algorithm for the implementation of the traffic estimator. Finally, protocol operation is illustrated with some simulation results in section 4 and the paper concludes with section 5.

2 Architecture overview

The proposed architecture uses two protocols to manage reservations: a reservation protocol to establish and maintain them, and a feedback protocol to inform the sender about the reservation status.

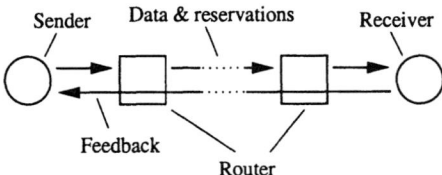

Fig. 1. Overview of the components in SRP.

Figure 1 illustrates the operation of the two protocols:

- Data packets with reservation information are sent from the sender to the receiver. The reservation information consists in a packet type which can take three values, one of them being ordinary best-effort (section 2.2). It is processed by routers, and may be modified by routers. Routers may also discard packets (section 2.1).
- The receiver sends feedback information back to the sender. Routers only forward this information; they don't need to process it (section 2.3).

Routers monitor the reserved traffic which is effectively present and adjust their global state information accordingly. Sections 2.1 to 2.3 illustrate the reservation and feedback protocol.

2.1   Reservation protocol

The reservation protocol is used in the direction from the sender to the receiver. It is implemented by the sender, the receiver, and the routers between them. As mentioned earlier, the reservation information is a packet type which may take three values:

Request This packet is part of a flow which is trying to gain *reserved* status. Routers may accept, degrade or reject such packets. When routers accept some *request* packets, then they commit to accept in the future a flow of reserved packets at the same rate. The exact definition of the rate is part of the estimator module.

Reserved This label identifies packets which are inside the source's profile and are allowed to make use of the reservation previously established by *request* packets. Given a correct estimation, routers should never discard *reserved* packets because of resource shortage.

Best effort No reservation is attempted by this packet.

Packet types are initially assigned by the sender, as shown in figure 2. A traffic source (i.e. the application) specifies for each packet if that packet needs a reservation. If no reservation is necessary, the packet is simply sent as *best-effort*. If a reservation is needed, the protocol entity checks if an already established reservation at the source covers the current packet. If so, the packet is sent as *reserved*, otherwise an additional reservation is requested by sending the packet as *request*.

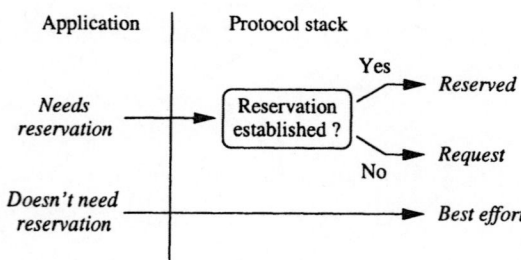

Fig. 2. Initial packet type assignment by sender.

Each router performs two processing steps (see also figure 3). First, for each *request* and *reserved* packet the estimator updates its current estimate of the resources used by the aggregate flows and decides whether to accept the packet (packet admission control). Then, packets are processed by various schedulers and queue managers inside the router.

- When a *reserved* packet is received, the estimator updates the resource estimation. The packet is automatically forwarded unchanged to the scheduler where it will have priority over best-effort traffic and normally is not discarded.

- When a *request* packet is received, then the estimator checks whether accepting the packet will not exceed the available resources. If the packet can be accepted, its *request* label is not modified. If the packet cannot be accepted, then it is degraded to best-effort
- If a scheduler or queue manager cannot accept a reserved or request packet, then the packet is either discarded or downgraded to *best-effort*.

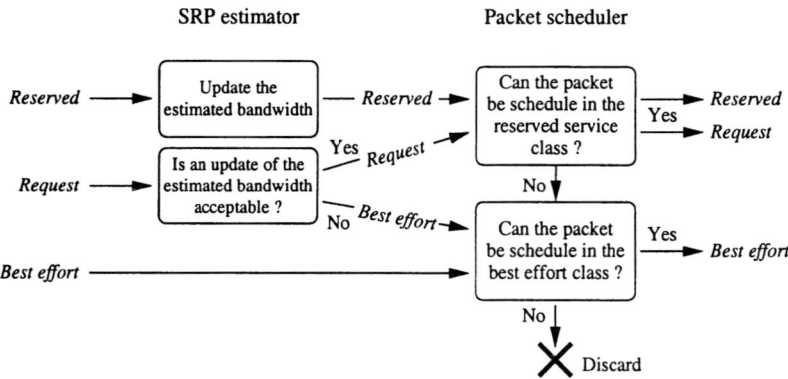

Fig. 3. Packet processing by routers.

Note that the reservation protocol may "tunnel" through routers that don't implement reservations. This allows the use of unmodified equipment in parts of the network which are dimensioned such that congestion is not a problem.

2.2 Packet type encoding

RFC2474 [9] defines the use of an octet in the IPv4 and IPv6 header for Differentiated Services (DS). This field contains the DS Code Point (DSCP), which determines how the respective packet is to be treated by routers (Per-Hop Behaviour, PHB). Routers are allowed to change the content of a packet's DS field (e.g. to select a different PHB).

As illustrated in figure 4, SRP packet types can be expressed by introducing two new PHBs (for *request* and for *reserved*), and by using the pre-defined DSCP value 0 for best-effort. DSCP values for *request* and *reserved* can be allocated locally in each DS domain.

2.3 Feedback protocol

The feedback protocol is used to convey information on the success of reservations and on the network status from the receiver to the sender. Unlike the

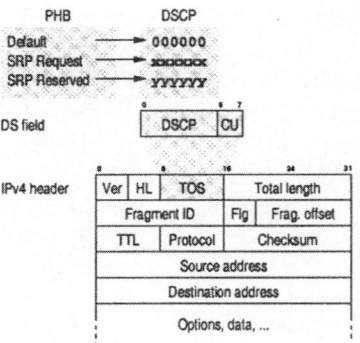

Fig. 4. Packet type encoding using Differentiated Services (IPv4 example).

reservation protocol, the feedback protocol does not need to be interpreted by routers, because they can determine the reservation status from the sender's choice of packet types.

Feedback information is collected by the receiver and it is periodically sent to the sender. The feedback consists of the number of bytes in *request* and *reserved* packets that have reached the receiver, and the local time at the receiver at which the feedback message was generated.

Receivers collect feedback information independently for each sender and senders maintain the reservation state independently for each receiver. Note that, if more than one flow to the same destination exists, attribution of reservations is a local decision at the source.

0	1	2	3	4	5	6	7	
Version	Reserved							
t0				t				
Reserved				Num REQ (t0)				
Reserved				Num REQ (t)				
Reserved				Num RSV (t0)				
Reserved				Num RSV (t)				

Fig. 5. Feedback message format.

Figure 5 illustrates the content of a feedback message: the time when the message was generated (t), and the number of bytes in *request* and *reserved* packets received at the destination (REQ and RSV). All counters wrap back to zero when they overflow.

In order to improve tolerance to packet loss, also the information sent in the previous feedback message (at time t0) is repeated. Portions of the message are reserved to allow for future extensions.

2.4 Shaping at the sender

The sender decides whether packets are sent as *reserved* or *request* based on its own estimate of the reservation it has requested and on the level of reservation along the path that has been confirmed via the feedback protocol. A source always uses the minimum of these two parameters to determine the appropriate output traffic profile.

Furthermore, the sender needs to filter out small differences between the actual reservation and the feedback in order to avoid reservations from drifting, and it must also ensure that *request* packets do not interfere with congestion-controlled traffic (e.g. TCP) in an unfair way [10].

2.5 Example

Figure 6 provides the overall picture of the reservation and feedback protocols for two end-systems connected through routers $R1$ and $R2$. The initial resource acquisition phase is followed by the generation of request packets after the first feedback message arrives. Dotted arrows correspond to degraded *request* packets, which passed the admission control test at router $R1$ but could not be accepted at router $R2$ because of resource shortage. Degradation of *requests* is taken into account by the feedback protocol. After receiving the feedback information the source sends *reserved* packets at an appropriate rate, which is smaller than the one at which *request* packets were generated.

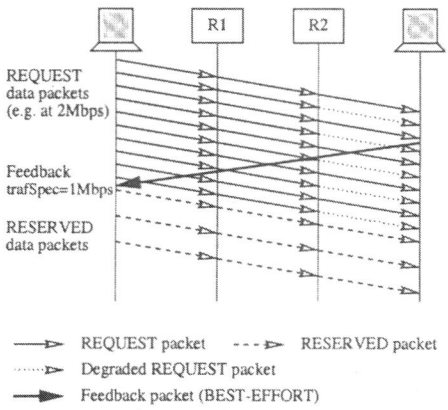

Fig. 6. Reservation and feedback protocol diagram.

2.6 Multicast

In order to support multicast traffic, we have proposed a design that slightly extends the reservation mechanism described in this sections. Refinement of this design is still the subject of ongoing work. A detailed description of the proposed mechanism can be found in [11].

3 Estimation modules

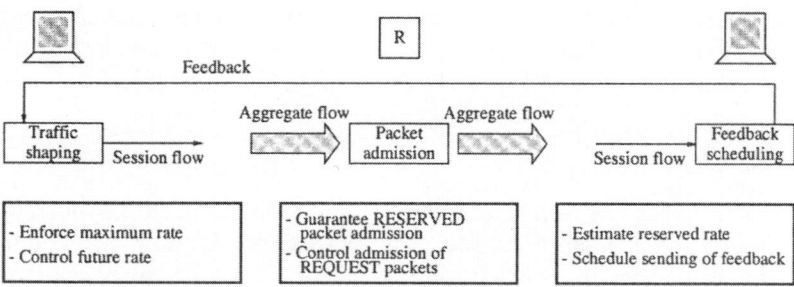

Fig. 7. Use of estimators at senders, routers, and receivers

We call *estimator* the algorithm which attempts to calculate the amount of resources that need to be reserved. The estimation measures the number of *requests* sent by sources and the number of *reserved* packets which actually make use of the reservation.

Estimators are used for several functions.

- Senders use the estimator for an optimistic prediction of the reservation the network will perform for the traffic they emit. This, in conjunction with feedback received from the receiver, is used to decide whether to send *request* or *reserved* packets.
- Routers use the estimator for packet-wise admission control and perhaps also to detect anomalies.
- In receivers, the estimator is fed with the received traffic and it generates an estimate of the reservation at the last router. This is used to schedule the sending of feedback messages to the source.

Figure 7 shows how the estimator algorithm is used in all network elements.

As described in section 2.1, a sender keeps on sending *requests* until successful reservation setup is indicated with a feedback packet, i.e. even until after the desired amount of resources has been reserved in the network. It's the feedback that is returned to the sender, which indicates the right allocation obtained on the path. When the source is feedback-compliant, the routers on the path start releasing a part of the over-estimated reservation already allocated. The feedback that is returned to the sender may also show an increased number of requests. The sender must not interpret those requests as a direct increase of the reservation. Instead, the sender estimator must correct the feedback information accordingly, which is achieved through the computation of the minimum of the feedback and of the resource amount requested by the source.

Our architecture is independent of the specific algorithm used to implement the estimator. Sections 3.1 and 3.2 describe two different solutions. The definition and evaluation of algorithms for reservation calculation in hosts and routers is still ongoing work. A detailed analysis of the estimation algorithms and additional improvements can be found in [12].

3.1 Basic estimation algorithm

The basic algorithm we present here is suitable for sources and destinations, and could be used as a rough estimator by routers. This estimator counts the number of requests it receives (and accepts) during a certain *observation interval* and uses this as an estimate for the bandwidth that will be used in future intervals of the same duration.

In addition to requests for new reservations, the use of existing reservations needs to be measured too. This way, reservations of sources that stop sending or that decrease their sending rate can automatically be removed. For this purpose the use of reservations can be simply measured by counting the number of *reserved* packets that are received in a certain interval.

To compensate for deviations caused by delay variations, spurious packet loss (e.g. in a best-effort part of the network), etc., reservations can be "held" for more than one observation interval. This can be accomplished by remembering the observed traffic over several intervals and using the maximum of these values (step 3 of the following algorithm). Given a hold time of h observation intervals, the maximum amount of resources which can be allocated Max, res and req (the total number of *reserved* and *request bytes* received in a given observation interval), the reservation R (in bytes) is computed by a router as follows. Given a packet of n bytes:

```
if (packet_type == REQ)
    if (R + req + n < Max) {
        accept;
        req = req + n;   // step 1
    }
    else degrade;

if (packet_type == RES)
    if (res + n < R) {
        accept;
        res = res + n;   // step 2
    }
    else degrade;
```

where initially $R, res, req = 0$. At the end of each observation cycle the following steps are computed:

```
for (i = h; i > 1; i--) R[i] = R[i-1];
R[1] = res + req;
R = max(R[h],R[h-1],...,R[1]); // step 3
res = req = 0;
```

The same algorithm can be run by the destination with the only difference that no admission checks are needed.

Examples of the operation of the basic algorithm are shown in section 4.1.

This easy algorithm presents several problems. First of all, the choice of the right value of the observation interval is critical and difficult. Small values make the estimation dependent on bursts of *reserved* or *request* packets and cause an overestimation of the resources needed. On the other hand, large intervals make the estimator react slowly to changes in the traffic profile. Then, the strictness of traffic acceptance control is fixed, while adaptivity would be highly desirable in order to make the allocation of new resources stricter as the amount of resources reserved gets closer to the maximum. These problems can be solved by devising an adaptive enhanced algorithm like the one described in the following section.

3.2 Enhanced estimation algorithm

Instead of using the same estimator in every network component, we can enhance the previous approach so that senders and receivers still run the simple algorithm described above, while routers implement an improved estimator.

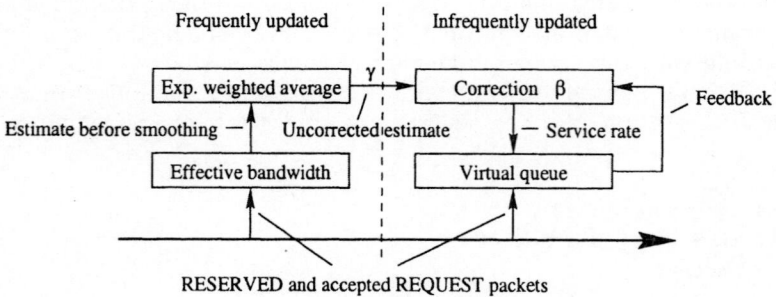

Fig. 8. Schematic design of an adaptive estimator.

We describe an example algorithm in detail in [11]. It consists of the principal components illustrated in figure 8: the effective bandwidth used by *reserved* and accepted *request* packets is measured and then smoothed by calculating an exponentially weighted average (γ). This calculation is performed for every single packet.

The estimate γ is multiplied with a correction factor β in order to correct for systematic errors in the estimation. Packets are added to a virtual queue (i.e. a counter), which is emptied at the estimated rate. If the estimate is too high, the virtual queue shrinks. If the estimate is too low, the virtual queue grows. Based on the size of the virtual queue, β can be adjusted.

4 Simulation

Section 4.1 provides a theoretic description of the behavior of the reservation mechanism in a very simple example, while section 4.2 shows the simulated behavior of the proposed architecture.

4.1 Reservation example

The network we use to illustrate the operation of the reservation mechanism, is shown in figure 9: the sender sends over a delay-less link to the router, which performs the reservation and forwards the traffic over a link with a delay of two time units to the receiver. The receiver periodically returns feedback to the sender.

The sender and the receiver both use the basic estimator algorithm described in section 3.1. The router may – and typically will – use a different algorithm (e.g. the one described in section 3.2).

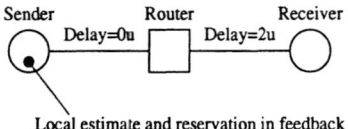

Fig. 9. Example network configuration.

The bandwidth estimate at the source and the reservation that has been acknowledged in a feedback message from the receiver are measured. In figure 10, they are shown with a thin continuous line and a thick dashed line, respectively. The packets emitted by the source are indicated by arrows on the reservation line. A full arrow head corresponds to *request* packets, an empty arrow head corresponds to *reserved* packets. For simplicity, the sender and the receiver use exactly the same observation interval in this example, and the feedback rate is constant.

The source sends one packet per time unit. First, the source can only send requests and the router reserves some resources for each of them. At point (1), the estimator discovers that it has established a reservation for six packets in four time units, but that the source has only sent four packets in this interval. Therefore, it corrects its estimate and proceeds. The first

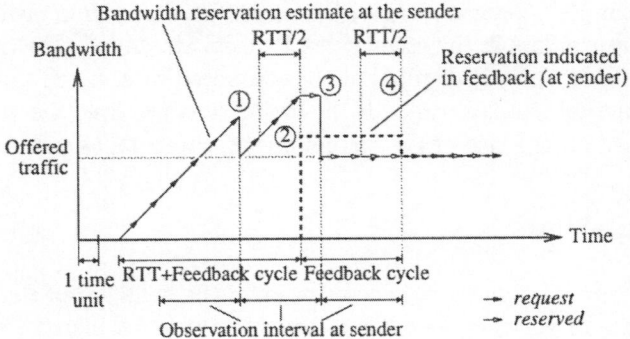

Fig. 10. Basic estimator example.

feedback message reaches the sender at point (2). It indicates a reservation level of five packets in four time units (i.e. the estimate at the receiver at the time when the feedback was sent), so the sender can now send *reserved* packets instead of *requests*. At point (3), the next observation interval ends and the estimate is corrected once more. Finally, the second feedback arrives at point (4), indicating the final rate of four packets in four time units. The reservation does not change after that.

4.2 Simulation results

The network configuration used for the simulation is shown in figure 11.[1] The grey paths mark flows we examine below.

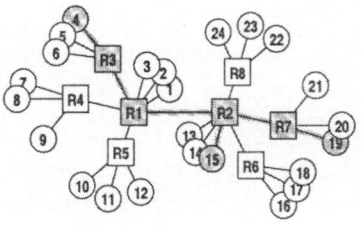

Fig. 11. Configuration of the simulated network.

There are eight routers (labeled **R1**...**R8**) and 24 hosts (labeled **1**...**24**). Each of the hosts **1**...**12** tries occasionally to send to any of the hosts **13**...**24**. Connection parameters are chosen such that the average number of concurrently active sources sending via the **R1**–**R2** link is approximately fifty. Flows

---

[1] The programs and configuration files used for the simulation are available on http://lrcwww.epfl.ch/srp/

have an on-off behaviour, where the on and off times are randomly chosen from the intervals $[5, 15]$ and $[0, 30]$ seconds, respectively. The bandwidth of a flow remains constant while the flow is active and is chosen randomly from the interval $[1, 200]$ packets per second.

All links in the network have a bandwidth of 4000 packets per second and a delay of 15 ms.[2] We allow up to 90% of the link capacity to be allocated to reserved traffic. The link between **R1** and **R2** is a bottleneck, which can only handle about 72% of the offered traffic. The delay objective D of each queue is 10 ms. The queue size per link is limited to 75 packets.

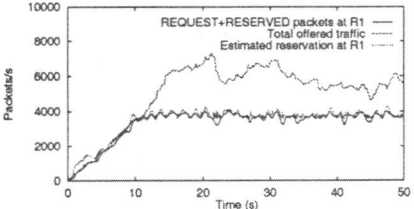

Fig. 12. Estimation and actual traffic at **R1** towards **R2**.

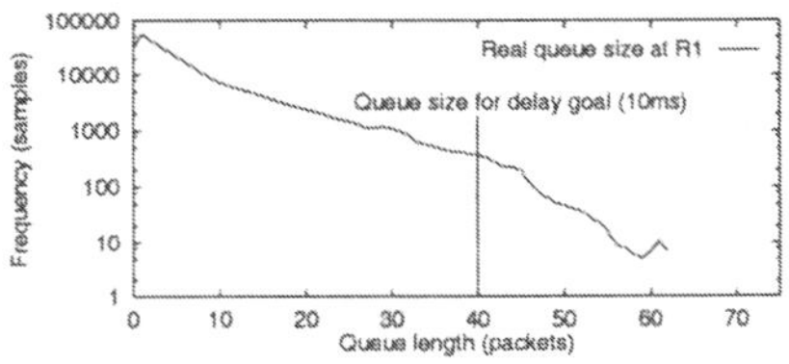

Fig. 13. Queue length at **R1** on the link towards **R2**.

Figure 12 shows the **R1–R2** link as seen from **R1**. We show the total offered rate, the estimated reservation ($\gamma\beta$) and the smoothed actual rates of *request* and *reserved* packets. Figure 13 shows the behaviour of the real queue. The system succeeds in limiting queuing delays to approximately the delay goal of 10 ms, which corresponds to a queue size of 40 packets. The queue limit of 75 packets is never reached.

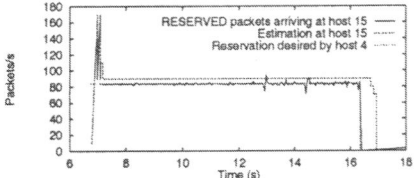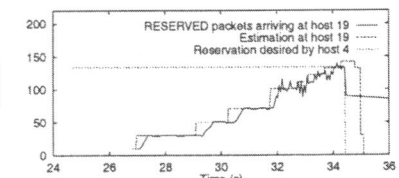

Fig. 14. End-to-end reservation from host **4** to host **15**.

Fig. 15. End-to-end reservation from host **4** to host **19**.

Finally, we examine some end-to-end flows. Figure 14 shows a successful reservation of 84 packets per second from host **4** to **15**. The requested rate, the estimation at the destination, and the (smoothed) rate of *reserved* packets

---

[2] Small random variations were added to link bandwidth and delay to avoid the entire network from being perfectly synchronized.

are shown. Similarly, figure 15 shows the same data for a less successful reservation host 4 attempts later to **19**, at a time when the offered traffic is almost twice a high as the bandwidth available at the bottleneck.[3]

During the entire simulated interval of 50 seconds, 3'368 *request* packets and 164'723 *reserved* packets were sent from **R1** to **R2**. This is 83% of the bandwidth of that link.

5 Conclusion

We have proposed a new scalable resource reservation architecture for the Internet. Our architecture achieves scalability for a large number of concurrent flows by aggregating flows at each link. This aggregation is made possible by delegating certain traffic control decisions to end systems – an idea borrowed from TCP. Reservations are controlled with estimation algorithms, which predict future resource usage based on previously observed traffic. Furthermore, protocol processing is simplified by attaching the reservation control information directly to data packets.

We did not present a conclusive specification but rather described the general concepts, gave examples for implementations of core elements, including the design of estimator algorithms for sources, destinations and routers, and showed some illustrative simulation results. Further work will focus on completing the specification, on evaluating and improving the algorithms described in this paper, and finally on the implementation of a prototype.

References

1. Diot, Christophe; Huitema, Christian; Turletti, Thierry. *Multimedia Applications should be Adaptive*, ftp://www.inria.fr/rodeo/diot/nca-hpcs.ps.gz, HPCS'95 Workshop, August 1995.
2. RFC2205; Braden, Bob (Ed.); Zhang, Lixia; Berson, Steve; Herzog, Shai; Jamin, Sugih. *Resource ReSerVation Protocol (RSVP) – Version 1 Functional Specification*, IETF, September 1997.
3. RFC1819; Delgrossi, Luca; Berger, Louis. *ST2+ Protocol Specification*, IETF, August 1995.
4. Ferrari, Domenico; Banerjea, Anindo; Zhang, Hui. *Network Support for Multimedia - A Discussion of the Tenet Approach*, Computer Networks and ISDN Systems, vol. 26, pp. 1267-1280, 1994.
5. The ATM Forum, Technical Committee. *ATM User-Network Interface (UNI) Signalling Specification, Version 4.0*, ftp://ftp.atmforum.com/pub/approved-specs/af-sig-0061.000.ps, The ATM Forum, July 1996.
6. The ATM Forum, Technical Committee. *ATM Forum Traffic Management Specification, Version 4.0*, ftp://ftp.atmforum.com/pub/approved-specs/af-tm-0056.000.ps, April 1996.

---

[3] In this simulation, sources did not back off if a reservation progressed too slowly.

7. RFC2475; Blake, Steven; Black, David; Carlson, Mark; Davies, Elwyn; Wang, Zheng; Weiss, Walter. *An Architecture for Differentiated Services*, IETF, December 1998.
8. Bernet, Yoram; Yavatkar, Raj; Ford, Peter; Baker, Fred; Zhang, Lixia; Speer, Michael; Braden, Bob; Davie, Bruce. *Integrated Services Operation Over Diffserv Networks* (work in progress), Internet Draft draft-ietf-issll-diffserv-rsvp-02.txt, June, 1999.
9. RFC2474; Nichols, Kathleen; Blake, Steven; Baker, Fred; Black, David. *Definition of the Differentiated Services Field (DS Field) in the IPv4 and IPv6 Headers*, IETF, December 1998.
10. Floyd, Sally; Mahdavi, Jamshid. *TCP-Friendly Unicast Rate-Based Flow Control*, http://www.psc.edu/networking/papers/tcp_friendly.html, Technical note, January 1997.
11. Almesberger, Werner; Ferrari, Tiziana; Le Boudec, Jean-Yves. *SRP: a Scalable Resource Reservation Protocol for the Internet*, Proceedings of IWQoS'98, pp. 107-116, IEEE, May 1998.
12. Ferrari, Tiziana. *QoS Support for Integrated Networks*, http://www.cnaf.infn.it/~ferrari/tesidot.html, Ph.D. thesis, November 1998.

Differentiated Internet Services

Florian Baumgartner, Torsten Braun, Hans Joachim Einsiedler and
Ibrahim Khalil

Institute of Computer Science and Applied Mathematics
University of Berne, CH-3012 Bern, Switzerland,
Tel +41 31 631 8681 / Fax +41 31 631 39 65
http://www.iam.unibe.ch/~rvs/

Abstract. With the grown popularity of the Internet and the increasing use of
business and multimedia applications the users' demand for higher and more predictable quality of service has risen. A first improvement to offer better than best-
effort services was made by the development of the integrated services architecture
and the RSVP protocol. But this approach proved only suitable for smaller IP
networks and not for Internet backbone networks. In order to solve this problem
the concept of differentiated services has been discussed in the IETF, setting up a
working group in 1997. The Differentiated Services Working Group of the IETF has
developed a new concept which is better scalable than the RSVP-based approach.
Differentiated Services are based on service level agreements (SLAs) that are negotiated between users and Internet service providers. With these SLAs users describe
the packets which should be transferred over the Internet with higher priority than
best-effort packets. The SLAs also define parameters such as the desired bandwidth
for these higher priority packets. The implementation of this concept requires additional functionality such as classification, metering, marking, shaping, policing etc.
within routers at the domain boundaries. This paper describes the Differentiated
Service architecture currently being defined by the IETF DiffServ working group
and the required components to implement the DiffServ architecture.

1 Introduction

The Internet, currently based on the best-effort model, delivers only one
type of service. With this model and FIFO queuing deployed in the network,
any non-adaptive sources can take advantage to grab high bandwidth while
depriving others. One can always run multiple web browsers or start multiple
FTP connections and grab substantial amount of bandwidth by exploiting the
best effort model. The Internet is also unable to support real time applications
like audio or video.

Incredible rapid growth of Internet has resulted in massive increases in
demand for network bandwidth performance guarantees to support both existing and new applications. In order to meet these demands, new Quality
of Service (QoS) functionalities need to be introduced to satisfy customer
requirements including efficient handling of both mission critical and bandwidth hungry web applications. QoS, therefore, is needed for various reasons:

- Better control and efficient use of networks resources (e.g. bandwidth).
- Enable users to enjoy multiple levels of service differentiation.
- Special treatment to mission critical applications while letting others to get fair treatment without interfering with mission sensitive traffic.
- Business Communication.
- Virtual Private Networks (VPN) over IP.

1.1 A Pragmatic Approach to QoS

A pragmatic approach to achieve good quality of service (QoS) is an adaptive design of the applications to react to changes of the network characteristics (e.g. congestion). Immediately after detecting a congestion situation the transmission rate may be reduced by increasing the compression ratio or by modifying the A/V coding algorithm. For this purpose functions to monitor quality of service are needed. For example, such functions are provided by the Real-Time Transport Protocol (RTP) [SCFJ96] and the Real-Time Control Protocol (RTCP). A receiver measures the delay and the rate of the packets received. This information is transmitted to the sender via RTCP. With this information the sender can detect if there is congestion in the network and adjust the transmission rate accordingly. This may affect the coding of the audio or video data. If only a low data rate is achieved, a coding algorithm with lower quality has to be chosen. Without adaptation the packet loss would increase, making the transmission completely useless. However, rate adaptation is limited since many applications need a minimum rate to work reasonably.

1.2 Reservation-based Approach

To achieve the QoS objective as mentioned in the earlier section, basically two approaches can be offered in a heterogeneous network like the Internet :

Integrated Service Approach: The Integrated Services Architecture based on the Resource Reservation Setup Protocol (RSVP) is based on absolute network reservation for specific flows. This can be supported in small LANs, where routers can store a small number of flow states. In the backbone, however, it would be extremely difficult, if not impossible, to store millions of flow states even with very powerful processors. Moreover, for short-lived HTTP connections, it is probably not practical to reserve resources in advance.

Differentiated Service (DiffServ): To avoid the scaling problem of RSVP, a differentiated service is provided for an aggregated stream of packets by marking the packets and invoking some differentiation mechanism (e.g. forwarding treatment to treat packets differently) for each marked packet on the nodes along the stream's path. A very general approach of this mechanism is to define a service profile (a contract between a user and the ISP) for each user (or group of users), and to design other mechanisms in the router that favors traffic conforming to those service profiles. These mechanisms might be classification, prioritization and resource allocation to allow the service provider to provision the network for each of the offered classes of service in order to meet the application (user) requirements.

2 DiffServ Basics and Terminology

The idea of differentiated services is based on the aggregation of flows, i.e. reservations have to be made for a set of related flows (e.g. for all flows between two subnets). Furthermore, these reservations are rather static since no dynamic reservations for a single connection are possible. Therefore, one reservation may exist for several, possibly consecutive connections.

IP packets are marked with different priorities by the user (either in an end system or at a router) or by the service provider. According to the different priority classes the routers reserve corresponding shares of resources, in particular bandwidth. This concept enables a service provider to offer different classes of QoS at different costs to his customers.

The differentiated services approach allows customers to set a fixed rate or a relative share of packets which have to be transmitted by the ISP with high priority. The probability of providing the requested quality of service depends essentially on the dimensions and configuration of the network and its links, i.e. whether individual links or routers can be overloaded by high priority data traffic. Though this concept cannot guarantee any QoS parameters as a rule it is more straightforward to be implemented than continuous resource reservations and it offers a better QoS than mere best-effort services.

2.1 Popular Services of the DiffServ Approach

At present, several proposals exist for the realization of differentiated services. Examples are:

Assured and Premium Services: The approach allowing the combination of different services like Premium and Assured Service seems to be very promising. In both approaches absolute bandwidth is allocated for aggregated flows. They are based on packet tagging indicating the service to be provided for a packet. Actually, assured service does not provide absolute bandwidth guarantee but offers soft guarantee with high probability that traffic marked with high priority tagging will be transmitted with high probability.

User Share Differentiation and Olympic Service: An alternative approach called User-Share Differentiation (USD) assigns bandwidth proportionally to aggregated flows in the routers (for example all flows from or to an IP address or a set of addresses). A similar service is provided by the Olympic service. Here, three priority levels are distinguished assigning different fractions of bandwidth to the three priority levels gold, silver and bronze, for example 60% for gold, 30% for silver and 10% for bronze.

2.2 DS byte marking

In differentiated services networks where service differentiation is the main objective, the differentiation mechanisms are triggered by the so-called DS byte (or ToS byte) marking of the IP packet header. Various service differentiation mechanisms (queuing disciplines), as we will study them in section 3, can be invoked dependent on the DS byte marking. Therefore, marking is one of most vital DS boundary enabling component and all DS routers must implement this facility.

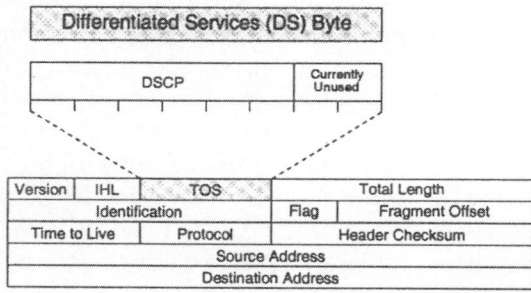

Fig. 1. DS byte in IPv4 [NBBB98]

In the latest proposal for packet marking the the first bit for IN or OUT-of-Profile traffic, the first 6 bits, called Differentiated Services Code point (DSCP), are used to invoke PHBs (see Figure 1). Router implementation should support recommended code point-to-PHB mappings. The default PHB, for example, is 000000. Since the DSCP field has 6 bits, the number of code points that can defined is $2^6 = 64$. This proposal will be the basis of future DiffServ development.

Many existing routers already use IP precedence field to invoke various PHB treatment similar to the fashion of DSCP. To remain compatible, routers can be configured to ignore bit 3,4 and 5. Code point 101000 and 101010 would, therefore, map to the same PHB. Router designers must consider the semantics described above in their implementation and do necessary and appropriate mapping in order to remain compatible with old systems.

2.3 Per Hop Behavior (PHB)

An introduction of PHB has already been given while discussing DS byte marking 2.2. Further [BW98] writes: *"Every PHB is the externally observable forwarding behavior applied at a DS capable node to a stream of packets that have a particular value in the bits of the DS field (DS code point). PHBs can also be grouped when it is necessary to describe the several forwarding behaviors simultaneously with respect to some common constraints."*

However, there is no rigid assignments of PHBs to DSCP bit patterns. These has several reasons:

- There are (or will be) a lot of more PHBs defined, than DSCPs available, making a static mapping impossible.
- The understanding of good choices of PHBs is at the beginning.
- It is desirable to have complete flexibility in the correspondence of PHB values and behaviors.
- Every ISP shall be able to create/map PHBs in his DiffServ domain.

For these reasons there are no static mappings between DS code points and PHBs. The PHBs are enumerated as they become defined and can be mapped to every DSCP within a DiffServ domain. As long as the enumeration space contains a large number of values (2^{32}), there is no danger of running out of space to list the PHB values. This list can be made public for maximum interoperability. Because of this interoperability, mappings between PHBs and DSCPs are proposed, even when every ISP can choose other mappings for the PHBs in his DiffServ domain.

Until now, two PHBs and corresponding DSCPs have been defined.

Table 1. The 12 different AF code points

Drop Precedences	AF Code points			
	Class 1	Class 2	Class 3	Class 4
Low Drop Precedence	001010	010010	011010	100010
Medium Drop Precedence	001100	010100	011100	100100
High Drop Precedence	001110	010110	011110	100110

Assured Forwarding PHB: Based on the current Assured Forwarding PHB (AF) group [HBWW99], a provider can provide four independent AF classes where each class can have one of three drop precedence values. These classes are not aggregated in a DS node and Random Early Detection (RED) [FJ93] is considered to be the preferred discarding mechanism. This required altogether 12 different AF code points as given in table 1.

In a Differentiated Service (DS) Domain each AF class receives a certain amount of bandwidth and buffer space in each DS node. Drop precedence indicates relative importance of the packet within an AF class. During congestion, packets with higher drop precedence values are discarded first to protect packets with lower drop precedence values. By having multiple classes and multiple drop precedences for each class, various levels of forwarding assurances can be offered. For example, Olympic Service can be achieved by mapping three AF classes to it's gold, silver and bronze classes. A low loss, low delay, low jitter service can also be achieved by using AF PHB group if packet arrival rate is known in advance. AF doesn't give any delay related service guarantees. However, it is still possible to say that packets in one AF class have smaller or larger probability of timely delivery than packets in another AF class. The Assured Service can be realized with AF PHBs.

Expedited Forwarding PHB: The forwarding treatment of the Expedited Forwarding (EF) PHB [JNP98] offers to provide higher or equal departure rate than the configurable rate for aggregated traffic. Services which need end-to-end assured bandwidth and low loss, low latency and low low jitter can use EF PHB to meet the desired requirements. One good example is premium service (or virtual leased line) which has such requirements. Various mechanisms like Priority Queuing, Weighted Fair Queuing (WFQ), Class Based Queuing (CBQ) are suggested to implement this PHB since they can preempt other traffic and the queue serving EF packets can be allocated bandwidth equal to the configured rate. The recommended code point for the EF PHB is 101110.

2.4 Service Profile

A service profile expresses an expectation of a service received by a user or group of users or behavior aggregate from an ISP. It is, therefore, a contract between a user and provider and also includes rules and regulations a user is supposed to obey. All these profile parameters are settled in an agreement called Service Level Agreement (SLA). It also contains Traffic Conditioning Agreement (TCA) as a subset, to perform traffic conditioning actions (described in the next subsection) and rules for traffic classification, traffic re-marking, shaping, policing etc. In general, a SLA might include performance parameters like peak rate, burst size, average rate, delay and jitter parameters, drop probability and other throughput characteristics. An Example is:

Service Profile 1: Code point: X, Peak rate= 2Mbps, Burst size=1200 bytes, avg. rate = 1.8 Mbps

Only a static SLA, which usually changes weekly or monthly, is possible with today's router implementation. The profile parameters are set in the router manually to take appropriate action. Dynamic SLAs change frequently and need to be deployed by some automated tool which can renegotiate resources between any two nodes.

2.5   Traffic Conditioner

Traffic conditioners [BBC+98] are required to instantiate services in DS capable routers and to enforce service allocation policies. These conditioners are, in general, composed of one or more of the followings: classifiers, markers, meters, policers, and shapers. When a traffic stream at the input port of a router is classified, it then might have to travel through a meter (used where appropriate) to measure the traffic behavior against a traffic profile which is a subset of SLA. The meter classifies particular packets as IN or OUT-of-profile depending on SLA conformance or violation. Based on the state of the meter further marking, dropping, or shaping action is activated.

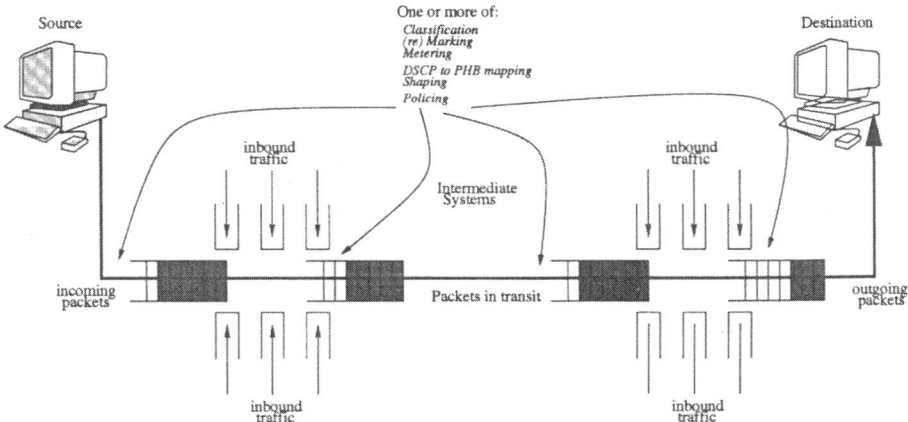

Fig. 2. DS Traffic Conditioning in Enterprise Network (as a set of queues)

Traffic Conditioners can be applied at any congested network node (Figure 2) when the total amount of inbound traffic exceeds the output capacity of the switch (or router). In Figure 2 routers between source and destination are modeled as queues in an enterprise network to show when and where traffic conditioners are needed. For example, routers may buffer traffic (i.e. shape them by delaying) or mark them to be discarded later during medium network congestion, but might require to discard packets (i.e. police traffic) during heavy network congestion when queue buffers fill up. As the number of routers grows in a network, congestion increases due to expanded volume of traffic and hence proper traffic conditioning becomes more important.

Traffic conditioners might not need all four elements. If no traffic profile exists then packets may only pass through a classifier and a marker.

Classifier: Classifiers categorize packets from a traffic stream based on the content of some portion of the packet header. It matches received packets to statically or dynamically allocated service profiles and pass those packets to an element of a traffic conditioner for further processing. Classifiers must be configured by some management procedures in accordance with the appropriate TCA.

Two types of classifiers exist:

BA Classifier: classifies packets based on patterns of DS byte (DS code point) only.

MF classifier: classifies packets based on any combination of DS field, protocol ID, source address, destination address, source port, destination port or even application level protocol information.

Markers: Packet markers set the DS field of a packet to a particular code point, adding the marked packet to a particular DS behavior aggregate. The marker can (i) mark all packets which are mapped to a single code point, or (ii) mark a packet to one of a set of code points to select a PHB in a PHB group, according to the state of a meter.

Meters: After being classified at the input of the boundary router, traffic from each class is typically passed to a meter. The meter is used to measure the rate (temporal properties) at which traffic of each class is being submitted for transmission which is then compared against a traffic profile specified in TCA (negotiated between the DiffServ provider and the DiffServ customer). Based on the the comparison some particular packets are considered conforming to the negotiated profile (IN-profile) or non-conforming (OUT-of-profile). When a meter passes this state information to other conditioning functions, an appropriate action is triggered for each packet which is either IN or OUT-of-profile (see Table 1).

Shapers: Shapers delay some packets in a traffic stream using a token bucket in order to force the stream into compliance with a traffic profile. A shaper usually has a finite-size buffer and packets are discarded if there is not sufficient buffer space to hold the delayed packets. Shapers are generally placed after either type of classifier.For example, shaping for EF traffic at the interior nodes helps to improve end to end performance and also prevents the other classes from being starved by a big EF burst. Only either a policer or a shaper is supposed to appear in the same traffic conditioner.

Policer: When classified packets arrive at the policer it monitors the dynamic behavior of the packets and discard or re-mark some or all of the packets in order to force the stream into compliance (i.e. force them to comply with configured properties like rate and burst size) with a traffic profile. By setting the shaper buffer size to zero (or a few packets) a policer can be implemented as a special case of a shaper. Like shapers policers can also be placed after either type of classifier. Policers, in general, are considered suitable to police traffic between a site and a provider(edge router) and after BA classifiers (backbone router). However, most researchers agree that policing should not be done at the interior nodes since it unavoidably involves flow classification. Policers are usually present in ingress nodes and could be based on simple token bucket filters.

3 Realizing PHBs: The Queuing Components

Since differentiated service is a kind of service discrimination, some traffic need to be handled with priority, some of the traffic needs to be discarded earlier than other traffic, some traffic needs to be serviced faster, and in general, one type of traffic always needs to better than the other. In earlier sections we have discussed about service profile and PHBs. It was made clear that in order to conform to the contracted profile and implement the PHBs, queuing disciplines play a crucial role. The queuing mechanisms typically need to be deployed at the output port of a router.

Since we need different kinds of differentiation under specific situations, the right queuing component (i.e PHB) needs to be invoked by the use of a particular code point. In this section, therefore, we will describe some of the most promising mechanisms which have already been or deserve to be considered for implementation in varieties of DS routers.

3.1 Absolute Priority Queuing

In absolute priority queuing (Figure 3), the scheduler gives higher-priority queues absolute preferential treatment over lower priority queues. Therefore, the highest priority queue receives the fastest service, and the lowest priority queue experiences slowest service among the queues.

The basic working mechanism is as follows: the scheduler would always scan the priority queues from highest to lowest to find the highest priority packet and then transmit it. When that packet has been completely served, the scheduler would start scanning again. If any of the queues overflows, packets are dropped and an indication is sent to the sender.

While this queuing mechanism is useful for mission critical traffic (since this kind of traffic is very delay sensitive) this would definitely starve the lower priority packets of the needed bandwidth.

3.2 WFQ

WFQ [Kes91](Figure 4)is a discipline that assigns a queue for each flow. A weight can be assigned to each queue to give a different proportion of the network capacity. As a result, WFQ can provide protection against other flows.

WFQ can be configured to give low-volume traffic flows preferential treatment to reduce response time and fairly share the remaining bandwidth between high volume traffic flows. With this approach bandwidth hungry flows are prevented from consuming much of network resources while depriving other smaller flows.

WFQ does the job of dynamic configuration since it adapts automatically to the changing network conditions. TCP congestion control and slow-start

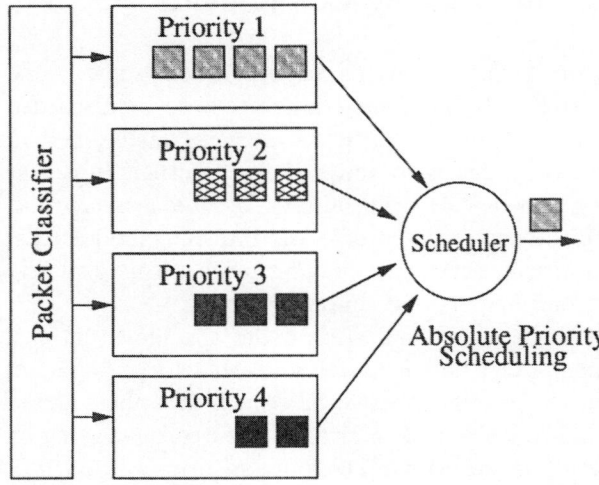

Fig. 3. Absolute Priority Queuing. The queue with the highest priority is served at first

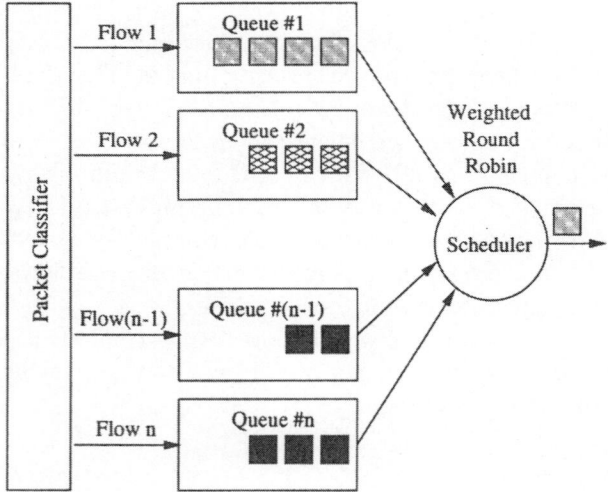

Fig. 4. Weighted Fair Queuing (WFQ)

features are also enhanced by WFQ, resulting in predictable throughput and response time for each active flow.

The weighted aspect can be related to values in the DS byte of the IP header. A flow can be allocated more access to queue resources if it has a higher precedence value.

3.3 Class Based Queuing (CBQ)

In an environment where bandwidth must be shared proportionally between users, CBQ [FJ95] (Figure 6) provides a very flexible and efficient approach to first classifying user traffic and then assigning a specified amount of resources to each class of packets and serving those queues in a round robin fashion.

A class can be an individual flow or aggregation of flows representing different applications, users, departments, or servers. Each CBQ traffic class has a bandwidth allocation and a priority. In CBQ, a hierarchy of classes (Figure 5) is constructed for link sharing between organizations, protocol families, and traffic types. Different links in the network will have different link-sharing structures. The link sharing goals are:

- Each interior or leaf class should receive roughly its allocated link-sharing bandwidth over appropriate time intervals, given the sufficient demand.
- If all leaf and interior classes with sufficient demand have received at least their allocated link-sharing bandwidth, the distribution of any excess bandwidth should not be arbitrary, but should follow some set of reasonable guidelines.

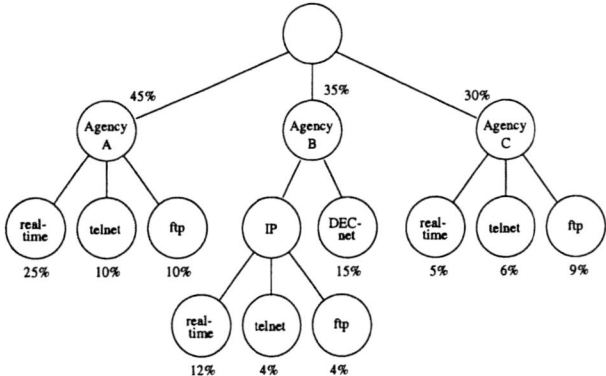

Fig. 5. Hierarchical Link-Sharing

The granular level of control in CBQ can be used to manage the allocation of IP access bandwidth across the departments of an enterprise, to provision bandwidth to the individual tenants of a multi-tenant facility.

Other than the classifier that assigns arriving packets to an appropriate class, there are three other main components that are needed in this CBQ mechanism: scheduler, rate-limiter (delayer) and estimator.

Scheduler: In a CBQ implementation, the packet scheduler can be implemented with either a packet-by-packet round robin (PRR) or weighted round robin (WRR) scheduler. By using priority scheduling the scheduler uses priorities, first scheduling packets from the highest priority level. Round-robin scheduling is used to arbitrate between traffic classes within the same priority level. In weighted round robin scheduling the scheduler uses weights proportional to a traffic class's bandwidth allocation. This weight finally allocates the number of bytes a traffic class is allowed to

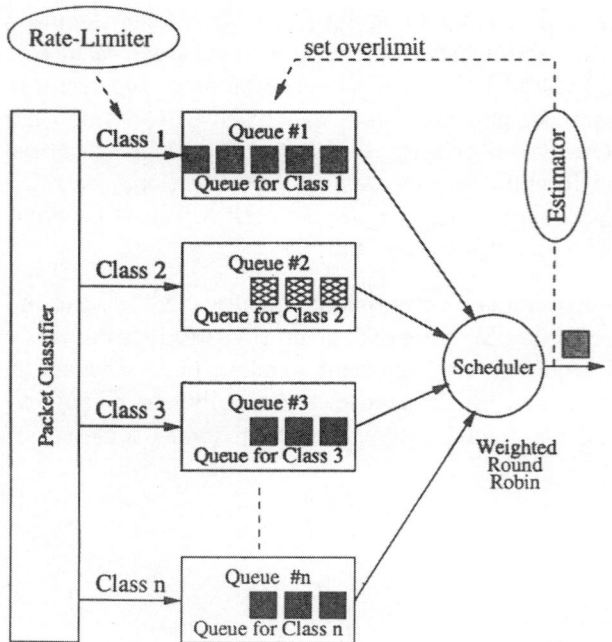

Fig. 6. Class Based Queuing: Main Components

send during a round of the scheduler. Each class at each round gets to send its weighted share in bytes, including finishing sending the current packet. That class's weighted share for the next round is decremented by the appropriate number of bytes. When a packet to be transmitted by a WRR traffic class is larger than the traffic class's weight but that class is underlimit[1], the packet is still sent, allowing the traffic class to borrow ahead from its weighted allotment for future rounds of the round-robin.

Rate-Limiter: If a traffic class is overlimit[2] and is unable to borrow from it's parent classes, the scheduler starts the overlimit action which might include simply dropping arriving packets for such a class or rate-limit overlimit classes to their allocated bandwidth. The rate-limiter computes the next time that an overlimit class is allowed to send traffic. Unless this future time has arrived, this class will not be allowed to send another packet until .

Estimator: The estimator estimates the bandwidth used by each traffic class over the appropriate time interval and determines whether each class is over or under its allocated bandwidth.

---

[1] If a class has used less than a specified fraction of its link sharing bandwidth (in bytes/sec, as averaged over a specified time interval)

[2] If a class has recently used more than its allocated link sharing bandwidth (in bytes/sec, as averaged over a specified time interval)

3.4 Random Early Detection (RED)

Random Early Detection (RED) [FJ93] is designed to avoid congestion by monitoring traffic load at points in the network and stochastically discarding packets when congestion starts increasing. By dropping some packets early rather than waiting until the buffer is full, RED keeps the average queue size low and avoids dropping large numbers of packets at once to minimize the chances of global synchronization.Thus, RED reduces the chances of tail drop and allows the transmission line to be used fully at all times. This approach has certain advantages:

- bursts can be handled better, as always a certain queue capacity can be reserved for incoming packets.
- by the lower average queue length real-time applications are better supported.

The working mechanism of RED is quite simple. It has two thresholds, minimum threshold $X1$ and a maximum threshold $X2$ for packet discarding or admission decision which is done by a dropper. Referring to Figure 7, when a packet arrives at the queue, the average queue (av_queue) is computed. If, $av\_queue < X1$, the packet is admitted to the queue; if $av\_queue \geq X2$, the packet is dropped. In the case, when the average queue size falls between the thresholds $X1 < av\_queue < X2$, the arriving packet is either dropped or queued, mathematically saying, it is dropped with linearly increasing probability.

When congestion occurs, the probability that the RED notifies a particular connection to reduce its window size is approximately proportional to that connection's share of the bandwidth. The RED congestion control mechanism monitors the average queue size for each output queue and using randomization choose connections to notify of that congestion.

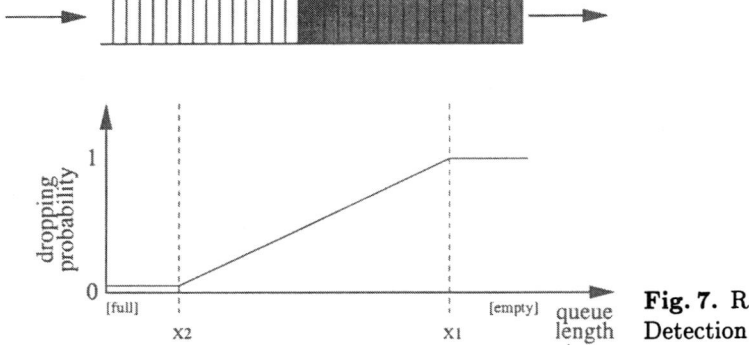

Fig. 7. Random Early Detection

It is very useful to the network since it has the ability to flexibly specify traffic handling policies to maximize throughput under congestion conditions.

RED is especially able to split bandwidth between TCP data flows in a fair way as lost packets automatically cause a reduction to a TCP data flow's packet rate. More problematic is the situation if non TCP conforming data flows (e.g. UDP based real-time or multicast applications) are involved. Flows not reacting to packet loss have to be handled by reducing their data rate specially to avoid an overloading of the network.

In general, RED statistically drops more packets from large users than from small ones. Therefore, traffic sources that generate the most traffic are more likely to be slowed down than traffic sources that generate little traffic.

3.5   RED with In and Out (RIO)

The queuing algorithm proposed for assured service RIO (RED with In and Out) [CW97] is an extension of the RED mechanism. This procedure shall make sure, that during overload primarily packets with high drop precedence (e.g. best-effort instead of assured service packets) are dropped. A data flow can consist of packets with various drop precedences, which can arrive at a common output queue. So changes to the packet order can be avoided affecting positively the TCP performance.

For in and out-of-profile packets a common queue using different dropping techniques for the different packet types is provided. The dropper for out of profile packets discards packets much earlier (e.g. a lower queue length) than the dropper for in profile packets. Further more the dropping probability for out of profile packets increases more than the probability for in packets. So, it shall be achieved that the probability for dropping in profile packets is kept very low. While the out-dropper used the number of all packets in the queue for the calculation of his probability, the in-dropper only uses the number of in profile packets (see figure 8). Using the same queue both types of packets will have the same delay. This might be a disadvantage of this concept. By dropping all out-of-profile packets at a quite small queue length this effect can be reduced but not eliminated.

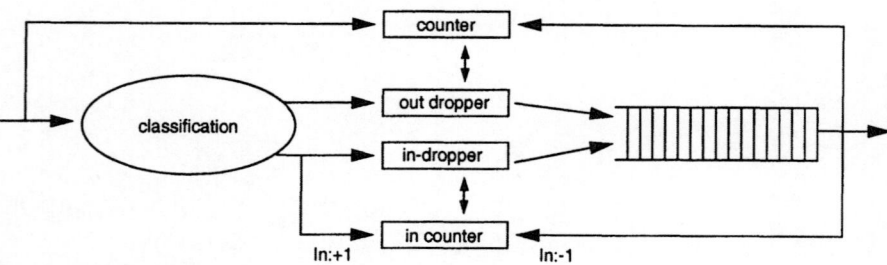

Fig. 8. RIO-Queuing

4 Differentiated Services in End-to-End Scenarios

4.1 Premium Service and Expedited Forwarding

With Premium Service the user negotiates with his ISP a maximum bandwidth for sending packets through the ISP network. Furthermore, the aggregated flow is described by the packets' source and destination addresses or address prefixes. In Figure 9 users and ISPs have agreed on a rate of three packets/s for traffic from A to B. The user configures the first-hop router in the individual subnet accordingly. In the example above a packet rate of two packets/s is allowed in every first-hop router as it can be expected that no two end systems will use the full bandwidth of two packets/s at the same time.

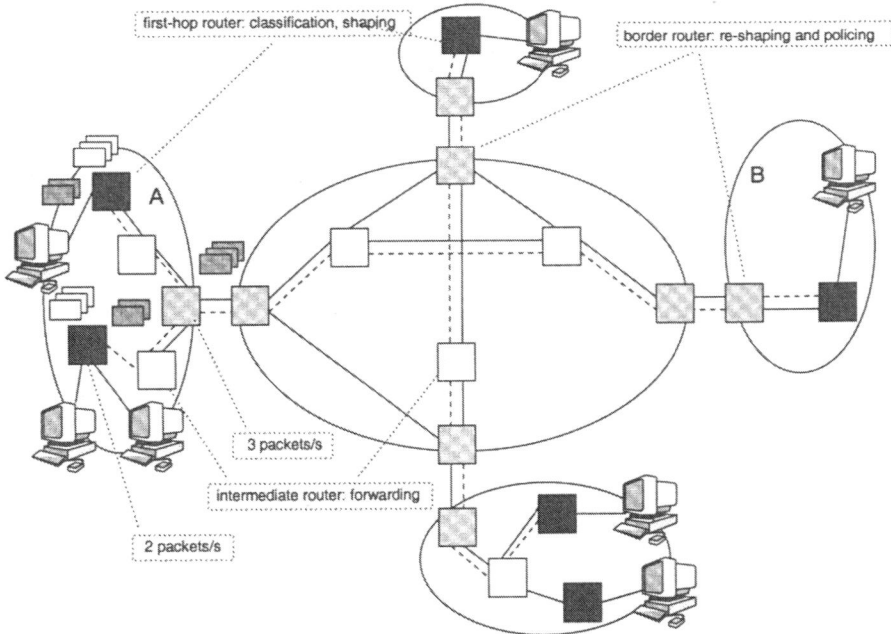

Fig. 9. Premium Service

First-hop routers have the task to classify the packets received from the end systems, i.e. to analyze if the Premium Service shall be provided to the packets or not. If yes, the packets are tagged as Premium Service and the data stream is shaped according to the maximum bandwidth. The user's border router re-shapes the stream (e.g. three packets per second) and transmits the packets to the ISP's border router, which performs policing functions, i.e. it checks whether the user's border router remains below the negotiated bandwidth of three packets/s. If each of the two first-hop routers allows two packets/s, one packet per second will be dropped by shaping or policing at the border routers. All first-hop and border-routers own two queues, one for EF-packets and one for all other (see Figure 9). If the EF-queue contains packets these are transmitted prior to others. The implementation of two queues in every router of the network (ISP and user network) equals to the realization of a virtual network for Premium Service traffic.

Premium Service offers a service corresponding to a private leased line, with the advantage of making free network capacities available to other tasks, resulting in lower fees for the users.

4.2 Assured Service

A potential disadvantage of Premium Service is the weak support for bursts and the fact that a user has to pay even if he is not using the whole bandwidth. The Assured Service tries to offer a service which cannot guarantee bandwidth but provides a high probability that the ISP transfers high-priority-tagged packets reliably. The definition of concrete services has not yet happened, but it is obvious to offer services similar to the IntServ controlled load service. The probability for packets to be transported reliably depends on the network capacity. An ISP may choose the sum of all bandwidths for Assured Service to remain below the bandwidth of the weakest link. In this case, only a small portion of the available capacity may be allocated in the ISP network. An advantage of the Assured Service is that users do not have to establish a reservation for a relative long time. With ISDN or ATM, users might be unable to use the reserved bandwidth because of the burstiness of their traffic, whereas Assured Service allows the transmission of short time bursts.

With the Assured Service the user negotiates a service profile with his service provider, e.g. the maximum amount or rate of high priority, i.e. Assured Service, packets. The user may then tag his packets as high priority within the end system or the first-hop router, i.e. assign them a tag for assured forwarding (AF) (see Figure 10). To avoid modifications in the end systems the first-hop router may analyze the packets with respect to their IP addresses and UDP-/TCP-Port and then assign them the according priority, i.e. set the AF-DSCP for conforming Assured Service packets. The maximum rate of high-priority (AF-DSCP) packets must not be exceeded. This is done by (re-)classification in the first-hop routers and in the user's border routers at the border to the ISP network. Nevertheless, the service provider has to check if the user remains below the maximum rate for high priority packets and apply corrective actions such as policing if necessary.

For example, the border router at the network entrance will tag the non-conforming packet as low priority (out of service, out of profile). An alternative would be to charge higher fees for non-conforming packets by the ISP. The tagging of low and high priority packets is done by use of the DS byte.

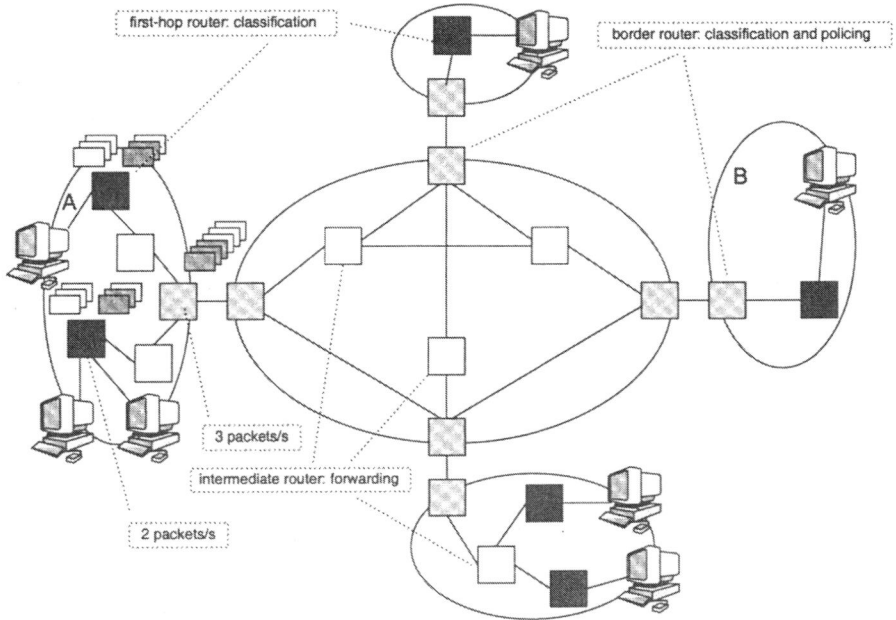

Fig. 10. Assured Service

Bursts are supported by making buffer capacity available for buffering bursty traffic. Inside the network, especially in backbone networks bursts can be expected to be compensated statistically.

4.3 Traffic Conditioning for Assured and Premium Service

The implementation of Assured and Premium Service requires several modifications of the routers. Mainly classification, shaping, and policing functions have to be performed to the router. These functions are necessary at the border between two networks, for example at the transition of the customer network to the ISP or between the ISPs. Service profiles have to be negotiated between the ISPs similar to the transition to the user.

First-hop router Figure 11 shows the first-hop router function for Premium and Assured Service. Received packets are classified and according to this the AF or EF-DSCP is set if the packet should be supported with Assured or Premium Service. As a parameter for the classification, source and destination addresses or information of higher protocols (e.g. port numbers) may be used. There are separate queues for each AF class, for EF and best effort traffic. So, a pure best-effort packet will be forwarded directly to the best-effort RED queue and the Assured Service packets get to their RED queues. The Assured Service packets are checked whether they conform to the service profile. The drop precedence will only be kept unchanged if the Assured Service bucket contains a token. Otherwise the drop precedence will be increased. The RED-based queuing shall guarantee that AF packets with higher drop precedence are dropped prior to AF packets with lower drop precedence, if the capacity is exceeded.

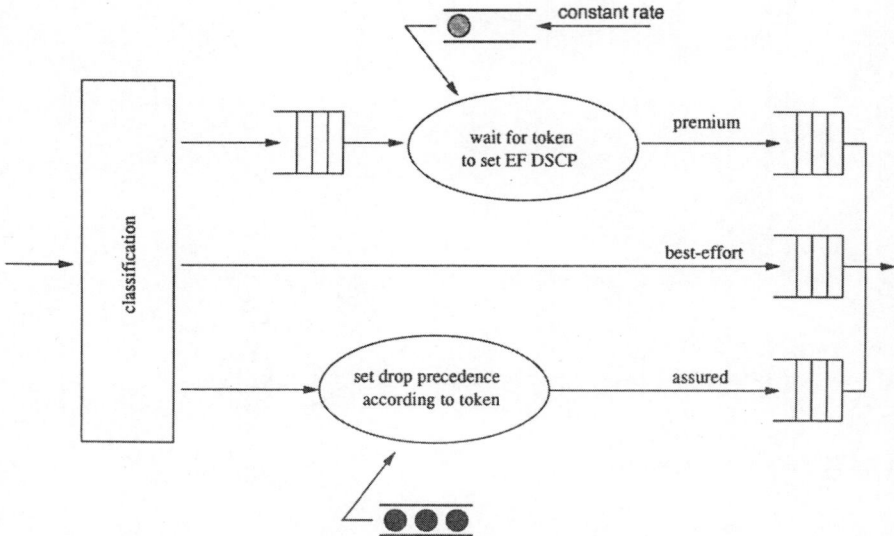

Fig. 11. First-hop router for Premium, Assured and best effort services

Border router Similar to the first-hop router an intermediate router has to perform shaping functions in order to guarantee that not more than the allowed packet rate is transmitted to the ISP. This is important since the ISP will check whether the user remains within the negotiated service profile. The border router in Figure 12 will therefore drop non conforming Premium service packets and increase the drop priority of non conforming Assured Service packets. Packets within an AF class but with different precedence values share the same queue since both types of packets may belong to the same source. A common queue avoids re-ordering of packets. This is especially important for TCP performance reasons.

First-Hop and Egress Border Routers Figure 13 shows the working principle of a first hop and an egress router for assured service. An egress border router is the border router, at which the packets are leaving the differentiated service domain. Received packets are classified and the AF DSCP is set, if assured service should be given to the packet. Source and destination addresses and information of higher protocols (e.g. port numbers) may

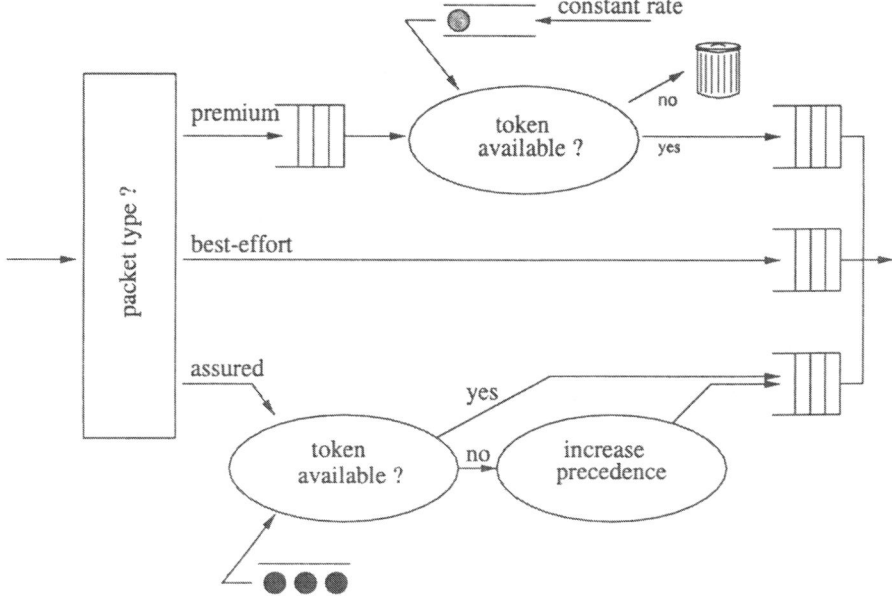

Fig. 12. Policing in a border router

be used as classification parameters. A pure best effort packet will directly be pushed to the output queue.

The AF-DSCP is set according to the availability of a token and then written to the AF output queue. Normal best effort traffic is directly pushed to the best effort queue.

The token buckets are configured according to the SLAs consisting of bit rates and the burst parameter. The bucket may be capable of keeping several tokens to support short time bursts. The bucket's depth depends on the arranged burst properties.

The difference between a first hop and an egress border router is the fact, that at the first hop router a packet is classified for the first time for this task information of higher protocols (TCP ports, type of the application) may be used, whereas the egress border router is capable of changing the drop precedence to meet the negotiated service profile.

Ingress Border Router The ISP has to ensure that the user meets the negotiated traffic characteristics. To achieve this, the ISP has to check in his ingress border router, which transmits the packets into his DS domain whether the user keeps the SLA. So the ingress border router of Figure 14 will change the drop precedence of non conforming packets.

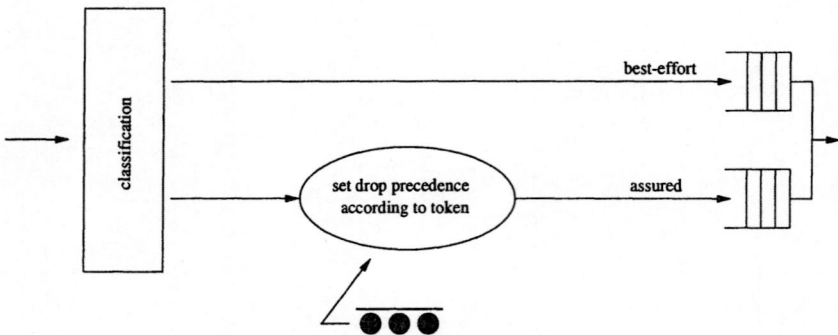

Fig. 13. First hop and egress border router for Assured Service

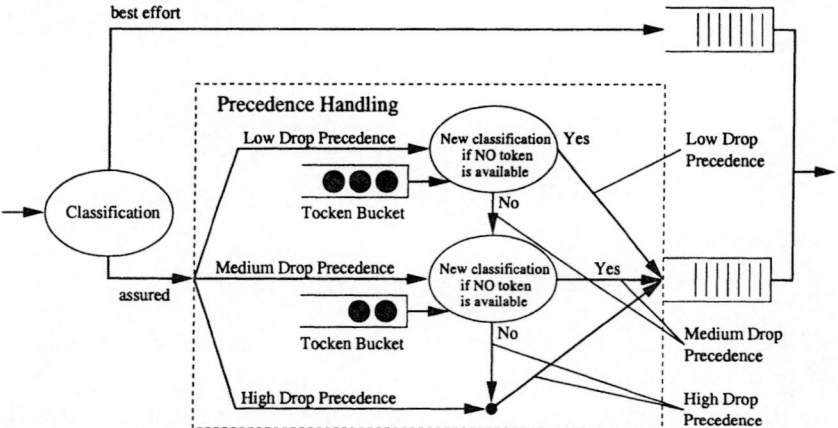

Fig. 14. Ingress border router with three drop precedences for Assured Service

4.4 User-Share Differentiation

Based upon packet tagging Premium and Assured Service models can fulfill the stipulated service parameters like bit rates with a high degree of probability only if the ISP network is dimensioned appropriately and non best-effort traffic is transmitted between certain known networks only.

If for instance two users have contracted a bit rate of 1 Mbps for Assured Service packets with an ISP and both wish to receive data simultaneously at a rate of 1 Mbps each from a WWW server which is connected to the network with a 1.5 Mbps link, the requested quality of service cannot be provided.

The User-Share Differentiation approach [Wan97] avoids this problem by contracting not absolute bandwidth parameters but relative bandwidth shares. A user will be guaranteed only a certain relative amount of the available bandwidth in an ISP network. In practice, the size of this share will be in direct relation to the charged costs.

In Figure 15, user A has allocated only half of the bandwidth of user B and one third of the bandwidth of user C. If A and B access the network on

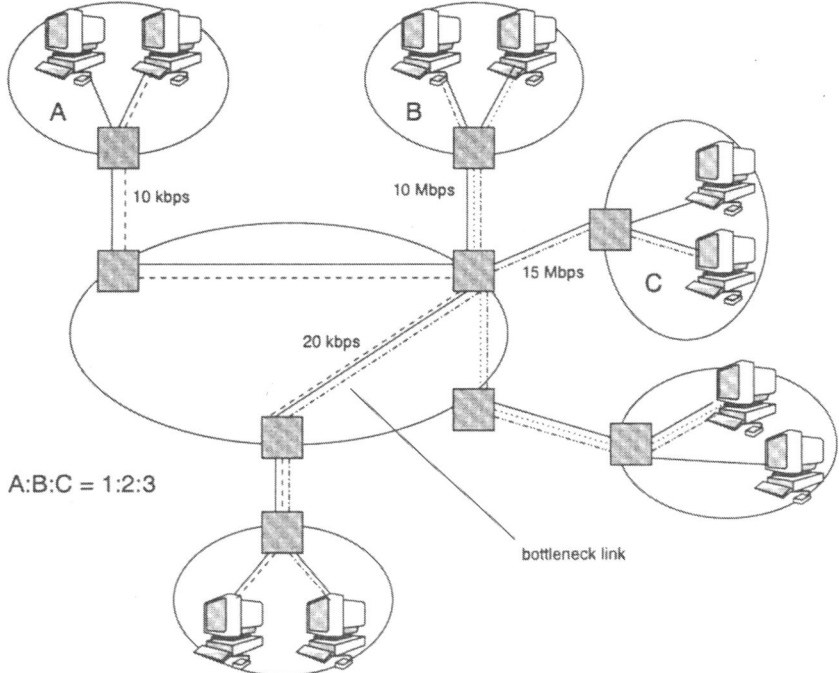

Fig. 15. User Share Differentiation (USD)

low bandwidth links with a capacity of 30 kbps at the same time, e.g. user B will receive a bandwidth of 20 kbps but user C will get merely 10 kbps. If B and C access the same or possibly a different network via a common high bandwidth link with a capacity of 25 Mbps, B will receive 10 Mbps and C only 15 Mbps.

Simpler router configuration is an important advantage of the USD approach. However, absolute bandwidth guarantees cannot supported. An additional drawback is that not only edge routers must be configured (as in the case of Premium or Assured Service) but also interior routers must be configured with the bandwidth shares.

5 Conclusion and Outlook

Standardization of Differentiated services is still under discussion. So far most of discussions have been centered around RED and Assured Service. Virtual Leased Line (or Premium Service) and it's implementations by EF PHB has been recently been discussed in [JNP98] which would require implementation of Priority Queuing, WFQ, CBQ etc. It is not clear where the policing and shaping should take place. Although, both AF and EF PHBs have been proposed, interaction between these two is a debatable issue.

RED and it's variants are complimentary to different scheduling algorithms, and fit very nicely with CBQ. RED is designed to keep queue sizes small (smaller than their maximum in a given implementation), and thus avoid tail drop and global TCP resynchronization. It is, therefore, expected that in router implementation all these service discipline need to coexist and some of those be complementary to each other. Nevertheless, new proposals for both AF and EF PHB strongly suggests that Class Based Queuing (CBQ), WFQ, and their variants will play stronger roles in the implementation of DiffServ.

Regarding interaction between the PHBs the EF draft says that other PHBs can coexist at the same DS node given that the requirements of AF classes are not violated. These requirements include timely forwarding which is at the heart of EF. On the other end, the AF PHB group distinguishes between the classes based on timely forwarding. The AF draft also says that "any other PHB groups may coexist with the AF group within the same DS domain provided that the other PHB groups do not preempt the resources allocated to the AF classes". The question here is: If they coexist should EF have more timely forwarding than the highest timely forwarded AF class by preempting any AF class as the EF document basically states?

What is needed here is EF must leave AF whatever has been allocated for AF. This would mean EF can actually preempt forwarding resources for AF. For example, one could take a 1.5 Mbps link and allow for 64 Kbps of it to be available to EF, with the remaining capacity available to AF. One could also state that EF has absolute priority over AF (up to the 64 Kbps allocated). In this case, EF would preempt AF (so long as it conforms to the 64 Kbps limit) and AF would always be assured that it has 1.5 Mbps - 64 Kbps of the link throughput.

There are lot more issues which are debatable and need attention for further research. However, we should always keep in mind that the whole point of DiffServ is to allow service providers to implement QoS pricing strategies in the first place.

References

[BBC+98] S. Blake, D. Black, M. Carlson, E. Davies, Z. Wang, and W. Weis. An architecture for differentiated services. Request for Comments 2475, December 1998.

[BW98]   Marty Borden and Christoph White. Management of phbs. Internet Draft `draft-ietf-diffserv-phb-mgmt-00.txt`, August 1998. work in progress.

[CW97]   D. Clark and J. Wroclawski. An approach to service allocation in the internet, work in progress. Internet Draft `draft-clark-diff-svc-alloc-00.txt`, Juli 1997. work in progress.

[FJ93]   Sally Floyd and Van Jacobson. Random early detection gateways for congestion avoidance. *IEEE/ACM Transactions on Networking*, August 1993.

[FJ95]   Sally Floyd and Van Jacobson. Link-sharing and resource management
            models for packet networks. *IEEE/ACM Transactions on Networking*,
            3(4), August 1995.

[HBWW99]   Juha Heinanen, Fred Baker, Walter Weiss, and John Wro-
            clawski.   Assured forwarding phb group.   Internet Draft
            `draft-ietf-diffserv-af-06.txt`, February 1999. work in progress.

[JNP98]   Van Jacobson, K. Nichols, and K. Poduri. An expedited forwarding
            phb. Internet Draft `draft-ietf-diffserv-af-02.txt`, October 1998.
            work in progress.

[Kes91]   S. Keshav. *Congestion Control in Computer Networks*. PhD thesis,
            Berkeley, September 1991.

[NBBB98]   K. Nichols, S. Blake, F. Baker, and D. Black. Definition of the differ-
            entiated services field (ds field) in the ipv4 and ipv6 headers. Request
            for Comments 2474, December 1998.

[SCFJ96]   H. Schulzrinne, S. Casner, R. Frederick, and V. Jacobson. Rtp: A
            transport protocol for real-time applications. Request for Comments
            1889, January 1996.

[Wan97]   Z. Wang.   User-share differentiation (usd) scalable band-
            width allocation for differentiated services.   Internet Draft
            `draft-wang-diff-serv-usd-00.txt`, November 1997.   work in
            progress.

A Portable Subroutine Library for Solving Linear Control Problems on Distributed Memory Computers[*]

Peter Benner[1], Enrique S. Quintana-Ortí[2], and Gregorio Quintana-Ortí[3]

[1] Zentrum für Technomathematik, Fachbereich 3 – Mathematik und Informatik, Universität Bremen, D–28334 Bremen, Germany; benner@math.uni-bremen.de.
[2] Departamento de Informática, Universidad Jaime I, 12080 Castellón, Spain; quintana@inf.uji.es.
[3] Same address as second author; gquintan@inf.uji.es.

Abstract. This paper describes the design of a software library for solving the basic computational problems that arise in analysis and synthesis of linear control systems. The library is intended for use in high performance computing environments based on parallel distributed memory architectures. The portability of the library is ensured by using the BLACS, PBLAS, and ScaLAPACK as the basic layer of communication and computational routines. Preliminary numerical results demonstrate the performance of the developed codes on parallel computers.

1 Introduction

In recent years, many new and reliable numerical methods have been developed for analysis and synthesis of moderate size linear time-invariant (LTI) systems. In generalized state-space form, such systems are described by the following models.

Continuous-time LTI system:

$$E\dot{x}(t) = Ax(t) + Bu(t), \qquad t > 0, \qquad x(0) = x^{(0)},$$
$$y(t) = Cx(t), \qquad t \geq 0. \tag{1}$$

Discrete-time LTI system:

$$Ex_{k+1} = Ax_k + Bu_k, \qquad k = 0, 1, 2, \ldots, \qquad x_0 = x^{(0)},$$
$$y_k = Cx_k, \qquad k = 0, 1, 2, \ldots. \tag{2}$$

In both cases, $A, E \in \mathbb{R}^{n \times n}$, $B \in \mathbb{R}^{n \times m}$, and $C \in \mathbb{R}^{p \times n}$. Here we assume that E is nonsingular. Descriptor systems with singular E also lead —after

---

[*] Partially supported by the DAAD programme *Acciones Integradas Hispano-Alemanas*. Enrique S. Quintana-Ortí and Gregorio Quintana-Ortí were also supported by the Spanish CICYT Project TIC96-1062-C03-03.

appropriate transformations— to the above problem formulation with a non-singular (and usually, diagonal or triangular) matrix E; see, e.g., [42,55].

The traditional approach to design a regulator for the above LTI systems involves the minimization of a cost functional of the form

$$J_c(x_0, u) \;=\; \frac{1}{2} \int_0^\infty \left(y(t)^T Q y(t) + 2 y(t)^T S u(t) + u(t)^T R u(t) \right) dt \qquad (3)$$

in the continuous-time case, and

$$J_d(x_0, u) \;=\; \frac{1}{2} \sum_{k=0}^\infty \left(y_k^T Q y_k + 2 y_k^T S u_k + u_k^T R u_k \right) \qquad (4)$$

in the discrete-time case. The matrices $Q \in \mathbb{R}^{p \times p}$, $S \in \mathbb{R}^{p \times m}$, and $R \in \mathbb{R}^{m \times m}$ are chosen in order to weight inputs and outputs of the system. The *linear-quadratic regulator problem* consists in minimizing (3) or (4) subject to the dynamics (1) or (2), respectively. It is well-known that the solution of this optimal control problem is given by the closed-loop control

$$u^*(t) = -R^{-1}(B^T X_c E + S^T C) x(t) =: F_c x(t), \qquad t \geq 0, \qquad (5)$$

in the continuous-time case, and

$$u_k^* = -(R + B^T X_d B)^{-1}(B^T X_d A + S^T C) x_k =: F_d x_k, \qquad k = 0, 1, 2, \ldots, \quad (6)$$

in the discrete-time case. See, e.g., [1,42,38] for details and further references. The matrices X_c and X_d in (5) and (6) denote particular solutions of the *(generalized) continuous-time algebraic Riccati equation* (CARE)

$$\begin{aligned} 0 \;=\; \mathcal{R}_c(X) &:= C^T Q C + A^T X E + E^T X A - \\ &\quad - (E^T X B + C^T S) R^{-1} (B^T X E + S^T C), \end{aligned} \qquad (7)$$

and the *(generalized) discrete-time algebraic Riccati equation* (DARE)

$$\begin{aligned} 0 \;=\; \mathcal{R}_d(X) &:= C^T Q C + A^T X A - E^T X E - \\ &\quad - (A^T X B + C^T S)(R + B^T X B)^{-1}(B^T X A + S^T C). \end{aligned} \qquad (8)$$

The optimal control in (5) and (6) is obtained from the *stabilizing* solutions of (7) and (8). That is, we need to compute X_c and X_d such that the resulting *closed-loop* matrices

$$A_c := E^{-1}(A + B F_c), \qquad F_c := -R^{-1}(B^T X_c E + S^T C), \qquad (9)$$

and

$$A_d := E^{-1}(A + B F_d), \qquad F_d := -(R + B^T X_d B)^{-1}(B^T X_d A + S^T C) \quad (10)$$

have stable spectra. In the continuous-time case this means that A_c has all its eigenvalues in the open left half plane while in the discrete-time case all the eigenvalues of A_d are of modulus less than one. We will call matrices (matrix pencils) with all eigenvalues in the open left half plane *c-stable* and those with spectra inside the open unit disk will be called *d-stable*. The matrices F_c and F_d are called the *optimal feedback gain matrices*. Under standard assumptions on LTI systems and the weighting matrices in the cost functionals $\mathcal{J}_c$ and $\mathcal{J}_d$, the stabilizing solutions of the CARE and DARE exist, are unique, and symmetric. See [38] for a detailed account on conditions for existence and uniqueness of solutions to (7) and (8).

The algebraic Riccati equations in (7) and (8) can be formulated in the more general forms

$$0 \; = \; \mathcal{R}_c(X) \; = \; \tilde{Q} + \tilde{A}^T X \tilde{E} + \tilde{E}^T X \tilde{A} - X \tilde{G} X \tag{11}$$

for the CARE, and

$$\begin{aligned} 0 \; = \; \mathcal{R}_d(X) = \tilde{Q} + \tilde{A}^T X \tilde{A} - \tilde{E}^T X \tilde{E} - \\ - \; (\tilde{A}^T X \tilde{B} + \tilde{S})(\tilde{R} + \tilde{B}^T X \tilde{B})^{-1}(\tilde{B}^T X \tilde{A} + \tilde{S}^T) \end{aligned} \tag{12}$$

for the DARE. Written in this form, they also include the algebraic Riccati equations arising in many areas of modern control theory like robust control, H_2- and H_∞-control, model reduction, etc.; see, e.g., [27,47,48,56]. The algorithms used here for the numerical solution of equations of the forms (11) and (12) do not depend on the particular form given in (7) and (8) and hence can be used to solve any algebraic Riccati equations given as in (11) and (12). Throughout this paper we will assume that stabilizing solutions of the CARE (11) and the DARE (12) exist and hence that the closed-loop matrices A_c and A_d defined above are c–stable and d–stable, respectively.

In the course of solving the above nonlinear systems of equations via Newton's method and in many other analysis and synthesis problems for LTI control problems, linear matrix equations of the form

$$\hat{A} X \hat{B} + \hat{C} X \hat{D} + \hat{E} = 0 \tag{13}$$

have to be solved. Here $\hat{A}, \hat{C} \in \mathbb{R}^{n \times n}$, $\hat{B}, \hat{D} \in \mathbb{R}^{m \times m}$, and $\hat{E}, X \in \mathbb{R}^{n \times m}$. Linear systems of equations as in (13) are called *generalized Sylvester equations*. Some particular instances of (13) are given below:

$$\hat{A} X + X \hat{D} + \hat{E} = 0, \qquad \textit{(Sylvester equation)} \tag{14}$$
$$\hat{A} X \hat{B} - X + \hat{E} = 0, \qquad \textit{(``discrete'' Sylvester equation)} \tag{15}$$

and for $\hat{E} = \hat{E}^T$,

$$\hat{A} X + X \hat{A}^T + \hat{E} = 0, \qquad \textit{(Lyapunov equation)} \tag{16}$$
$$\hat{A} X \hat{C}^T + \hat{C} X \hat{A}^T + \hat{E} = 0, \qquad \textit{(generalized Lyapunov equation)} \tag{17}$$
$$\hat{A} X \hat{A}^T - X + \hat{E} = 0, \qquad \textit{(Stein equation)} \tag{18}$$
$$\hat{A} X \hat{A}^T - \hat{C} X \hat{C}^T + \hat{E} = 0, \qquad \textit{(generalized Stein equation)} \tag{19}$$

Stein equations are often also referred to as *discrete Lyapunov equations*.

In addition to the above, we will consider special cases of (generalized) Lyapunov and Stein equations where $\hat{E}$ is semidefinite and factored as $\hat{E} = \pm\hat{E}_1\hat{E}_1^T$. In this case, if $\hat{A} - \lambda\hat{C}$ is a stable matrix pencil (the generalized eigenvalues of the matrix pencil are stable), then the solution of the corresponding Lyapunov or Stein equation is also semidefinite and can be factored as $X = \pm X_1 X_1^T$. This is the case, e.g., when computing the *controllability Gramian W_c* and *observability Gramian W_o* of a continuous-time LTI system via the Lyapunov equations

$$AW_cE^T + EW_cA^T + BB^T = 0, \tag{20}$$

$$A^TW_oE + E^TW_oA + C^TC = 0. \tag{21}$$

In the discrete-time case these Gramians are given by the corresponding Stein equations

$$AW_cA^T + EW_cE^T + BB^T = 0, \tag{22}$$

$$A^TW_oA - E^TW_oE + C^TC = 0. \tag{23}$$

The Gramians of LTI systems play a fundamental role in many analysis and design problems of LTI systems as computing balanced, minimal, or partial realizations, the Hankel singular values and Hankel norm of LTI systems, and model reduction. Often, the *Cholesky factors X_1* of the solutions to the above equations are needed. Hence, special algorithms are designed to compute these factors without ever forming the solution matrix explicitly.

We consider special algorithms for all the above equations. The subroutines resulting from implementing these algorithms will be used in order to tackle some computational problems for LTI systems:

C1 stabilize an LTI system, i.e., find $F \in \mathbb{R}^{m \times n}$ such that $E - \lambda(A + BF)$ is a stable matrix pencil;

C2 model reduction, i.e., find low-order matrices (E_r, A_r, B_r, C_r) such that the LTI system defined by these matrices approximates the input-output behavior of the original system;

C3 solve the linear-quadratic optimization problems discussed above using (5) and (6);

C4 compute the optimal H_2 controller;

C5 compute a suboptimal H_∞ controller.

In addition to the computational subroutines provided by the PBLAS and ScaLAPACK [15] and the solvers for the above linear and nonlinear matrix equations we will also need tools for the spectral decomposition of matrices and matrix pencils in order to accomplish Task **C1**.

The need for parallel computing in this area can be seen from the fact that already for a system with state-space dimension $n = 1000$, the corresponding Sylvester, Lyapunov, Stein, or Riccati equations represent a set of linear or nonlinear equations with one million unknowns. Systems of such a dimension driven by ordinary differential(-algebraic) equations are not uncommon in chemical engineering applications and are standard for second order systems arising from modeling mechanical multibody systems or large flexible space structures. We assume here that the coefficient matrices are dense and $n < 6000$. Larger systems, as those arising from the discretization of partial differential equations, usually involve sparse matrices. If sparsity is to be exploited, other computational techniques have to be employed [34,45]. The algorithms considered here are implemented in Fortran 77 using the kernels in libraries BLACS, PBLAS, and ScaLAPACK. The resulting subroutines will form a subroutine library with tentative name **PLILCO**, **P**arallel Software **Library** for **linear Control** theory.

This prospectus of the future PLILCO is organized as follows. In Section 2 we will review the basic numerical algorithms that can be employed in order to solve the computational problems needed to accomplish the required tasks. In order to obtain a high portability of the subroutines to be implemented, we will follow the guidelines and computation model used in ScaLAPACK [15] as well as the implementation and documentation standards given in [16]. A short review of the parallel computing paradigms used and a survey of the design and contents of the prospective library will be given in Section 3. Preliminary results in Section 4 will demonstrate the performance of the developed subroutines in several parallel computing environments with shared/distributed memory. An outlook on future activities is given in Section 5.

2 Numerical Algorithms

2.1 The QR and QZ algorithms

The traditional approaches to solving the computational problems introduced in the preceding section involve the computation of invariant/deflating subspaces by means of the QR/QZ algorithms; see, e.g., [26,49].

The QR algorithm consists of an initial reduction step which transforms a given matrix $A \in \mathbb{R}^{n \times n}$ to upper Hessenberg form, i.e.,

$$A_0 := U_0^T A U_0 = \left[\diagdown \right], \qquad (24)$$

where U_0 is orthogonal. Afterwards, a sequence of similarity transformations $A_{j+1} := U_{j+1}^T A_j U_{j+1}$ for $j = 0, 1, 2, \ldots$ is performed. The transformation matrices U_j are chosen such that all iterates A_j are upper Hessenberg matrices and converge to upper quasi-triangular form. That is, if $\tilde{A}_* = \lim_{j \to \infty} A_j$, then $\tilde{A}_*$ is upper triangular with 1×1 and 2×2 blocks on the diagonal. The 1×1 blocks correspond to real eigenvalues of A while 2×2 blocks represent pairs of complex conjugate eigenvalues of A. Usually, convergence takes place in

$\mathcal{O}(n)$ iterations. The similarity transformations with U_j can be implemented at a computational cost of $\mathcal{O}(n^2)$ such that the overall computational cost of this algorithm is $\mathcal{O}(n^3)$. If we set $\tilde{U} := \lim_{j \to \infty} \prod_{k=0}^{j} U_j$, then $\tilde{A}_* = \tilde{U}^T A \tilde{U}$. The upper quasi-triangular matrix $\tilde{A}_*$ is called the *(real) Schur form* of A. Applying a finite sequence of orthogonal similarity transformations to $\tilde{A}_*$, the diagonal blocks can be swapped such that the upper $k \times k$ block of the transformed matrix contains those eigenvalues of A that are inside some subset of the complex plain which is closed under complex conjugation. If we denote the accumulated transformation matrices that achieve this re-ordering by $\hat{U}$ and set $U := \hat{U}\tilde{U}$ then the first k columns of U span the A-invariant subspace corresponding to these eigenvalues.

The QZ algorithm applied to matrix pencil $A - \lambda E$ computes orthogonal matrices $\tilde{U}, \tilde{Z} \in \mathbb{R}^{n \times n}$ such that $\tilde{U}^T(A - \lambda E)\tilde{Z} = A_* - \lambda E_*$, where A_* is upper quasi-triangular and E_* is upper triangular. Again the matrices $\hat{U}, \hat{Z}$ c an be chosen such that the first k columns of $Z := \hat{Z}\tilde{Z}$ span a particular right deflating subspace of $A - \lambda E$ corresponding to some desired subset of eigenvalues of $A - \lambda E$. The QZ algorithm is equivalent to applying the QR algorithm to AE^{-1} without ever forming the product or the inverse explicitly. The matrix pencil $A_* - \lambda E_*$ is called the *generalized (real) Schur form* of $A - \lambda E$.

In order to compute a spectral decomposition of a matrix or matrix pencil as required, e.g., in Task **C1**, the QR (QZ) algorithm can be applied to the matrix (pencil). The re-ordering must then be performed such that the spectrum of the leading $k \times k$ diagonal block of A_* $(A_* - \lambda B_*)$ corresponds to the eigenvalues on the one side of the line dividing the spectrum while the trailing diagonal block corresponds to the eigenvalues of the other side of this line. For continuous-time systems, usually a spectral division along the imaginary axis is needed while for discrete-time systems, the usual spectral division line is the unit circle.

When solving the symmetric linear matrix equations (16)–(19) with the most widely used method, the *Bartels-Stewart method*, the QR and QZ algorithms are used for initial reductions of the involved matrix $\hat{A}$ or matrix pencil $\hat{A} - \lambda \hat{C}$ to upper quasi-triangular form. This initial stage is followed by a backsubstitution process in order to solve the resulting triangular systems. Note that the main computational work is done during the initial reduction. This approach is used, e.g., in [5,22,23,45] for the equations (16)–(19) and also when solving semidefinite Lyapunov and Stein equations of the form (20)–(23) in [30,54,45].

For the nonsymmetric equations (14) and (15), it is usually sufficient to transform one of the coefficient matrices to upper quasi-triangular form and the other one to Hessenberg form. This approach is called the *Hessenberg-Schur method* following [25] and is extended to (13) in [22,23].

The algebraic Riccati equations (11) and (12) can be solved via the QR/QZ algorithms using the relation to certain invariant or deflating subspaces of the corresponding matrices/matrix pencils. If the stable right deflating subspace of

$$H - \lambda K := \begin{bmatrix} \tilde{A} & \tilde{G} \\ \tilde{Q} & -\tilde{A}^T \end{bmatrix} - \lambda \begin{bmatrix} \tilde{E} & 0 \\ 0 & \tilde{E}^T \end{bmatrix} \tag{25}$$

is spanned by $\begin{bmatrix} Z_{11} \\ Z_{21} \end{bmatrix}$, $Z_{11}, Z_{21} \in \mathbb{R}^{n \times n}$, and Z_{11} is invertible, the stabilizing solution of (11) is given by $X_c = -Z_{12}Z_{11}^{-1}\tilde{E}^{-1}$. Hence the CARE (11) can be solved applying the QZ algorithm to $H - \lambda K$ and re-ordering the eigenvalues such that the stable eigenvalues (i.e., those with negative real parts) appear in the upper $n \times n$ diagonal block of the generalized Schur form of $H - \lambda K$. Then the first n columns of the matrix Z computed by the QZ algorithm span the required stable right deflating subspace of $H - \lambda K$. Note that the optimal control $u_*(t)$ can be computed using $F_c = R^{-1}(B^T Z_{12}Z_{11} - S^T C)$ without solving the CARE explicitly. In case $E = I_n$, it is sufficient to apply the QR algorithm to the *Hamiltonian matrix* H from (25) and to order the Schur form of H accordingly. This approach was first suggested in [39] and outlined in [3] for $E \neq I_n$. The resulting methods are called the *(generalized) Schur vector methods*.

Similar observations as in the continuous-time case lead to Schur vector methods for DAREs as given in (12). Here the QZ algorithm and an appropriate re-ordering are to be applied to

$$M - \lambda L = \begin{bmatrix} \tilde{A} & 0 & \tilde{B} \\ \tilde{Q} & -\tilde{E}^T & \tilde{S} \\ \tilde{S}^T & 0 & \tilde{R} \end{bmatrix} - \lambda \begin{bmatrix} \tilde{E} & 0 & 0 \\ 0 & -\tilde{A}^T & 0 \\ 0 & -\tilde{B}^T & 0 \end{bmatrix}. \tag{26}$$

If the generalized Schur form of $M - \lambda L$ is ordered such that the leading $n \times n$ diagonal block contains the eigenvalues inside the unit disk, then the first n columns of the Z-matrix computed by the QZ algorithm span the stable (with respect to the unit circle) right deflating subspace of $M - \lambda L$. Partitioning these n columns of Z as $[Z_{11}^T, Z_{21}^T, Z_{31}^T]^T$, where $Z_{11}, Z_{21} \in \mathbb{R}^{n \times n}$, $Z_{31} \in \mathbb{R}^{m \times n}$, and assuming Z_{11} nonsingular, $X_d = Z_{21}Z_{11}^{-1}\tilde{E}^{-1}$ and $F_d = Z_{31}Z_{11}^{-1}$ [42]. Note that using this approach it is possible to compute the optimal control u_k^* directly without solving the DARE explicitly. The computational cost of this approach can be lowered if R is invertible and well-conditioned by applying the QZ algorithm to

$$\tilde{M} - \lambda \tilde{L} = \begin{bmatrix} \tilde{A} - \tilde{R}^{-1}\tilde{S} & 0 \\ \tilde{Q} - \tilde{S}^T\tilde{R}^{-1}\tilde{S} & \tilde{E}^T \end{bmatrix} - \lambda \begin{bmatrix} \tilde{E} & -\tilde{B}\tilde{R}^{-1}\tilde{B}^T \\ 0 & (\tilde{A} - \tilde{R}^{-1}\tilde{S})^T \end{bmatrix}. \tag{27}$$

If $E = I_n$, $\tilde{M} - \lambda \tilde{L}$ is a symplectic matrix pencil. These Schur vector methods for the discrete-time case have been proposed in [3,44,53].

If the standard approaches to the spectral division problem and to the solution of the linear and nonlinear matrix equations described above are to be used for computations on parallel distributed memory computers, we will need efficient implementations of the QR and QZ algorithms for these computing environments. In ScaLAPACK, only the QR algorithm is available so far. However, in order to solve the linear matrix equations considered here, the QR algorithm can only be used for (14)–(16) and (18). In all other cases, the QZ algorithm has to be employed in the initial stage when solving these equations via the most frequently used Hessenberg-Schur and Bartels-Stewart methods as described above. Solving (11) and (12) by the (generalized) Schur vector methods, again the QR algorithm can only be used in the CARE case with $E \neq I_n$; for all other cases, the QZ algorithm is needed.

A different approach to solving the algebraic Riccati equations (11) and (12) is to consider these equations as nonlinear sets of equations. From this perspective, the most obvious choice to solve algebraic Riccati equations is Newton's method. In each iteration step of Newton's method applied to CAREs or DAREs [3,33,37,38,42], a (generalized) Lyapunov or Stein equation of the form (16)–(19) has to be solved; see Section 2.4 below. Thus a parallel implementation of Newton's method also depends heavily on the parallel performance of the Lyapunov or Stein solver employed, i.e., if the Bartels-Stewart method is to be used, once more on the efficiency of the parallelized QR/QZ algorithms.

¿From the above considerations we can conclude that in order to use the traditional algorithms for solving linear and algebraic Riccati matrix equations, it is necessary to have efficient parallelizations of the QR and QZ algorithms. However, several experimental studies report the difficulties in parallelizing the double implicit shifted QR algorithm on parallel distributed multiprocessors (see, e.g., [17,24,31,51]). The algorithm presents a fine granularity which introduces performance losses due to communication start-up overhead (latency). Besides, traditional data layouts (column/row block scattered) lead to an unbalanced distribution of the computational load. A different approach relies on a block Hankel distribution, which improves the balancing of the computational load [31]. Attempts to increase the granularity by employing multishift techniques have been recently proposed in [32]. Nevertheless, the parallelism and scalability of these algorithms are still far from those of matrix multiplications, matrix factorizations, triangular linear systems solvers, etc.; see, e.g., [15] and the references given therein.

Although the parallelization of the QR algorithm has been thoroughly studied, in contrast, the parallelization of the QZ algorithm remains unexplored to the best of our knowledge. Moreover, since both the QR and the QZ algorithms are composed of the same type of fine-grain computations, similar or even worse parallelism and scalability results are to be expected from the QZ algorithm.

In order to avoid the problems arising from the difficult parallelization of the QR and QZ algorithms, we will use a different computational approach here. It is well-known that under suitable assumptions, the above matrix equations can be solved via the sign function method. It has long been acknowledged that algorithms based on the sign function are relatively easy to parallelize. The methods that will be employed in the PLILCO will be considered in the next sections.

2.2 The Sign Function Method and the Smith Iteration

The sign function method was first introduced in 1971 by Roberts [46] for solving algebraic Riccati equations of the form (11) with $E = I_n$. Roberts also shows how to solve stable Sylvester and Lyapunov equations via the matrix sign function. The application to CAREs and DAREs with $E \neq I_n$ is investigated in [20,21] while the application to (16) with $E \neq I_n$ is examined in [13].

The computation of the sign function requires basic numerical linear algebra tools like matrix multiplication, inversion and/or solving linear systems. These computations are implemented efficiently on most parallel architectures and, in particular, ScaLAPACK [15] provides easy to use and portable computational kernels for these operations. Hence, the sign function method is an appropriate tool to design and implement efficient and portable numerical software for distributed memory parallel computers.

Let $Z \in \mathbb{R}^{n \times n}$ have no eigenvalues on the imaginary axis and denote by $Z = S \begin{bmatrix} J^- & 0 \\ 0 & J^+ \end{bmatrix} S^{-1}$ its Jordan decomposition with $J^- \in \mathbb{C}^{k \times k}$, $J^+ \in \mathbb{C}^{(n-k) \times (n-k)}$ containing the Jordan blocks corresponding to the eigenvalues in the open left and right half planes, respectively. Then the *matrix sign function* of Z is defined as

$$\text{sign}(Z) := S \begin{bmatrix} -I_k & 0 \\ 0 & I_{n-k} \end{bmatrix} S^{-1}. \qquad (28)$$

Note that $\text{sign}(Z)$ is unique and independent of the order of the eigenvalues in the Jordan decomposition of Z (see, e.g., [38, Section 22.1]). Many other equivalent definitions for $\text{sign}(Z)$ can be given; see, e.g., the recent survey paper [35].

The application of the matrix sign function method to a matrix pencil $Z - \lambda Y$ as given in [20] in case Z and Y are nonsingular can be presented as

$$Z_0 := Z, \qquad Z_{k+1} := \frac{1}{2c_k} \left(Z_k + c_k^2 Y Z_k^{-1} Y \right), \qquad k = 0, 1, 2, \ldots, \qquad (29)$$

where c_k is a scaling parameter. E.g., for determinantal scaling, c_k is given as $c_k = (|\det(Z_k)|/|\det(Y)|)^{\frac{1}{n}}$ [20]. This iteration is equivalent to computing the sign function of the matrix $Y^{-1}Z$ via the standard Newton iteration as proposed in [46]. The property needed here is that if $Z_\infty := \lim_{k \to \infty} Z_k$, then $(Z_\infty - Y)/2$ (or $(Z_\infty + Y)/2$) defines the skew projection onto the stable (or anti-stable) right deflating subspace of $Z - \lambda Y$ parallel to the anti-stable (or stable) deflating subspace.

In [20] the iteration (29) is used to compute the stabilizing solution of the CARE (11) and the DARE (12) using the matrix pencils (25) and (27). The algebraic Riccati equations (11) can be solved by applying (29) to $Z - \lambda Y = H - \lambda K$ and then forming the resulting projector $Z_\infty - Y$ onto the stable deflating subspace of $H - \lambda K$. A basis of this subspace is then given by the range of that projector. This subspace is usually not computed explicitly as $X_E := X_c \tilde{E}$ can be obtained by solving the overdetermined but consistent set of linear equations

$$\begin{bmatrix} Z_{12} \\ Z_{22} + \tilde{E}^T \end{bmatrix} X_E = \begin{bmatrix} Z_{11} + \tilde{E} \\ Z_{21} \end{bmatrix}, \tag{30}$$

where $Z_\infty = \begin{bmatrix} Z_{11} & Z_{12} \\ Z_{21} & Z_{22} \end{bmatrix}$; see [18,20,38,46]. The matrix X_c can be obtained by solving $X_E = X_c \tilde{E}$ while the optimal gain matrix and therefore the optimal control is obtained directly using $F_c = -R^{-1}(B^T X_E + S^T C)$.

The DARE (12) can not be solved directly using the sign function method as we need the d–stable deflating subspace of $\tilde{M} - \tilde{L}$ from (27). One possibility to switch back-and-forth between c– and d–stable matrix pencils $A - \lambda B$ (or c– and d–stable deflating subspaces) is the *Cayley transformation*

$$C_\mu(A - \lambda B) = (\mu A + B) - \lambda(A - \mu B), \qquad |\mu| = 1, \; \det(A - \mu B) \neq 0.$$

In order to keep computations real, one has to choose $\mu = \pm 1$; here we restrict ourselves to $\mu = 1$. It is well-known (see, e.g., [40,43]) that if $A - \lambda B$ is c–stable (d–stable), then $C_\mu(A - \lambda B)$ is d–stable (c–stable) and the c–stable (d–stable) right deflating subspace of $A - \lambda B$ is the d–stable (c–stable) right deflating subspace of $C_\mu(A - \lambda B)$. Hence, the DARE (12) can be solved with the sign function method applied to $C_\mu(\tilde{M} - \lambda \tilde{L})$. The solution X_d is then obtained from (30) replacing X_c by X_d.

Note that none of the methods considered so far can be used to solve (12) via (26): as we need the d–stable deflating subspace of $M - \lambda L$, the sign function method can not be applied directly. Though this subspace is given by the c–stable right deflating subspace of the Cayley transformed matrix pencil $C_\mu(M - \lambda L)$, the sign function method can in general not be used here as $M + L$ and $M - L$ may be singular.

A different approach to solve the spectral division problem and the considered matrix equations is reviewed in Section 2.3. This approach will also overcome the problems for the DARE (12) mentioned above.

The (generalized) Lyapunov and Stein equations (16) are special instances of the CARE (11) and DARE (12), respectively. This implies that one can solve (16) and (17) by means of the sign function method applied to the matrix pencil in (25) which then takes the form

$$H - \lambda K = \begin{bmatrix} \hat{A} & 0 \\ \hat{E} & -\hat{A}^T \end{bmatrix} - \lambda \begin{bmatrix} \hat{C} & 0 \\ 0 & \hat{C}^T \end{bmatrix}. \tag{31}$$

For stable matrix pencils $\hat{A} - \lambda\hat{C}$, $H - \lambda K$ is regular and has an n-dimensional stable deflating subspace such that the solution of (16) can be obtained analogously to that of (11).

In [13] it is observed that applying the generalized Newton iteration (29) to the matrix pencil $H - \lambda K$ in (31) and exploiting the block-triangular structure of all matrices involved, (29) boils down to

$$
\begin{aligned}
A_0 &:= \hat{A}, & A_{k+1} &:= \frac{1}{2}\left(A_k + \hat{C}A_k^{-1}\hat{C}\right), \\
E_0 &:= \hat{E}, & E_{k+1} &:= \frac{1}{2}\left(E_k + \hat{C}^T A_k^{-T} E_k A_k^{-1}\hat{C}\right),
\end{aligned}
\qquad k = 0, 1, 2, \ldots
$$

$$(32)$$

and that $X = \frac{1}{2}\hat{C}^{-T}\left(\lim_{k\to\infty} E_k\right)\hat{C}^{-1}$. In case $\hat{C} = I_n$, the iteration in (32) has already been derived by Roberts [46]. The semidefinite Lyapunov equations as in (20)–(23) can be solved using a factored version of the iteration for the $\hat{E}_k$'s in (32), i.e., the iteration is performed starting with the factor $\hat{F}$ of $\hat{E} = \hat{F}^T\hat{F}$. This iteration then converges to $2X_1\hat{C}$ if the solution is factored as $X = X_1^T X_1$. Details of this algorithm can be found in [13] and its application to computing the system Gramians for continuous-time LTI systems as given in (20), (21) is described in [11].

In case the spectra of $\hat{A} - \lambda\hat{C}$ and $\hat{B} - \lambda\hat{D}$ satisfy $\sigma(\hat{A}, \hat{C}) \subset \mathbb{C}^-$ and $\sigma(\hat{B}, \hat{D}) \subset \mathbb{C}^-$, the Sylvester equation (14) can also be solved using the sign function method applied to

$$
H - \lambda K := \begin{bmatrix} \hat{D} & 0 \\ \hat{E} & \hat{A} \end{bmatrix} - \lambda \begin{bmatrix} \hat{B} & 0 \\ 0 & -\hat{C} \end{bmatrix}.
\tag{33}
$$

Using again the block-triangular structure of the matrix pencil $H - \lambda K$, the iteration can be performed on the blocks as follows:

$$
\begin{aligned}
A_0 &:= \hat{A}, \quad D_0 := \hat{D}, \quad E_0 := \hat{E}, \\
A_{k+1} &:= \tfrac{1}{2}\left(A_k + CA_k^{-1}C\right), \\
D_{k+1} &:= \tfrac{1}{2}\left(D_k + BD_k^{-1}D\right), \\
E_{k+1} &:= \tfrac{1}{2}\left(E_k + CA_k^{-1}E_k D_k^{-1}B\right),
\end{aligned}
\qquad k = 0, 1, 2, \ldots.
\tag{34}
$$

The solution of (13) is then given by the solution of the linear system of equations $2\hat{C}X\hat{B} = \lim_{k\to\infty} E_k$.

In case $\hat{C} = I_n$ and $\hat{D} = I_m$, other iterative schemes for computing the sign function like the Newton-Schulz iteration or Halley's method can also be implemented efficiently to solve the corresponding Lyapunov and Sylvester equations (16) and (14); details of the resulting algorithms will be reported in [14].

So far we have only considered the linear matrix equations for continuous-time control problems. That is, we have assumed stability with respect to the imaginary axis. In discrete-time control problems, stability properties are given with respect to the unit circle. The linear matrix equations encountered in discrete-time control problems are (15), (18), and (19). Let us first consider (15) of which (18) is a special instance. If we rewrite the equation in fixed point form, $X = \hat{A}X\hat{B} + \hat{E}$ and form the fixed point iteration

$$X_0 := \hat{E}, \qquad X_{k+1} = \hat{E} + \hat{A}X_k\hat{B}, \quad k = 0, 1, 2, \ldots.$$

then this iteration converges to X if A and B are d-stable. The convergence rate of this iteration is linear. A quadratically convergent version of the fixed point iteration is suggested in [19,50],

$$
\begin{aligned}
A_0 &:= \hat{A}, \quad B_0 := \hat{B}, \quad X_0 := \hat{E}, \\
X_{k+1} &:= A_k X_k B_k + X_k, \\
A_{k+1} &:= A_k^2, \quad B_{k+1} := B_k^2,
\end{aligned}
\qquad k = 0, 1, 2, \ldots. \tag{35}
$$

The above iteration is referred to as the *Smith iteration*. We employ it to solve (18) and (15). In case (19) is to be solved with the Smith iteration, one has to apply (35) to $(\hat{A}\hat{C}^{-1})^T X(\hat{A}\hat{C}^{-1}) - X + \hat{C}^{-T}\hat{E}\hat{C}^{-1} = 0$. This has the disadvantage that the iteration is started with data that is already corrupted by roundoff errors basically determined by $\mathrm{cond}\left(\hat{C}\right)$, i.e., the condition of $\hat{C}$ with respect to matrix inversion defined by $\mathrm{cond}\left(\hat{C}\right) = \|\hat{C}\|\|\hat{C}^{-1}\|$.

One possibility to avoid the initial inversion of $\hat{C}$ when solving (19) by the Smith iteration is to transform (19) to a generalized Lyapunov equation without inverting any matrices using the Cayley transformation and then applying (32) to the transformed equation

$$(\hat{A} + \hat{C})^T X(\hat{A} - \hat{C}) + (\hat{A} - \hat{C})^T X(\hat{A} + \hat{C}) + 2Q = 0 \tag{36}$$

which has the same solution as (19). Of course, the same approach can be used for (18) setting $\hat{C} = I_n$. But this yields a generalized Lyapunov equation. In order to obtain a standard Lyapunov equation of the form (16) one has to multiply (36) from the left by $(\hat{A} - \hat{C})^{-T}$ and $(\hat{A} - \hat{C})^{-1}$ from the right. This introduces again unnecessary rounding errors and we will therefore not follow this approach here.

2.3 The Disk Function Method

Let $Z - \lambda Y$, $Z, Y \in \mathbb{R}^{n\times n}$, be a regular matrix pencil having no eigenvalues on the unit circle. Suppose the *Weierstraß (Kronecker) canonical form* of $Z - \lambda Y$ is given by

$$Z - \lambda Y = T\begin{bmatrix} J^0 - \lambda I & 0 \\ 0 & J^\infty - \lambda N \end{bmatrix} S^{-1}$$

where Jordan blocks corresponding to eigenvalues inside the unit disk are collected in J^0 while J^∞ corresponds to eigenvalues outside the unit disk and N contains nilpotent blocks corresponding to infinite eigenvalues. The *matrix pencil disk function* is defined in [6] as

$$\text{disk}\,(Z,Y) \; := \; S\left(\begin{bmatrix} I_k & 0 \\ 0 & 0 \end{bmatrix} - \lambda \begin{bmatrix} 0 & 0 \\ 0 & I_{n-k} \end{bmatrix}\right) S^{-1} \; =: \; D_Z - \lambda D_Y. \quad (37)$$

A *matrix disc function* was also introduced in [46] using a different approach. In [6] it is shown that this is a special case of the above definition using $Y = I_n$. ¿From the disk function, we can obtain the d–stable deflating subspace of $Z - \lambda Y$ as D_Z is a skew projector onto this subspace. Hence, a basis for this subspace is given by a basis of the column space of D_Z.

The disk function has received some interest in recent years as it provides the mathematical framework for an algorithm proposed in [41] and made feasible for practical computations in [4] for solving the spectral division problem. This *inverse free spectral division algorithm* can be given as follows:

$$Z_0 \; := \; Z, \qquad Y_0 \; := \; Y,$$

$$\begin{bmatrix} Y_k \\ -Z_k \end{bmatrix} \; := \; \begin{bmatrix} U_{11} & U_{12} \\ U_{21} & U_{22} \end{bmatrix} \begin{bmatrix} R_k \\ 0 \end{bmatrix} \quad \text{(QR decomposition)}, \qquad k = 0,1,2,\ldots.$$

$$Z_{k+1} \; := \; U_{12}^T Z_k, \qquad Y_{k+1} \; := \; U_{22}^T Y_k,$$

$$(38)$$

It follows that $\text{disk}\,(Z,Y) = (Z_\infty + Y_\infty)^{-1}(Y_\infty - \lambda Z_\infty)$. Hence a basis for the d–stable right deflating subspace of $Z - \lambda Y$ can be computed via a rank-revealing QR decomposition of $(Z_\infty + Y_\infty)^{-1}Y_\infty$, where $\lim_{k\to\infty}(Z_k, Y_k) =: (Z_\infty, Y_\infty)$. Note that this QR decomposition can be computed without explicitly inverting $(Z_\infty + Y_\infty)$; see [4] for details. Moreover, a complete spectral decomposition of $Z - \lambda Y$ along the unit circle can be computed using only one iteration of the form (38); see [52].

¿From the above considerations we can conclude that the DARE (12) can be solved applying iteration (38) to $M - \lambda L$ from (26). It is shown in [6] that it is not necessary to compute a basis for the d–stable deflating subspace explicitly. Using the relation between the nullspaces $\text{Ker}(D_Z) = \text{Ker}(Z_\infty)$, and the fact that if the stabilizing solution X_d of (12) exists, then $[\, I_n \; (X_d \tilde{E})^T \; F_d^T \,] \in \text{Ker}(D_Z)$, one can show (see [6]) that the solution of the DARE and the optimal gain matrix of the discrete-time optimal control problem can be obtained from the solution of the overdetermined but consistent set of linear equations

$$\begin{bmatrix} Z_{12} & Z_{13} \\ Z_{22} & Z_{23} \\ Z_{32} & Z_{33} \end{bmatrix} \begin{bmatrix} X_E \\ F_d \end{bmatrix} = - \begin{bmatrix} Z_{11} \\ Z_{21} \\ Z_{31} \end{bmatrix}. \quad (39)$$

Here, the Z_{kj}, $k,j = 1,2,3$, define a block partitioning of Z_∞ conformal to the partitioning in (26). Solving (12) with this approach will be referred to as the *disk function method*.

Note that the CARE (11) can also be solved with the disk function method by applying the iteration (38) to $C_\mu(H - \lambda K)$ with $H - \lambda K$ as in (25). The solution X_c of (11) as well as the optimal gain matrix F_c can be obtained from the overdetermined but consistent set of linear equations

$$\begin{bmatrix} Z_{12} \\ Z_{22} \end{bmatrix} X_E = \begin{bmatrix} Z_{11} \\ Z_{21} \end{bmatrix}, \qquad Z_\infty := \begin{bmatrix} Z_{11} & Z_{12} \\ Z_{21} & Z_{22} \end{bmatrix},$$

and $X_c = X_E \tilde{E}^{-1}$, $F_c = -(B^T X_E + S^T C)$.

The linear matrix equations (18) and (19) can also be solved via the disk function method applied to (27) noting that (18), (19) are special instances of (12). The corresponding matrix pencil then takes the form

$$\tilde{M} - \lambda\tilde{L} = \begin{bmatrix} \hat{A} & 0 \\ \hat{E} & \hat{C}^T \end{bmatrix} - \lambda \begin{bmatrix} \hat{C} & 0 \\ 0 & \hat{A}^T \end{bmatrix}. \tag{40}$$

Equations (16) and (17) are special instances of (11) and hence can be solved via the disk function method applied to $C_\mu(H - \lambda K)$ for $H - \lambda K$ as in (31).

Unfortunately, the iteration in (38) can not be decomposed into iterations on the matrix blocks such that no computational savings is obtained compared to the solution of the DARE. Hence, the computational cost for solving linear matrix equations with the disk function method is in general prohibitive.

More details and computational aspects of the disk function method can be found in [4,6,7,52]. Though a general scaling strategy in order to accelerate convergence in (38) is yet not known, an initial scaling of $H - \lambda K$ in (25) can significantly improve the disk function methods for CAREs; see [6].

2.4   Newton's Method

The methods presented so far have addressed the algebraic Riccati equations by their relation to eigenproblems. By nature, they are systems of nonlinear equations. It is therefore straightforward to apply methods for solving nonlinear equations. In [37], Kleinman shows that Newton's method, applied to the CARE (11) with $\hat{E} = I_n$ and properly initialized, converges to the desired stabilizing solution of the CARE. The application to the generalized equation (11) is considered in [3,42]. Given some initial guess X_0, the resulting algorithm can be stated in different ways. We have chosen here the variant that is most robust with respect to accumulation of rounding errors.

> FOR $k = 0, 1, 2, \ldots$ "until convergence"
>
> 1. $A_k := \tilde{A} - \tilde{G}X_k$.
>
> 2. Solve for N_k in the generalized Lyapunov equation
>
> $$0 = \mathcal{R}_c(X_k) + A_k^T N_k \tilde{E} + \tilde{E}N_k A_k.$$
>
> 3. $X_{k+1} := X_k + N_k$.

The main computational cost in this algorithm comes from the solution of the (generalized) Lyapunov equation in each iteration step. If $\tilde{G}$ is positive semidefinite and under the assumption used throughout this paper, i.e., X_c exists such that $\sigma\,(\tilde{E}^{-1}(\tilde{A} - \tilde{G}X_c)) \subset \mathbb{C}^-$, it can be shown that convergence to X_c is globally quadratic if X_0 is chosen such that $\tilde{E}^{-1}A_0$ is c–stable.

Remark 1. Convergence of Newton's method for algebraic Riccati equations can also be proved under slightly more general assumptions than used here [29,28]: suppose there exists a gain matrix $\tilde{F}_c$ such that $\tilde{A}_c$ in (9) is c–stable, that is, the underlying LTI system (1) is *c–stabilizable*. Furthermore, assume that $\tilde{G}$ is positive semidefinite and there exists a maximal symmetric solution X_+ of (11), i.e., $X_+ \geq X$ for any other symmetric solution of (11). Then $A_+ := \tilde{E}^{-1}(\tilde{A} - \tilde{G}X_+)$ has all its eigenvalues in the closed left half plane. The matrix $X+$ is therefore called *almost stabilizing*. It is the unique solution of (11) with this property and coincides with X_c if the latter exists [38, Chapter 7]. Then the Newton iteration converges to X_+ from any stabilizing initial guess X_0 [29,38]. The convergence rate is usually linear if X_+ is not stabilizing, but quadratic convergence may still occur. A simple trick presented in [29] can improve the convergence in the linear case significantly.

Analogous observations hold in the discrete-time case [28].

Finding a stabilizing X_0 usually is a difficult task and requires the stabilization of an LTI system, i.e., the solution of Task **C1**. The computational cost is equivalent to one iteration step of Newton's method; see, e.g., [49] and the references therein. Moreover, X_0 determined by a stabilization procedure may lie far from X_c. Though ultimately quadratic convergent, Newton's method may initially converge slowly. This can be due to a large error $\|X_0 - X_c\|$ or to a disastrously bad first step, leading to a large error $\|X_1 - X_c\|$; see, e.g., [36,6,8]. Due to the initial slow convergence, Newton's method often requires too many iterations to be competitive with other Riccati solvers. Therefore it is most frequently only used to refine an approximate CARE solution computed by any other method.

Recently an exact line search procedure was suggested that accelerates the initial convergence and avoids "bad" first steps [6,8]. Specifically, Step 3. of Newton's method given above is modified to $X_{k+1} = X_k + t_k N_k$, where t_k is chosen in order to minimize the Frobenius norm of the residual $\mathcal{R}_c(X_k + t N_k)$. As computing the exact minimizer is very cheap compared to a Newton step and usually accelerates the initial convergence significantly while benefiting from the quadratic convergence of Newton's method close to the solution, this method becomes attractive, even as a solver for CAREs (at least in some cases), see [6,8,9] for details. Moreover, for some ill-conditioned CAREs, exact line search improves Newton's method also when used only for iterative refinement. Note that the line search strategy discussed in [6,8,9] also includes the trick described in Remark (1) for accelerating the linear convergence in case X_c does not exist.

Similarly, Newton's method can be applied to the DARE (12). The resulting algorithm is described in [33] for $\tilde{E} = I_n$ and in [3,42] for $\tilde{E} \neq I_n$. The main computation there is again the solution of a linear matrix equation which is in this case a (generalized) Stein equation. Again, line searches can be employed to (partially) overcome the difficulties mentioned above as they apply analogously to DAREs [6].

In both the continuous- and discrete-time case, in each iteration step of Newton's method, a linear matrix equation has to be solved. Hence the key to an efficient parallelization of Newton's method is an efficient solver for the linear matrix equation in question. Hence we employ the iterative schemes discussed above (sign function method or Smith iteration). Note that all other computations required by Newton's method apart from solving Lyapunov equations basically consist of matrix multiplications and can therefore be implemented efficiently on parallel computers. The parallelization of Newton's method with exact line search based upon solving the generalized Lyapunov equations via (32) is discussed in [9] where also several numerical experiments are reported.

3   Prospectus of the PLILCO

In this section we first describe the ScaLAPACK library [15] which is used as the parallel infrastructure for our PLILCO routines. We then describe the specific routines in PLILCO, including both the available routines and those that will be included in a near future to extend the functionality of the library.

3.1   The ScaLAPACK library

The ScaLAPACK (Scalable LAPACK) library [15] is designed as an extension of the successful LAPACK library [2] for parallel distributed memory multiprocessors. ScaLAPACK mimics the LAPACK, both in structure and notation. The parallel kernels in this library rely on the use of those in the PBLAS (Parallel BLAS) library and the BLACS (Basic Linear Algebra Communication Subroutines). The serial computations are performed by calls to routines from the BLAS and LAPACK libraries; the communication routines in BLACS are usually implemented on top of a standard communication library as MPI or PVM.

This structured hierarchy of dependences (see Figure 2) enhances the portability of the codes. Basically, a parallel algorithm that uses ScaLAPACK routines can be migrated to any vector processor, superscalar processor, shared memory multiprocessor, or distributed memory multicomputer where the BLAS and the MPI (or PVM) are available.

ScaLAPACK implements parallel routines for solving linear systems, linear least squares problems, eigenvalue problems, and singular value problems.

The performance of these routines depends on those of the serial BLAS and the communication library (MPI or PVM).

ScaLAPACK employs the so-called message-passing paradigm. That is, the processes collaborate in solving the problem and explicit communication requests are performed whenever a process requires a datum that is not stored in its local memory.

In ScaLAPACK the computations are performed by a logical grid of $P_r \times P_c$ processes. The processes are mapped onto the physical processors, depending on the available number of these. All data (matrices) have to be distributed among the process grid prior to the invocation of a ScaLA-PACK routine. It is the user's responsibility to perform this data distribution. Specifically, in ScaLAPACK the matrices are partitioned into $nb \times nb$ square blocks and these blocks are distributed (and stored) among the processes in column-major order. A graphical representation of the data layout is given in Figure 1 for a logical grid of 2×3 processes.

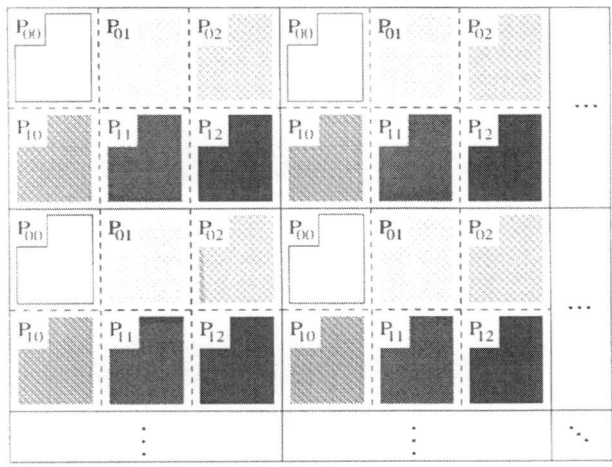

Fig. 1. Data layout in a logical grid of 2×3 processes.

Although not strictly part of ScaLAPACK, the library also provides routines for distributing a matrix among the process grid. The communication overhead of this initial distribution is well balanced in most medium and large-scale applications by the improvements in performance achieved with parallel computation.

3.2 Structure of PLILCO

PLILCO heavily relies on the use of the available parallel infrastructure in ScaLAPACK (see Figure 2). Although ScaLAPACK is incomplete, the kernels available in the current version (1.6) allows us to implement most of our PLILCO routines. PLILCO will benefit from future extensions and developments in the ScaLAPACK project. Improvements in performance of the PBLAS kernels will also be specially welcome.

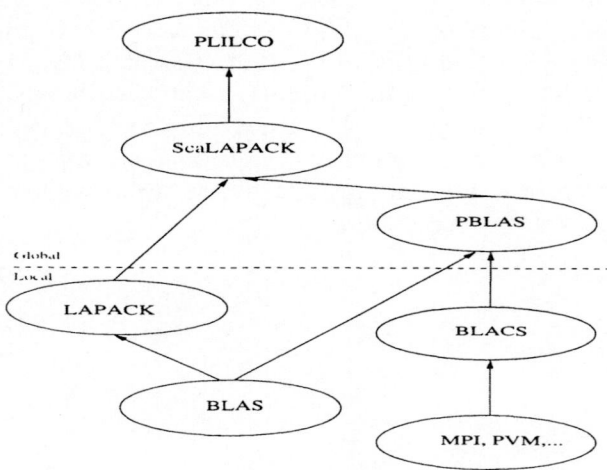

Fig. 2. Structure of PLILCO.

In PLILCO the routines are named, following the convention in LAPACK and ScaLAPACK, as PDxxyyzz. The PD- prefix in each name indicates that this is a Parallel routine with Double-precision arithmetic. The following two letters, xx indicate the type of LTI system addressed by the routine. Thus, GE or GG indicate, respectively, a standard LTI system ($E = I_n$) or a generalized LTI system ($E \neq I_n$). The last four letters in the name indicate the specific problem (yy), and the method employed for that problem (zz). In most cases, a sequence Cyzz indicates that the routine deals with a continuous-time problem while a sequence Dyzz indicates a discrete-time problem.

The PLILCO routines can be classified in 4 groups according to their functionality (computation of basic matrix functions, linear matrix equation solvers, optimal control, and feedback stabilization). Two more groups will be included in the near future. We next review the routines in each of these 4 groups.

Group MTF: Basic matrix functions.
The routines in this group implement iterative schemes to compute functions

of matrices or matrix pencils. For instance, three routines are available for computing the sign function of a matrix. These routines employ different variants for the matrix sign function iteration:

- PDGESGNW. The Newton iteration.
- PDGESGNS. The Newton-Schulz iteration.
- PDGESGHA. The Halley iteration.

Two more routines are designed for computing the sign function or the disk function of matrix pencils:

- PDGGSGNW. The generalized Newton iteration for the matrix sign function.
- PDGGDKMA. The iteration (38) for computing the disk function.

Note that the iteration (38) for the disk function only deals with matrix pencils and therefore PLILCO does not provide any routine for the standard problem. The disk function of a matrix Z is obtained by applying routine PDGGDKMA to $Z - \lambda I$.

Table 1 lists the PLILCO routines in the MTF group.

Type of Function	Problem	
	Standard	Generalized
Matrix sign function	PDGESGNW PDGESGNS PDGESGHA	PDGGSGNW
Matrix pencil disk function		PDGGDKMA

Table 1. PLILCO routines in MTF group.

Group LME: Linear matrix equation solvers.

The routines in this group are solvers for several particular instances of generalized Sylvester matrix equations, see (13)–(19).

All the solvers for those equations arising in continuous-time LTI systems are based on the matrix sign function and require that the coefficient matrices of the equation are stable.

PLILCO includes three solvers for stable Sylvester equations that differ in the iteration used for the computation of the matrix sign function:

- PDGECSNW. The Newton iteration.
- PDGECSNS. The Newton-Schulz iteration.
- PDGECSHA. The Halley iteration.

In the generalized problem a solver for generalized Sylvester equations is also included:

– PDGGCSNW. The generalized Newton iteration as in (34).

All these solvers have their analogous routines for stable Lyapunov equations (three routines) and stable generalized Lyapunov equations (one routine):

– PDGECLNW. The Newton iteration.
– PDGECLNS. The Newton-Schulz iteration.
– PDGECLHA. The Halley iteration.
– PDGGCLNW. The generalized Newton iteration as in (32).

Furthermore, in case the constant term $\hat{E}$ is semidefinite it is also possible to obtain the Cholesky factor of the solution directly by means of routines

– PDGECLNC. The Newton iteration for the Cholesky factor.
– PDGGCLNC. The generalized Newton iteration for the Cholesky factor.

In the discrete-time case, the iterative solvers in PLILCO are based on the Smith iteration and require the coefficient matrices to be stable (in the discrete-time sense). So far PLILCO only includes two solvers, for the discrete-time Sylvester equation (15) and the Stein (or discrete-time Lyapunov) equation (18), respectively:

– PDGEDSSM. The Smith iteration for (15).
– PDGEDLSM. The Smith iteration for (18).

Special versions of the Smith iteration for computing the Cholesky factors of semidefinite Stein equations as in (22) and (23) are to be developed in the future.

The solution of generalized discrete-time linear equations can be obtained by transforming this equation into an standard one, as there is no generalized version of the Smith iteration. Note that this transformation involves explicit inversion of the coefficient matrices in the generalized equation.

Table 2 summarizes the PLILCO routines in the LME group.

Group RIC: Riccati matrix equation solvers.

We include in this group solvers both for CARE and DARE.

In the continuous-time case, PLILCO provides solvers based on three different methods, these are, Newton's method, the matrix sign function, and the matrix disk function. Moreover, in the standard case, three different variants are proposed for Newton's method depending on the Lyapunov solver that is employed. Thus, we have the following CARE solvers:

– PDGECRNW. Newton's method with the Newton iteration for solving the Lyapunov equations.
– PDGECRNS. Newton's method with the Newton-Schulz iteration for solving the Lyapunov equations.
– PDGECRHA. Newton's method with the Halley iteration for solving the Lyapunov equations.

Type of equation	Problem	
	Standard	Generalized
Sylvester	PDGECSNW PDGECSNS PDGECSHA	PDGGCSNW
Lyapunov	PDGECLNW PDGECLNS PDGECLHA PDGECLNC	PDGGCLNW PDGGCLNC
Discrete-time Sylvester	PDGEDSNW	
Stein	PDGEDLSM	

Table 2. PLILCO routines in the LME group.

- PDGECRSG. The matrix sign function method.
- PDGECRDK. The matrix disk function method.

Similarly, we have the following generalized CARE solvers:

- PDGGCRNW. Newton's method with the generalized Newton iterative scheme for solving the generalized Lyapunov equations.
- PDGGCRSG. The generalized matrix sign function method.
- PDGGCRDK. The matrix disk function method.

PLILCO also includes the following solvers for DARE (two routines) and generalized DARE (two routines):

- PDGEDRSM. Newton's method with the Smith iteration for solving the discrete-time Lyapunov equations.
- PDGEDRDK. The matrix disk function method for DARE.
- PDGGDRSM. Newton's method with the Smith iteration for solving the discrete-time generalized Lyapunov equations.
- PDGGDRDK. The matrix disk function method for generalized DARE.

Table 3 lists the PLILCO routines in the RIC group.

Group STF: Feedback stabilization of LTI systems.

This group includes routines for partial and complete (state feedback) stabilization of LTI systems. The routines in this group use the linear matrix equations solvers in group LME to deal with the different equations arising in standard and generalized, continuous-time and discrete-time LTI systems. PLILCO thus includes several state feedback stabilizing routines, which differ in the linear matrix equation that has to be solved and, therefore, the iteration employed. The feedback stabilization of continuous-time LTI systems can be obtained by means of routines:

Type of Equation	Problem	
	Standard	Generalized
CARE	PDGECRNW	PDGGCRNW
	PDGECRNS	
	PDGECRHA	
	PDGECRSG	PDGGCRSG
	PDGECRDK	PDGGCRDK
DARE	PDGEDRSM	PDGGDRSM
	PDGEDRDK	PDGGDRDK

Table 3. PLILCO routines in the RIC group.

- PDGECFNW. The Newton iteration.
- PDGECFNS. The Newton-Schulz iteration.
- PDGECFHA. The Halley iteration.
- PDGGCFNW. The generalized Newton iteration.

In the discrete-time case, the unique routine available so far is the following:

- PDGEDFSM. The Smith iteration.

Table 4 lists the names of the routines in group STF.

Type of LTI system	Problem	
	Standard	Generalized
continuous-time	PDGECFNW	PDGGSTNW
	PDGECFNS	
	PDGECFHA	
discrete-time	PDGEDFSM	

Table 4. PLILCO routines in STF group.

Future extensions of PLILCO will include at least two more groups:

Group MRD: Model reduction of LTI systems.

Group H2I: Computation of H_2- and H_∞-controllers.

4   Preliminary Results

In this section we present some of the preliminary results obtained with the PLILCO routines on different parallel architectures. Specifically, we report results for the Lyapunov equation solver PDGECLNW and the generalized Lyapunov equation solver PDGGCLNW.

As target parallel distributed memory architectures we evaluate our algorithms on an IBM SP2 and a Cray T3E. In both cases we use the native BLAS, the MPI communication library, and the LAPACK, BLACS, and ScaLAPACK libraries [2,15] to ensure the portability of the algorithms.

The IBM SP2 that we used consists of 80 RS/6000 nodes at 120 MHz, and 256 MBytes RAM per processor. Internally, the nodes are connected by a TB3 high performance switch. The Cray T3E-600 has 60 DEC Alpha EV5 nodes at 300 MHz, and 128 Mbytes RAM per processor. The communication network has a bidimensional torus topology.

Table 5 reports the performance of the Level-3 BLAS matrix product (in Mflops, or millions of floating-point arithmetic operations per second), and the latency and bandwidth for the communication system of each platform.

	DGEMM (Mflops)	Latency (sec.)	Bandwith (Mbit/sec.)
IBM SP2	200	30×10^{-6}	90
Cray T3E	400	50×10^{-6}	166

Table 5. Basic performance parameters of the parallel architectures.

In both matrix equations, the coefficient matrix A is generated with random uniform entries. This matrix is stabilized by a shift of the eigenvalues ($A := A - \|A\|_F I_n$ in the continuous-time case and $A := A/\|A\|_F$ in the discrete-time case). In case a LTI system is required, $\hat{A} - \lambda \hat{E}$ is obtained from as $\hat{A} = R$, $\hat{E} = Q^H$ where Q and R are obtained from a QR factorization A. The solution matrix X is set to a matrix with all entries equal to one and matrix Q is then chosen to satisfy the corresponding linear matrix equation.

All experiments were performed using Fortran 77 and IEEE double-precision arithmetic ($\varepsilon \approx 2.2 \times 10^{-16}$). In our examples, the solution is obtained with the accuracy that could be expected from the conditioning of the problem. A more detailed study of the accuracy of these solvers is beyond the scope of this paper. For details and numerical examples demonstrating the performance and numerical reliability of the proposed equation solvers, see [9–14].

The figures show the Mflops ratio per node when the number of nodes is increased and the ratio n/p is kept constant. Thus, we are measuring the scalability of our parallel routines. The results in the figures are averaged for 5 executions on different randomly generated matrices. In these figures the solid line indicates the maximum attainable real performance (that of DGEMM) and the dashed line represents the performance of the corresponding linear matrix equation solver.

Figure 3 reports the Mflops ratio per node for routine PDGECLNW on the Cray T3E platform and routine PDGGCLNW on the IBM SP2 platform.

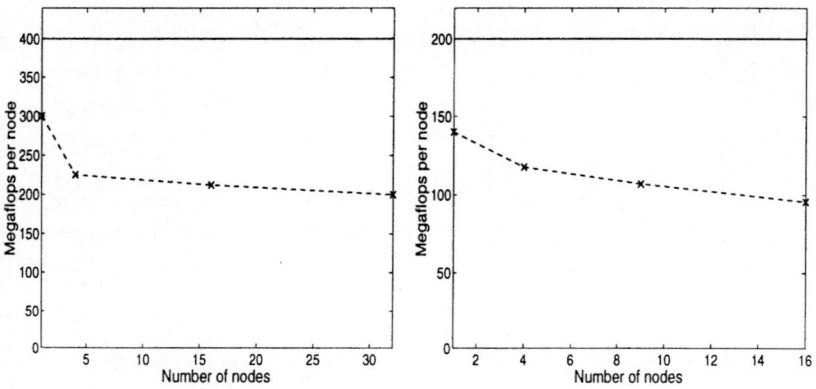

Fig. 3. Mflop ratio for routine PDGECLNW on the Cray T3E with $n/p = 750$ (left), and routine PDGGCLNW on the IBM SP2, with $n/p = 1000$ (right).

Both figures show similar results. The performance per node of the algorithms decreases when the number of processors is increased from 1 to 4 due to the communication overhead of the parallel algorithm. However, as the number of processors is further increased, the performance only decreases slightly showing the scalability of the solvers.

5   Concluding Remarks

We have described the development of a software library for solving the computational problems that arise in analysis and synthesis of linear control systems. The library is intended for solving medium-size and large-scale problems and the numerical results demonstrate its performance on shared and distributed memory parallel architectures. The portability of the library is ensured by using the PBLAS, BLACS, and ScaLAPACK. It is hoped that this high-performance computing approach will enable users to deal with large scale problems in linear control theory.

References

1. B.D.O. Anderson and J.B. Moore. *Optimal Control – Linear Quadratic Methods.* Prentice-Hall, Englewood Cliffs, NJ, 1990.

2. E. Anderson, Z. Bai, C. Bischof, J. Demmel, J. Dongarra, J. Du Croz, A. Greenbaum, S. Hammarling, A. McKenney, S. Ostrouchov, and D. Sorensen. *LAPACK Users' Guide.* SIAM, Philadelphia, PA, second edition, 1995.

3. W.F. Arnold, III and A.J. Laub. Generalized eigenproblem algorithms and software for algebraic Riccati equations. *Proc. IEEE*, 72:1746–1754, 1984.

4. Z. Bai, J. Demmel, and M. Gu. An inverse free parallel spectral divide and conquer algorithm for nonsymmetric eigenproblems. *Numer. Math.*, 76(3):279–308, 1997.

5. R.H. Bartels and G.W. Stewart. Solution of the matrix equation $AX + XB = C$: Algorithm 432. *Comm. ACM*, 15:820–826, 1972.

6. P. Benner. *Contributions to the Numerical Solution of Algebraic Riccati Equations and Related Eigenvalue Problems.* Logos–Verlag, Berlin, Germany, 1997. *Also:* Dissertation, Fakultät für Mathematik, TU Chemnitz–Zwickau, 1997.

7. P. Benner and R. Byers. Disk functions and their relationship to the matrix sign function. In *Proc. European Control Conf. ECC 97*, Paper 936. BELWARE Information Technology, Waterloo, Belgium, 1997. CD-ROM.

8. P. Benner and R. Byers. An exact line search method for solving generalized continuous-time algebraic Riccati equations. *IEEE Trans. Automat. Control*, 43(1):101–107, 1998.

9. P. Benner, R. Byers, E.S. Quintana-Ortí, and G. Quintana-Ortí. Solving algebraic Riccati equations on parallel computers using Newton's method with exact line search. Berichte aus der Technomathematik, Report 98–05, Universität Bremen, August 1998. Available from `http://www.math.uni-bremen.de/zetem/berichte.html`.

10. P. Benner, M. Castillo, V. Hernández, and E.S. Quintana-Ortí. Parallel partial stabilizing algorithms for large linear control systems. *J. Supercomputing*, to appear.

11. P. Benner, J.M. Claver, and E.S. Quintana-Ortí. Efficient solution of coupled Lyapunov equations via matrix sign function iteration. In A. Dourado et al., editor, *Proc. 3rd Portuguese Conf. on Automatic Control CONTROLO'98*, Coimbra, pages 205–210, 1998.

12. P. Benner, J.M. Claver, and E.S. Quintana-Ortí. Parallel distributed solvers for large stable generalized Lyapunov equations. *Parallel Processing Letters*, to appear.

13. P. Benner and E.S. Quintana-Ortí. Solving stable generalized Lyapunov equations with the matrix sign function. *Numer. Algorithms*, to appear.

14. P. Benner, E.S. Quintana-Ortí, and G. Quintana-Ortí. Solving linear matrix equations via rational iterative schemes. In preparation.

15. L.S. Blackford, J. Choi, A. Cleary, E. D'Azevedo, J. Demmel, I. Dhillon, J. Dongarra, S. Hammarling, G. Henry, A. Petitet, K. Stanley, D. Walker, and R.C. Whaley. *ScaLAPACK Users' Guide.* SIAM, Philadelphia, PA, 1997.

16. I. Blanquer, D. Guerrero, V. Hernandez, E. Quintana-Ortí, and P. Ruiz. Parallel-SLICOT implementation and documentation standards. SLICOT Working Note 1998-1, `http://www.win.tue.nl/niconet/`, September 1998.

17. D. Boley and R. Maier. A parallel QR algorithm for the unsymmetric eigenvalue problem. Technical Report TR-88-12, University of Minnesota at Minneapolis, Department of Computer Science, Minneapolis, MN, 1988.

18. R. Byers. Solving the algebraic Riccati equation with the matrix sign function. *Linear Algebra Appl.*, 85:267–279, 1987.

19. E.J. Davison and F.T. Man. The numerical solution of $A'Q + QA = -C$. *IEEE Trans. Automat. Control*, AC-13:448–449, 1968.

20. J.D. Gardiner and A.J. Laub. A generalization of the matrix-sign-function solution for algebraic Riccati equations. *Internat. J. Control*, 44:823–832, 1986.

21. J.D. Gardiner and A.J. Laub. Parallel algorithms for algebraic Riccati equations. *Internat. J. Control*, 54:1317–1333, 1991.

22. J.D. Gardiner, A.J. Laub, J.J. Amato, and C.B. Moler. Solution of the Sylvester matrix equation $AXB + CXD = E$. *ACM Trans. Math. Software*, 18:223–231, 1992.

23. J.D. Gardiner, M.R. Wette, A.J. Laub, J.J. Amato, and C.B. Moler. Algorithm 705: A Fortran-77 software package for solving the Sylvester matrix equation $AXB^T + CXD^T = E$. *ACM Trans. Math. Software*, 18:232–238, 1992.

24. G.A. Geist, R.C. Ward, G.J. Davis, and R.E. Funderlic. Finding eigenvalues and eigenvectors of unsymmetric matrices using a hypercube multiprocessor. In G. Fox, editor, *Proc. 3rd Conference on Hypercube Concurrent Computers and Appl.*, pages 1577–1582, 1988.

25. G. H. Golub, S. Nash, and C. F. Van Loan. A Hessenberg–Schur method for the problem $AX + XB = C$. *IEEE Trans. Automat. Control*, AC-24:909–913, 1979.

26. G.H. Golub and C.F. Van Loan. *Matrix Computations*. Johns Hopkins University Press, Baltimore, third edition, 1996.

27. M. Green and D.J.N Limebeer. *Linear Robust Control*. Prentice-Hall, Englewood Cliffs, NJ, 1995.

28. C.-H. Guo. Newton's method for discrete algebraic Riccati equations when the closed-loop matrix has eigenvalues on the unit circle. *SIAM J. Matrix Anal. Appl.*, 20:279–294, 1998.

29. C.-H. Guo and P. Lancaster. Analysis and modification of Newton's method for algebraic Riccati equations. *Math. Comp.*, 67:1089–1105, 1998.

30. S.J. Hammarling. Numerical solution of the stable, non-negative definite Lyapunov equation. *IMA J. Numer. Anal.*, 2:303–323, 1982.

31. G. Henry and R. van de Geijn. Parallelizing the QR algorithm for the unsymmetric algebraic eigenvalue problem: myths and reality. *SIAM J. Sci. Comput.*, 17:870–883, 1997.

32. G. Henry, D.S. Watkins, and J.J. Dongarra. A parallel implementation of the nonsymmetric QR algorithm for distributed memory architectures. LAPACK Working Note 121, University of Tennessee at Knoxville, 1997.

33. G.A. Hewer. An iterative technique for the computation of steady state gains for the discrete optimal regulator. *IEEE Trans. Automat. Control*, AC-16:382–384, 1971.

34. A.S. Hodel and K.R. Polla. Heuristic approaches to the solution of very large sparse Lyapunov and algebraic Riccati equations. In *Proc. 27th IEEE Conf. Decis. Cont.*, Austin, TX, pages 2217–2222, 1988.

35. C. Kenney and A.J. Laub. The matrix sign function. *IEEE Trans. Automat. Control*, 40(8):1330–1348, 1995.

36. C. Kenney, A.J. Laub, and M. Wette. A stability-enhancing scaling procedure for Schur-Riccati solvers. *Sys. Control Lett.*, 12:241–250, 1989.

37. D. L. Kleinman. On an iterative technique for Riccati equation computations. *IEEE Trans. Automat. Control*, AC-13:114–115, 1968.

38. P. Lancaster and L. Rodman. *The Algebraic Riccati Equation*. Oxford University Press, Oxford, 1995.

39. A.J. Laub. A Schur method for solving algebraic Riccati equations. *IEEE Trans. Automat. Control*, AC-24:913–921, 1979.

40. A.J. Laub. Algebraic aspects of generalized eigenvalue problems for solving Riccati equations. In C.I. Byrnes and A. Lindquist, editors, *Computational and Combinatorial Methods in Systems Theory*, pages 213–227. Elsevier (North-Holland), 1986.

41. A.N. Malyshev. Parallel algorithm for solving some spectral problems of linear algebra. *Linear Algebra Appl.*, 188/189:489–520, 1993.

42. V. Mehrmann. *The Autonomous Linear Quadratic Control Problem, Theory and Numerical Solution*. Number 163 in Lecture Notes in Control and Information Sciences. Springer-Verlag, Heidelberg, July 1991.

43. V. Mehrmann. A step toward a unified treatment of continuous and discrete time control problems. *Linear Algebra Appl.*, 241–243:749–779, 1996.

44. T. Pappas, A.J. Laub, and N.R. Sandell. On the numerical solution of the discrete-time algebraic Riccati equation. *IEEE Trans. Automat. Control*, AC-25:631–641, 1980.

45. T. Penzl. Numerical solution of generalized Lyapunov equations. *Adv. Comp. Math.*, 8:33–48, 1997.

46. J.D. Roberts. Linear model reduction and solution of the algebraic Riccati equation by use of the sign function. *Internat. J. Control*, 32:677–687, 1980. (Reprint of Technical Report No. TR-13, CUED/B-Control, Cambridge University, Engineering Department, 1971).

47. A. Saberi, P. Sannuti, and B.M. Chen. H_2 *Optimal Control*. Prentice-Hall, Hertfordshire, UK, 1995.

48. G. Schelfhout. *Model Reduction for Control Design*. PhD thesis, Dept. Electrical Engineering, KU Leuven, 3001 Leuven–Heverlee, Belgium, 1996.

49. V. Sima. *Algorithms for Linear-Quadratic Optimization*, volume 200 of *Pure and Applied Mathematics*. Marcel Dekker, Inc., New York, NY, 1996.

50. R.A. Smith. Matrix equation $XA+BX = C$. *SIAM J. Appl. Math.*, 16(1):198–201, 1968.

51. G.W. Stewart. A parallel implementation of the QR algorithm. *Parallel Computing*, 5:187–196, 1987.

52. X. Sun and E.S. Quintana-Ortí. Spectral division methods for block generalized Schur decompositions. PRISM Working Note #32, 1996. Available from http://www-c.mcs.anl.gov/Projects/PRISM.

53. P. Van Dooren. A generalized eigenvalue approach for solving Riccati equations. *SIAM J. Sci. Statist. Comput.*, 2:121–135, 1981.

54. A. Varga. A note on Hammarling's algorithm for the discrete Lyapunov equation. *Sys. Control Lett.*, 15(3):273–275, 1990.

55. A. Varga. Computation of Kronecker-like forms of a system pencil: Applications, algorithms and software. In *Proc. CACSD'96 Symposium, Dearborn, MI*, pages 77–82, 1996.

56. K. Zhou, J.C. Doyle, and K. Glover. *Robust and Optimal Control*. Prentice-Hall, Upper Saddle River, NJ, 1995.

ParaStation User Level Communication

Joachim M. Blum and Thomas M. Warschko and Walter F. Tichy

Institut für Programmstrukturen und Datenorganisation, Fakultät für Informatik, Am Fasanengarten 5, Universität Karlsruhe, D-76128 Karlsruhe, Germany

Summary. PULC (ParaStation User Level Communication) is a user-level communication library for workstation clusters. PULC provides a multi-user, multi-programming communication library for user-level communication on top of high-speed communication hardware. This paper describes the design of the communication subsystem, a first implementation on top of the ParaStation communication adapter, and benchmark results of this first implementation.

PULC removes the operating system from the communication path and offers a multi-process environment with user-space communication. Additionally, it moves some operating system functionality to the user-level to provide higher efficiency and flexibility. Message demultiplexing, protocol processing, hardware interfacing, and mutual exclusion of critical sections are all implemented in user-level. PULC offers the programmer multiple interfaces including TCP user-level sockets, MPI [CGH94], PVM [BDG+93], and Active Messages [CCHvE96]. Throughput and latency are close to the hardware performance (e.g., the TCP socket protocol has a latency of less than 9 μs).

Keywords: Workstation Cluster, Parallel and Distributed Computing, User-Level Communication, High-Speed Interconnects.

1. Introduction

Common network protocols are designed for general purpose communication in a LAN/ WAN environment. These protocols reside in the kernel of an operating system and are built to interact with divers communication hardware. To handle this diversity, many standardised layers exist. Each layer offers an interface through which the other layers can access its services. This layered architecture is useful for supporting divers hardware but leads to high and inefficient protocol stacks. Protocols which are using standardised interfaces of the operating system are unaware of superior hardware functionality and often reimplement features in software even if the hardware already provides them. Another inefficiency is due to copy operations between kernel- and user-space and within the kernel itself. To transmit a message the kernel has to copy the data from or to user-space. The copying between protected address space boundaries often adds more latency than the physical transmission of a message. In addition, the kernel copies the data several times from one buffer to another while traversing layers of the protocol stack. On the positive side, the traditional communication path with the kernel as single point of access to the hardware ensures correct interaction with the hardware and mutual exclusion of competing processes.

For parallel computation on clusters of workstations, many of the protocols which are designed for wide area networks are too inefficient. Therefore, cluster computing must take new approaches.

The most promising technique is to move protocol processing to user-level. This technique opens up the opportunity to investigate optimised protocols for parallel processing. With user-level protocols there is no need to use the standardised interfaces between the operating system and the device driver. Thus, the reimplementation of services in software which are already provided by the hardware can be avoided.

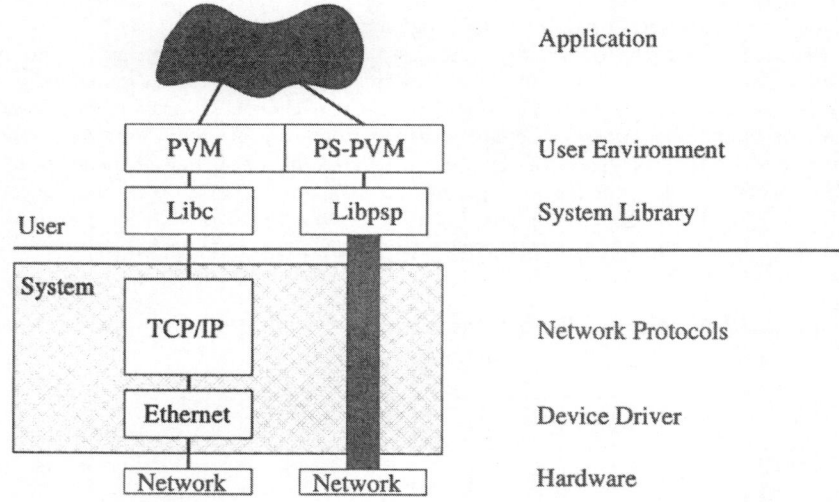

Fig. 1.1. User-level communication highway

User-level communication removes the kernel from the critical path of data transmission. Figure 1.1 shows how user-level communication shortcuts the access to the communication hardware. High-performance communication protocols are based on superior hardware features to speed up communication. Copying data between kernel- and user-space is avoided and the implementation of true zero-copy protocols is possible. These key issues minimise latency and lead to high throughput.

But user-level communication has also its drawbacks, because now the single point of access to the communication hardware, namely the kernel, is missing. Therefore many user-level communication libraries restrict the number of processes on a node to a single process. Enabling multiple processes on one node in user-level raises difficulties, but also offers a lot of benefits. Once problems, such as demultiplexing of messages and ensuring correct interaction between multiple processes are solved, the high-speed communication network can be used similar to a cluster with regular communication channels such as Unix sockets.

The goal of PULC is to provide a multi-user, multi-programming communication library for user-level communication on top of a high-speed communication hardware. The first implementation of PULC uses the ParaStation communication adapter, which is described in section 3.. Section 4. presents design alternatives and the optimisation techniques used. In section 5., this paper describes the implementation of PULC on top of ParaStation. Performance figures for two different hardware platforms are presented in section 6.. The last two sections present the conclusion and the plans for future work.

2. Related Work

There are several approaches targeting efficient parallel computing on workstation clusters. Some of them use custom hardware which support memory mapped communication. SHRIMP [DBDF97] builds a client server computing environment on top of a virtual shared memory. Similar to PULC, SHRIMP offers standardised interfaces such as Unix sockets. Digital's Memory Channel [FG97] is proprietary to DEC Alphas and uses address space mapping to transfer data from one process to another. On top of this low level mechanism Memory Channel offers MPI and PVM. Many recent parallel machines, e.g. IBM SP2, are a collection of regular workstations connected with a high speed interconnect.

Others use commodity hardware to implement communication subsystems. OSCAR (e.g. [JR97]) implements MPI on top of SCI cards. Fast Messages [CPL+97] and Active Messages [CCHvE96] are approaches for MPP systems ported to workstation clusters. Both offer low latency protocols which can be used to build other communication libraries on top. As an example the *Berkeley Fast Socket* protocol [SR97] is build on top of Active Messages. Similar to PULC, it provides an object code compatible socket interface. It's latency is about 75 μs and it's throughput reaches 33 MByte/s on Myrinet. But in contrast to PULC it has some restrictions in the use of fork() and exec() calls. Differently from the current PULC implementation, it provides interoperability between Fast Socket and other applications on the same cluster whereas PULC only provides it for out-of-cluster communication.

BIP [PT97] and Myricom GM [myr] implement low level interfaces to the Myrinet hardware. They are comparable with the PULC hardware abstraction layer but lack on higher protocols. Gamma [CC97] builds Active Messages on top of Fast Ethernet cards and gets nearly full performance by adding a system call and building a special protocol in the Linux kernel. UNet [WBvE97] uses Fast Ethernet and ATM to build an abstraction of the network interface. Dependent on the hardware support, they use kernel or user-level communication. They've even built a memory management system to enable DMA transfer to previously unpinned pages. Such a memory management is not implemented in PULC, but could be done as soon as hardware with DMA transfer and on-board processors are used.

In the Berkeley NOW project [ACP95], GLUnix offers a transparent global view of a cluster. As in PULC the network of workstations can be used similar to a single parallel machine. Their main focus is on Active Messages and therefore no other protocols are implemented.

3. ParaStation Hardware

The first implementation of PULC uses the ParaStation high-speed communication card as communication hardware. ParaStation is the reengineered MPP-network of Triton/1 [HWTP93], an MPP-system built at the University of Karlsruhe. Within a workstation cluster the ParaStation hardware is dedicated to parallel applications while the operating system continues to use standard hardware (e.g., Ethernet).

The network topology is based on a two-dimensional toroidal mesh. Table-based, self-routing packet switching transports data using virtual cut-through routing. The size of a packet can vary from 4 to 508 bytes. Packets are delivered in order and no packets are lost. Flow control is provided at link level and the unit of flow control is one packet. These features enable the software to use a simple fragmentation/defragmentation scheme. The communication processor used involves a routing delay of about $250ns$ per node and offers a maximum throughput of 16 MByte/s per link.

The ParaStation hardware resides on an interface card which plugs into the PCI-bus of the host system. Thus, it is possible to use ParaStation on a wide range of machines from different vendors. A more detailed description of the hardware is given in [WBT97].

4. Design of PULC

A new communication subsystem has to fulfil several issues to be helpful for parallel computing. First, parallel computing is highly dependent on very low latency and high throughput. The performance available for the user has to be close to the hardware limits. Therefore, deep protocol stacks are deadly for parallel computing.

Second, communication hardware is getting faster and more intelligent. New approaches, such as DMA transfers and communication processors on the interface cards enable high performance and flexible protocol processing. A new communication protocol has to be well-suited for these technologies.

Third, communication libraries offer different interfaces and semantics to the programmer. Not each communication library is well-suited for all users of a cluster of workstations. Therefore, a new communication subsystem has to offer different interfaces (communication libraries). It should also be extensible for new approaches in this field.

Fourth, workstation clusters are often used by several people for parallel computing. Having user-level access to the hardware usually prohibits simultaneous use of one node by several processes. A new approach should support a multi-process environment.

Therefore the main goal was that PULC supports fine grained parallel programming on workstation clusters while still providing the benefits of multi-process environments.

The most challenging problem in a multi-process environment is the demultiplexing of incoming messages. Generally there are three possible places where message demultiplexing can take place:

1. In the operating system: The operating system either checks periodically the hardware for pending messages or it is interrupted by the hardware when a message has arrived. The operating system unpacks the message header and stores the message data in a corresponding queue in kernel space. From the viewpoint of the kernel it doesn't matter if the message is for the currently running process or for any other process.

2. In the communication processor: Each communicating process has a memory area which is accessible by the communication hardware. The communication processor checks the header and decides where the message fragment should be stored. The number of accessible memory areas is limited, however. To solve this problem the communication system can either limit the number of communicating processes or it buffers the message intermediately, where the processes can access the data (in kernel space, common message area, or a trusted process' address space).

3. In the low level communication software in user-space: A user process periodically checks the hardware (or gets interrupted), and receives the message. If the message is not addressed to the receiving process, the process stores the message in a message pool accessible by the destination process.

In all cases the destination process executes a receive call and gets the data from the intermediate storage and stores it into the final destination. If the final destination is known and accessible at the time of message demultiplexing, the message can be stored directly in this area. This is known as *true zero copy* message reception [BBVvE95].

PULC divides the message demultiplexing and the message reception in two different modules. The *PULC message handler* demultiplexes incoming messages. This message handler can either run on the communication processor or it can be linked to each user process. The *PULC interface* receives the message for the process. It always runs in the address space of the communicating process. Both modules communicate by calling each other or by updating queues in a shared message area.

Another challenging task is resource management. Resources (buffers, sockets, etc.) are usually managed by the operating system. When moving the communication out of the kernel, this task can be accomplished by a regular user process. The resource manager has to control access to the hardware and cleans up after application shutdowns. In PULC, this task is performed by the *PULC resource manager*.

Figure 4.1 gives an overview of the major parts of PULC.

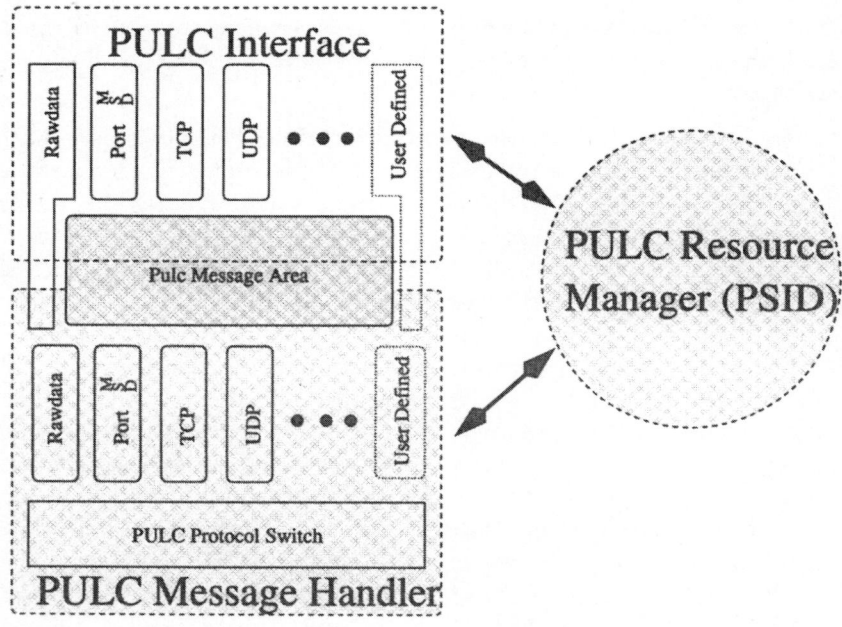

Fig. 4.1. PULC Architecture

PULC Programming Interface: This module acts as programming interface for any application. The design is not restricted to a particular interface definition such as Unix sockets. It is possible and reasonable to have several interfaces (or protocols) residing side by side, each accessible through its own API. Thus, different APIs and protocols can be implemented to support a different quality of service, ranging from standardised interfaces (i.e. TCP or UDP sockets), widely used programming environments (i.e. MPI or PVM), to specialised and proprietary APIs (ParaStation ports and a true zero copy protocol called Rawdata). All in all, the PULC interface is the programmer- visible interface to all implemented protocols.

PULC Message Handler: The message handler is responsible to handle all kind of (low level) data transfer, especially incoming and outgoing messages, and is the only part to interact directly with the hardware. It consists of a protocol-independent part and a specific implementation for each protocol defined within PULC. The protocol-independent part is the *protocol switch* which dispatches incoming messages and demultiplexes them to protocol specific *receive handlers*. To get high-speed communication, the protocols have to be as lean as possible. Thus, PULC protocols are not layered on top of each other; they reside side by side. Sending a message avoids any intermediate buffering. After checking the data buffer, the sender directly transfers the data to the hardware. The specific protocols inside the message handler are responsible for the coding of the protocol header information.

PULC Resource Manager: This module is implemented as a Unix daemon process (PSID) and supervises allocated resources, cleans up after application shutdowns, and controls access to common resources. Thus, it takes care of tasks usually managed by the operating system.

To be portable among different hardware platforms and operating systems, PULC implements all hardware and operating system specific parts in a module called hardware abstraction layer (HAL). Choosing an interconnection network with different quality of services would force the adoption of the PULC message handler to these services the communication hardware provides. E.g. if the hardware doesn't provide in-order delivery, the message handler has use the PULC functions which provide a reordering of fragments.

4.1 Resources provided by PULC

PULC supports the implementation of different protocols by offering a variety of resources together with associated interfaces to access them. The protocol independent resources are *message fragments, communication ports, semaphores,* and *process control blocks*. A *Message fragment* consists of a fragment control block and the message data and several fragments are concatenated to form a messages. Fragmentation is essential, because the underlying hardware has limited packet size. Therefore, PULC fragments have fixed sizes in memory and fragments are allocated as fixed sized memory blocks. This may waste memory, but allocating and managing variable sized chunks of memory is time consuming. Several messages together form a message queue of a *port*. The port is the basic addressable element in PULC communication. Different protocols use the ports as the channels to their communication partners. The resource manager frees a port and all fragments inside its message queue when no process is using it anymore. For the TCP/UDP protocol, another resource called *socket* is provided. A socket uses a port as its communication channel and stores additional socket specific information. To know about all the resources which are allocated by a specific process, PULC keeps information about a process in a *process control block*. These information are use to clean up the allocated resources when the process exits.

If the PULC message handler runs on the host processor, several processes can access common resources. To ensure mutual exclusion of processes to protect critical sections (manipulating queues or other resources), PULC provides user-level semaphores. Processor specific atomic operations, such as *test and set* or *load/store locked*, are used to implement them.

For an easy implementation of the protocols, PULC offers support functions to access the resources. E. g., PULC provides routines to store fragments into the message queue of a port. There are only three different strategies to store fragments in a message queue. PULC classifies the ports and the protocol calls its appropriate routine. In general, message queues of a port can be classified in the following way:

- Single stream: All fragments are stored in a single queue disregarding any message boundaries or message sources.
- Multiple Stream: All fragments of a the same source are stored in a queue. Fragments of different sources are stored in different queues.
- Datagrams: Fragments of different messages and different sources are stored in different queues. Each message has its own queue.

In addition to this classification these routines have to know if the hardware delivers the fragments of a message in order or if a reordering of the fragments is necessary. Fortunately the ParaStation hardware provides in-order delivery. The same holds for our HAL implementation for the Myrinet card.

4.2 PSID: The PULC Coordinator

Since PULC is fully implemented in user-space, the operating system does not manage the resources. This task is done by a resource manager (PSID: ParaStation Daemon). It cleans up resources of dead processes and organises access to the message area. Before a process can communicate with PULC, the process has to register with the PSID. The PSID can grant or deny access to the message area and the hardware.

The PSID also checks if the version used by the PULC interface and the PULC message handler are compatible. The version check makes corruption of data impossible. The PSID can restrict the access to the communication subsystem to a specific user or a maximum number of processes. This enables the cluster to run in an optimised way, since multiple processes slow down application execution due to scheduling overhead.

All PSIDs are connected to each other. They exchange local information and transmit demands of local processes to the PSID of the destination node. With this cooperation, PULC offers a distributed resource management. The single system semantic of PULC is ensured by the PSIDs. They spawn and kill client processes on demand of other processes. PULC transfers remote spawning or killing requests to the PSID of the destination node. PULC uses operating system functionality to spawn and kill the processes on the local node. The spawned process runs with same user id as the spawning process.

PULC redirect the output of spawned process on the terminal of the mother process. Therefore it offers a transparent view of the cluster.

The PSIDs periodically exchange load information. With this information PULC provides load balancing when spawning new tasks. There are several spawning strategies possible:

- Spawn a new task on the specified node: No selection is done by PULC. The spawn request is transfered to the remote PSID, which creates the new task. A new task identifier is returned in the result.
- Spawn a task on the next node: PULC keeps track of the node which was used to spawn the last task on. This strategy selects the next node by incrementing the node number.
- Spawn a task on a unloaded node: Before spawning, PULC orders the available nodes by their load. After that, PULC spawns on the nodes with the least heavy load.

These strategies allow a PULC cluster to run in a balanced fashion, while still allowing the programmer to specify the exact node, when problem solved requires a specific communication pattern.

4.3 The PULC Message Handler

The PULC message handler is responsible for receiving and sending messages.

4.3.1 Sending messages. Sending a message avoids any intermediate buffering. After checking the buffer, the sender directly transfers the data to the hardware. The specific protocols inside the message handler are responsible for the coding of the protocol header information. PULC doesn't restrict the length or form of the header. PULC just specifies the form of the hardware header with its protocol id. The rest of the message header must be interpretable by the protocol specific receive handler. If the receiver is on the local node, the receive handler optimises message transfer by directly calling the appropriate receive handler of the protocol.

4.3.2 Receiving a message. If the hardware supports a demultiplexing of messages, the PULC message handler runs on the communication processor of the hardware. It has some memory common with each receiving process. The data can directly be transferred to this memory area.

The first generation of the ParaStation card does not support any message demultiplexing at hardware level and so the PULC message handler has to be part of a process and runs in the address space of its host process. During reception of a message the PULC message handler can detect that it is not addressed to its own host process. It has to store the message in a commonly accessible message area (SHM) where the destination process can read the message. Whether a message is received with true zero copy, or it is stored intermediately, depends on the used protocol.

The *PULC protocol switch* reads only the hardware header of the message and the protocol identifier. After decoding the id, the protocol switch directly transfers control to the receive handler of the protocol, which reads the rest of the message. This *header forwarding* is extremely fast and does not do any unnecessary copy of the data. The protocols can store the data directly in user data structures, as it is done in the rawdata protocol, or queue the data in the a message queue (TCP,UDP,PORT-M/S/D). Other protocols can do it in their specific way.

PULC allows multiple processes to communicate concurrently since different processes can use different communication ports. The protocol interface and the protocol receive handler have to ensure the correct cooperation while receiving a message.

In a hardware-supported PULC message handler a shared port must reside in an area where both processes can access it. If both processes trust each other, the port can reside in a message area which is mapped in both processes. If they do not trust each other, the message handler has to protect the port in its own memory area. Both processes would have to access the message in the port through the message handler API. This is much slower than the solution with a direct access.

4.4 PULC Interface

Each protocol in the message handler can have its own interface. The interface is the counterpart of the message handler. The message handler receives a message and puts it in the message area whereas the interface functions get these messages as soon as they are received completely. The cooperation between the interface functions and the receive handler of the protocol includes correct locking of the port and its message queues. Correct interaction is necessary since PULC doesn't have control of the scheduling decisions of the Operating System. Thus the receive handler could be in a critical section while the Operating System switches to a process which conflicts with this critical section. This could destroy consistency.

A process can use several interfaces at the same time. E. g., it can use the sockets for regular communication and PULC's ability to spawn processes through the Port-M interface.

The socket interface to PULC is the same as for BSD Unix sockets. This interface allows easy porting of applications and libraries to the fast communication protocols. Destinations which are not reachable inside the PULC cluster are redirected to regular operating system calls. All communication in Unix is based on the socket interface. By providing a compatible interface, porting applications to PULC is just a relinking.

PULC sockets use specially tuned methods with caching of recently used structures. This allows an extremely fast communication with minimal protocol overhead. Each socket has a port as its communication channel. The socket receive handler only knows about the ports and uses different enqueuing strategies for UDP (datagram ports) and TCP sockets (single stream ports). The socket interface provides the interaction between the communication ports and the socket descriptor. Sockets can be shared among different processes due to a fork() call and can be inherited by a exec() call. During fork(), the socket is duplicated but both sockets share the same communication port (the count attribute of the port is incremented). Thus, both processes have access to the message queue of the socket. After an exec() and a reconnection to PULC the sockets and the ports of the message area are inserted into the private socket and port descriptor tables. Therefore the process has access to these abstractions again.

4.4.1 Communication Libraries on Top of PULC. There are several communication libraries built on top of PULC. Most of them are just the standard Unix distributions on top of sockets. The applications which use these libraries just have to be linked with the PULC sockets. These libraries include P4 [BL92] and tcgmsg [Har91]. Others such as PVM [BDG+93] have been changed [BWT96] in a way that they can be used simultaneously to the standard sockets. This enables a direct comparison of the operating system communication and PULC. The implementation shows that PVM adds a significant overhead to the regular socket communication. This isn't obvious when PVM is used with regular sockets (see section 6.). This lead to a new approach[OBWT97], which optimised PVM on top of the port-D interface. PULC already provides efficient and flexible buffer management and therefore this functionality could be eliminated in the PVM source. This PSPVM2 is still interoperable with other PVMs running on any other cluster or supercomputer. PSPVM2 views the whole PULC cluster as one single parallel system.

The PULC MPI implementation is based on MPICH. MPICH provides a *channel* interface which hardware manufacturers can use to port MPICH to their own communication subsystem. This channel interface is implemented on top of PULC's port-D protocol. MPICH on PULC uses PULC's dynamic process creation at startup. The implementation is well-suited for MPI-2, which is supporting dynamic process creation at run-time. It is possible to support MPI directly as an interface to PULC. Most of the functionality is already provided in the Port protocol.

5. Implementation

There exists two implementations of PULC, one for Intel-PCs running Linux and the other for DEC-Alpha workstations running Digital Unix. Both of them use the ParaStation high-speed communication card as communication hardware. As described in section 3., ParaStation offers many useful services to the software protocols, but unfortunately, it has no communication processor on board. Thus, the implementation uses a commonly accessible shared memory area (see figure 5.1) to store messages and control information. The PULC library itself, in particular the PULC message handler, acts as the trusted base within the whole system. The library is statically linked to each application and ensures correct interaction between all parts of the system. The operating system is only invoked at system and application startup.

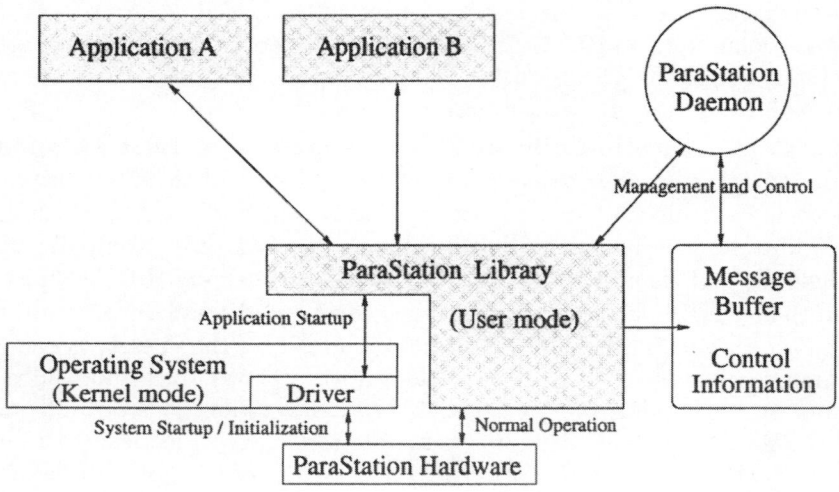

Fig. 5.1. ParaStation User-Level Communication

Operating system and hardware specific parts of the library are placed in a separate module (the HAL). Therefore only this module has to be change when porting PULC to another platform. This is currently done for the Myrinet communication card.

Since the message handler is part of each process, the message area is mapped into each communicating process. This enables the message handler to receive messages for different processes and to demultiplex them to the correct receiving port. The multi-process ability of this solution is quite expensive due to the locking of ports, as well as locking data transmission to and from the hardware.

Using a commonly accessible message area suffers from a (minimal) lack of protection. The implemented message demultiplexing implies that all communicating processes trust each other. A malfunctioning process accessing the common message area directly is able to corrupt data owned by another process and can possibly crash the system. But the risk is minimal since the address space of Alpha processors (64 bit addresses) is approximately 2^{52} times larger than the size of the message area (configuration dependent). If a wrong address is produced once a second, a corruption of data in the message area could happen approximately every 2^{27} years. On the other hand, the trusted system is open for malicious hackers with access to the cluster, but this is a tolerable disadvantage when compared to the performance benefits gained from this policy. If this lack of protection is considered harmful, PULC can be configured to allow only a specific number of process or only a specific user access to the communication system concurrently.

6. Performance Evaluation

This section shows the efficiency of the PULC implementation. The performance of the different protocols is presented and the results are explained. Performance is measured on a cluster where each node in the cluster is a fully configured workstation.

6.1 Communication Benchmark

Communication subsystems can be compared by evaluating the latency and throughput of the systems. PULC offers several interfaces and runs on several hardware/operating system environments. Our test clusters consist of two Pentium PCs (166MHz) running Linux 2.0 and two Alphas 21164 (500MHz) running Digital Unix 4.0.

PULC's results are compared with the operating system performance whenever possible. The test consists of a pairwise exchange program to measure the throughput and a ping-pong test to measure the latency, which is calculated by the round trip time divided by two.

```
/* Sender code */              /* Receiver Code */
StartTimer()                   StartTimer()
for(i=0;i<LOOPS;i++)           for(i=0;i<LOOPS;i++)
   SendMessage(buffer,size)       SendMessage(buffer,size)
   ReceiveMessage(buffer,size)    ReceiveMessage(buffer,size)
end                            end
EndTimer()                     EndTimer()
```

Fig. 6.1. Pairwise exchange test program

In the exchange program(see figure 6.1) both processes send a message to the other and wait for the receive of the other. Therefore both processes execute always the same command.

In the ping-pong test(see figure 6.2), one process sends and the other receives, after receiving the message, the receiver sends the message back to the sender.

Surprisingly, the slower Pentium system performs better than the Alpha system in both latency and throughput at lower layers (see Table 6.3). This

```
/* Sender code */              /* Receiver Code */
StartTimer()                   StartTimer()
for(i=0;i<LOOPS;i++)           for(i=0;i<LOOPS;i++)
   SendMessage(buffer,size)       ReceiveMessage(buffer,size)
   ReceiveMessage(buffer,size)    SendMessage(buffer,size)
end                            end
EndTimer()                     EndTimer()
```

Fig. 6.2. Pingpong test program

| protocol-layer | Alpha 21164, 500 MHz | | | | Pentium, 166 MHz | | | |
| | ParaStation | | OS/Ethernet | | ParaStation | | OS/Ethernet | |
	latency [μs]	band-width [MB/s]	latency [μs]	band-width [MB/s]	latency [μs]	band-width [MB/s]	latency [μs]	band-width [MB/s]
hardware	4.2	12.4			3.4	15.6		
rawdata	5.1	11.9			6.4	14.8		
port-M	8.9	9.5			14.6	11.8		
socket	9.0	9.6	115	1.1	13.9	11.9	308	1.0
PVM	78.0	8.7	289	1.0	158	7.8	776	0.8
PVM (port-M)	11.5	9.4			27.2	11.5		
socket (self)	3.2	318.8	390	33.0	19.0	107.0	578	30.0

Fig. 6.3. Communication Performance of the PULC system

is due to the architectural differences between the two systems. In particular Alpha's capability to combine writes to the same memory location requires additional synchronisation. As the ParaStation communication interface is implemented as a FIFO buffer, memory barrier instructions (MB) are inserted after each write to the FIFO. The MB instruction itself waits for all outstanding read and write operations and thus limits the performance. In addition to the write combining bottleneck, the semaphore mechanism which is used in the Alphas is not as fast as the semaphores on the Pentium. A lock operation on the Alphas takes about 1 μs whereas a Pentium provides mutual exclusion within 200 ns. The semaphore bottleneck is also visible in multi-process protocols .

The line titled *hardware* in the table above shows the performance of the hardware abstraction layer described in section 5. and reflects the maximum performance one can get using ParaStation on the stated workstation.

The additional latency of 0.9 μs on the Alpha (3 μs on the Pentium) introduced by the rawdata protocol is due to guarantee mutual exclusion and correct interaction between concurrent processes. Multiple ports are addressed by the port protocol. This multi-programming environment adds additional 3.8 μs (8.2 μs on Pentium) to the rawdata protocol. Providing full TCP socket functionality within 9 μs opens up a wide range of fine grained parallel programs on top of sockets. As reported in [BWT96] standard programming environments, such as PVM, add a huge amount of latency to the sockets. This is not noticeable when slow operating system sockets are used. When running PVM on top of PULC sockets 89 % (91% on a Pentium) of the latency is caused by these packages. These numbers show that these standard environments do not adopt well to high speed protocols. This lead to an optimisation of PVM on top of ports. As reported, the port-M protocol already provides most of the functionality that PVM has to implement on top of sockets, e.g. a very inefficiently implemented buffer management. Using the whole functionality of PULC, PVM only adds 2.5 μs (13.4 μs on the Pentium) to the port-M protocol latency. This shows that even with standardised interfaces, PULC offers great performance.

6.2 Application Benchmark

A user doesn't focus on the pure message passing numbers. The more important fact is how the system behaves with real applications. This section presents performance measurements of the system in two different areas with two different communication libraries. First, the PVM implementation is measured with a widely used linear algebra package and second, the NAS parallel benchmark is used to compare the system to the Cray T3E, a dedicated parallel system.

6.2.1 Linear Algebra Package on top of PVM. This test uses a linear equation solver for dense systems, called *xslu*, which is part of ScaLAPack [CDD+95], a popular linear algebra package. ScaLAPack uses BLACS as a communication interface to different communication libraries such as MPI or PVM. In this test PVM acts as the underlying subsystem. The test is run on up to 8 Alphas (500 MHz, 256 MB Ram, Digital Unix 4.0b) connected with the ParaStation hardware.

ScaLAPACK on 160 MBit ParaStation with PSPVM2							
Problem size (n)	1 work-station MFlop	2 workstations		4 workstations		8 workstations	
		Speed up	MFlop	Speed up	MFlop	Speed up	MFlop
3000	443	1.75	759	2.18	966	2.62	1161
4000		1.70	753	2.47	1093	3.23	1431
5000		1.85	821	2.68	1187	3.74	1656
6000				2.90	1285	4.04	1789
7000				3.11	1379	4.37	1939
8000						4.61	2044
9000						5.07	2247
10000						5.22	2312

The table shows that on top of ParaStation the application scales good in terms of problem size and number of processors. A maximum performance of 2.3 GFlops is achieved which compare quite well to dedicated parallel machines. Unfortunately, *xslu* depends on high bandwidth and thus ParaStation with about 10MByte/s throughput is the real bottleneck.

6.2.2 NAS Parallel Benchmark on top of MPI. The second test measures the performance of the system with the NAS Parallel Benchmark suite. This suite is widely used to compare different parallel platforms. It is based on top of MPI and it runs without any source code modifications.

Some tests require a power of two number and others a square number of processors. Therefore not all columns are filled in each test.

The FT benchmark is a 3-D FFT application. MG is a multi grid benchmark. The LU benchmark does a matrix decomposition. It is the only benchmark in the NPB 2.0 suite that sends large numbers of very small (40 byte) messages. Therefore it shows the performance of the communication subsystem for fine-grained applications. EP (embarrassing parallel) usually shows the performance of a single node. The communication subsystem is not used frequently. IS (integer sort) sorts a number of integers in parallel. CG (conjugate gradient) and IS exchange a lot of data in huge data chunks. All of these codes require a power-of-two number of processors. The SP(pentadiagonal solver) and BT (block diagonal solver) algorithms are more coarse grained implementations. They solve three sets of uncoupled systems of equation using multi partition schemes. Both the SP and BT codes require a square number of processors[1].

As a comparison the numbers achieved by a Cray T3E-900 are presented, which has similar processors per node. The communication subsystem is a highly optimised three dimensional torus. The third level cache is eliminated and therefore tests which are memory intensive run good on the T3E and test which can mostly run in the cache perform worse than on regular workstations.

The Cray T3E provides a bandwidth of about 300 MB/s and a latency of about 2 μs at hardware level . Therefore one could expect a comparable performance for tests which do not depend on bandwidth.

---

[1] For a detailed description of the tests please refer to http://science.nas.nasa.gov/Software/NPB/

NAS Parallel Benchmark on ParaStation and T3E				
Test on no. of nodes Class A	1	2	4	8
BT ParaStation	n/a	n/a	144.4	n/a
BT Cray T3E-900	n/a	n/a	226.7	n/a
CG ParaStation	19.7	44.5	55.15	75.72
CG Cray T3E-900		46.5	86.0	241.4
EP ParaStation	4			31.96
EP Cray T3E-900		5.2	10.4	20.8
IS ParaStation	1.46	2.23	2.15	3.72
IS Cray T3E-900		6.6	12.9	22.1
LU ParaStation				579.48
LU Cray T3E-900		134.4	270.4	531.1
MG ParaStation				299.77
MG Cray T3E-900		171.5	313.9	720.8
FT ParaStation				86.02
FT Cray T3E-900		85.3	169.5	330.4
SP ParaStation		n/a	106.55	
SP Cray T3E-900	n/a	n/a	172.4	n/a

The table shows the results measured on ParaStation and the results taken from the NAS homepage for the T3E. Higher numbers mean better performance.

In some test ParaStation behaves very good compared to the expensive dedicated system. Unsurprisingly, these are the test with minimal communication (EP) and the test with many small messages (LU), because the MPI latency is about the same on both systems.

During the other tests, which depend on high throughput, the ParaStation system began to swap received messages to user-space due to an overflow of the message storage. This effect limited the performance and a new version PULC will optimise this swapping. But even with this swapping effect, the resulting numbers are even better than other much more expensive dedicated machines, such as the IBM SP/2, SGI Origin, and Cray T3D[2].

7. Conclusion

PULC shows extremely good performance on all protocols. Many programs benefit from the high speed of the PULC library. PULC's design offers nearly the raw performance of high-speed communication cards to the user while still providing standardised interfaces. The design goal of a multi-user/multi-programming environment at full speed was reached. PULC is also easily

---

[2] See the performance numbers at http://science.nas.nasa.gov/Software/NPB/NPB2Results/index.html

adapted to new hardware and brings efficient parallel processing to workstations clusters. Presented performance results compare well with parallel systems. PULC is included in the ParaStation system, which was introduced into market in 1996[3] and is currently ported to the Myrinet communication adapter. First results show that TCP throughput will raise up to 60 MB/s while latency will increase to about 20 μs. These are first numbers, where the message handler still runs in low level software.

The pure user-level approach in the ParaStation system showed many drawbacks which could only be resolved by introducing some security holes and performance limitations. Especially the performance of multiple processes on one node is dependent on the coscheduling strategies used. Unfortunately, I couldn't find a coscheduling strategy which is good for multithreaded and interprocess communication at the same time. More research has to be done in this area.

8. Future Work

In future, the ParaStation team will work on next-generation ParaStation hardware. Current issues for a new network design are fiber optic links, optimised packet switching, and flexible DMA engines to reach an application-to-application bandwidth of about 100 MByte/s. Similar to Myrinet the new hardware will be able to run the message handler on board. Therefore any security whole will be eliminated.

PULC and the full ParaStation environment is going to be ported to other systems with PCI bus (e.g., Sun/Solaris, IBM-PowerPC/AIX, SGI/IRIX). PULC itself will be ported to other communication hardware. Additional interfaces and protocols, such as Active Messages, are considered to be implemented as protocols inside of PULC. This would give them a performance boost over the current implementation which are implemented on top of sockets or ports.

Furthermore, the analysis of the message demultiplexing showed that this task can be done in the OS, the communication hardware, and the low level software. All three cases will be implemented and evaluated.

References

[ACP95] Thomas E. Anderson, David E. Culler, and David A. Patterson. A Case for NOW (Network of Workstations). *IEEE Micro*, 15(1):54–64, February 1995.

---

[3] For further information, see http://wwwipd.ira.uka.de/ParaStation.

[BBVvE95] Anindya Basu, Vineet Buch, Werner Vogels, and Thorsten von Eicken. U-net: A user-level network interface for parallel and distributed computing. In *Proc. of the 15th ACM Symposium on Operating Systems Principles, Copper Mountain, Colorado*, December 3-6, 1995.

[BDG+93] A. Beguelin, J. Dongarra, Al Geist, W. Jiang, R. Manchek, and V. Sunderam. *PVM 3 User's Guide and Reference Manual*. ORNL/TM-12187, Oak Ridge National Laboratory, May 1993.

[BL92] Ralph Buttler and Ewing Lusk. *User's Guide to the p4 Parallel Programmimg System*. ANL-92/17, Argonne National Laboratory, October 1992.

[BWT96] Joachim M. Blum, Thomas M. Warschko, and Walter F. Tichy. PSPVM:Implementing PVM on a high-speed Interconnect for Workstation Clusters. In *Proc. of 3rd Euro PVM Users' Group Meeting*, Munich, Germany, Oct.7-9, 1996.

[CC97] G. Chiola and G. Ciaccio. Gamma: a low-cost network of workstations based on active messages. In *5th EUROMICRO workshop on Parallel and Distributed Processing*, 1997.

[CCHvE96] Chi-Chao Chang, Grzegorz Czajkowski, Chris Hawblitzel, and Thorsten von Eicken. Low-Latency Communication on the IBM RISC System/6000 SP. In *ACM/IEEE Supercomputing '96, Pittsburgh, PA*, November 1996.

[CDD+95] J. Choi, J. Demmel, I. Dhillon, J. Dongarra, S. Ostrouchov, A. Petitet, K. Stanley, D. Walker, and R. C. Whaley. ScaLAPACK: A Portable Linear Algebra Library for Distributed Memory Computers – Design Issues and Performance. Technical Report UT CS-95-283, LAPACK Working Note #95, University of Tennesee, 1995.

[CGH94] Lyndon Clarke, Ian Glendinning, and Rolf Hempel. The MPI Message Passing Interface Standard. Technical report, March 94.

[CPL+97] Chien, Pakin, Lauria, Buchanan, Hane, Giannini, and Prusakova. High Performance Virtual Machines (HPVM): Clusters with Supercomputing APIs and Performance. In *Eighth SIAM Conference on Parallel Processing for Scientific Computing (PP97)*, 1997.

[DBDF97] Stefanos N. Damianakis, Angelos Bilas, Cezarz Dubnicki, and Edward W. Felten. Client Server Computing on Shrimp. *IEEE Micro*, pages 8-17, January/February 1997.

[FG97] Marco Fillo and Richard B. Gillett. Architecture and implementation of memory channel 2. Technical report, Digital Equipment Coropration, 9 1997.

[Har91] R. J. Harrison. Portable tools and applications for parallel computers. *International Journal on Quantum Chem.*, 40:847–863, 1991.

[HWTP93] Christian G. Herter, Thomas M. Warschko, Walter F. Tichy, and Michael Philippsen. Triton/1: A massively-parallel mixed-mode computer designed to support high level languages. In *7th International Parallel Processing Symposium, Proc. of 2nd Workshop on Heterogeneous Processing*, pages 65–70, Newport Beach, CA, April 13–16, 1993.

[JR97] H. Jin and W. Rehm. Performance of message passing and shared memory on sci-based smp cluster. In *Proceedings of Fifth High Performance Computing Symposium, Atlanta, Georgia*, April 6-10 1997.

[myr] *The GM API.*

[OBWT97] Patrick Ohly, Joachim M. Blum, Thomas M. Warschko, and Walter F. Tichy. PSPVM2:PVM for ParaStation. In *Proc. of 1st Workshop on Cluster Computing*, Chemnitz, Germany, Nov.6-7, 1997.

[PT97] Loic Prylli and Bernard Tourancheau. New protocol design for high performance networking. Technical report, LIP-ENS Lyon, 69364 Lyon, France, 1997.

[SR97] David Culler Steve Rodrigues, Tom Anderson. High-performance local-area communication using fast sockets. In *USENIX '97*, 1997.

[WBT97] Thomas M. Warschko, Joachim M. Blum, and Walter F. Tichy. ParaStation: Efficient Parallel Computing by Clustering Workstations: Design and Evaluation. *Journal of Systems Architecture*, 1997. Elsevier Science Inc., New York, NY 10010. *To appear.*

[WBvE97] Matt Welsh, Anindya Basu, and Thorsten von Eicken. ATM and Fast Ethernet Network Interfaces for user-level communication. In *roceedings of the Third International Symposium on High Performance Computer Architecture (HPCA), San Antonio*, 1997.

TOP-C: Task-Oriented Parallel C for Distributed and Shared Memory

Gene Cooperman*

College of Computer Science
Northeastern University
Boston, MA 02115
gene@ccs.neu.edu

Summary. The "holy grail" of parallel software systems is a parallel programming language that will be as easy to use as a sequential one, while maintaining most of the potential efficiency of the underlying parallel hardware. TOP-C (Task-Oriented Parallel C) attempts such a model by presenting a task abstraction that hides much of the details of the underlying hardware. DSM (Distributed Shared Memory) also attemps such a model, but along an orthogonal direction. By presenting a shared memory model of memory, it hides much of the details of message-passing required by the underlying hardware. This article reviews the TOP-C model and then presents ongoing research on combining the advantages of both models in a single system.

1. Introduction

This paper proposes the TOP-C model as a way to easily organize computations on DSM systems with many processors, while maintaining high concurrency. The proposed model allows the application writer to implicitly declare segments of his environment that correspond to the program objects that he is using. The segments are implicit in that the application writer need only declare to TOP-C which segments are modified by a given routine.

TOP-C has been successful in executing many large, parallel applications [4, 8, 10, 11, 12, 17]. TOP-C is implemented as a C library, and does not require a modification of the programming language of the application. As with any C library, the TOP-C library can also be used by a C++ program. One can choose any of three TOP-C libraries to choose between: SMP (Symmetric MultiProcessing, or shared memory) architectures, distributed memory architectures, and a sequential architecture. The application writer may continue to use his or her favorite programming language as long as that language has an interface to C libraries.

It should be noted that current high-end SMP architectures (many processors) are quite similar to DSM systems with hardware support. Hence, there appears to be a gradual progression from low-latency SMP through medium-latency DSM systems, with no sharp dividing line. Accordingly, we talk about the SMP version of the TOP-C model with the intention that this also applies to DSM.

---

* Supported in part by NSF Grant CCR-9732330.

Section 2. describes the TOP-C model. Section 3. then motivates why the model needs to be extended when the environment uses a lot of memory. Section 4. then describes a natural way to enhance the TOP-C model by providing an application abstraction of *segments*. If the application program is an object-oriented C++ program, then each segment will often correspond to an object.

Section 5. then describes how the enhanced TOP-C model maps onto a DSM architecture. In particular, there is an important issue of how the multiple segments of the TOP-C environment map onto the multiple pages of a DSM system. We are still in the process of obtaining a suitable DSM, and so we have not had the opportunity to test TOP-C in this environment. Nevertheless, a paper analysis describes many of the DSM features that we expect will be necessary for TOP-C to run efficiently on top of DSM.

2. The TOP-C Model

The TOP-C model has been described in [7]. The model is sufficiently flexible to also be easily ported to interactive languages [5, 6]. The model has also been applied to metacomputing [9], due to the ease of checkpointing the current state and sending a copy of that state to a new process joining the computation. The model has been successfully used in a variety of applications [4, 8, 10, 11, 12, 17].

The model allows a single file of application code to be executed as a sequential, SMP, or distributed memory application, by simply linking with a different library. Portability is emphasized by building on top of a POSIX threads library (for SMP) or MPI [14] (for distributed memory). MPI was chosen as a widely available message-passing standard, with good efficiency. The TOP-C distribution also contains its own small, unoptimized subset implementation of MPI, allowing one to quickly set up a small, self-contained application. Further, the portability of TOP-C makes it easy to re-target to another message-passing platform, such as PVM. TOP-C is freely distributed at `ftp://ftp.ccs.neu.edu/pub/people/gene/top-c/`.

The programming style is SPMD (Single Program, Multiple Data). This is executed in the context of a master-slave architecture and an environment or global state. This environment receives lazy, incremental updates, in a fashion that will be made clear later.

The user interface has purposely been kept simple by restricting the user interface to a single, primary system call: `master_slave()`. That function requires as parameters, four application functions declared by the user: `set_task_input()`, `do_task()`, `get_task_output()` and `update_environment()`. The philosophy is to present the higher-level task abstraction to the application. This should be contrasted to lower level interfaces that present either a message-passing abstraction or a shared memory abstraction.

The *task* is the first abstraction. The first two application-defined functions, `set_task_input()` and `do_task()`, implicitly define the input-output behavior of the task. The third function, `get_task_output()`, returns an *action* to be taken, based upon the task output. The three primary actions are `NO_ACTION`, `REDO`, and `UPDATE`. When the application specifies the `UPDATE` action, the application-specific function, `update_environment()` is called on each process (including the master). The routine, `update_environment()` uses the task output to introduce an incremental update.

The figure below illustrates the flow of control between master and each of several slaves for a task.

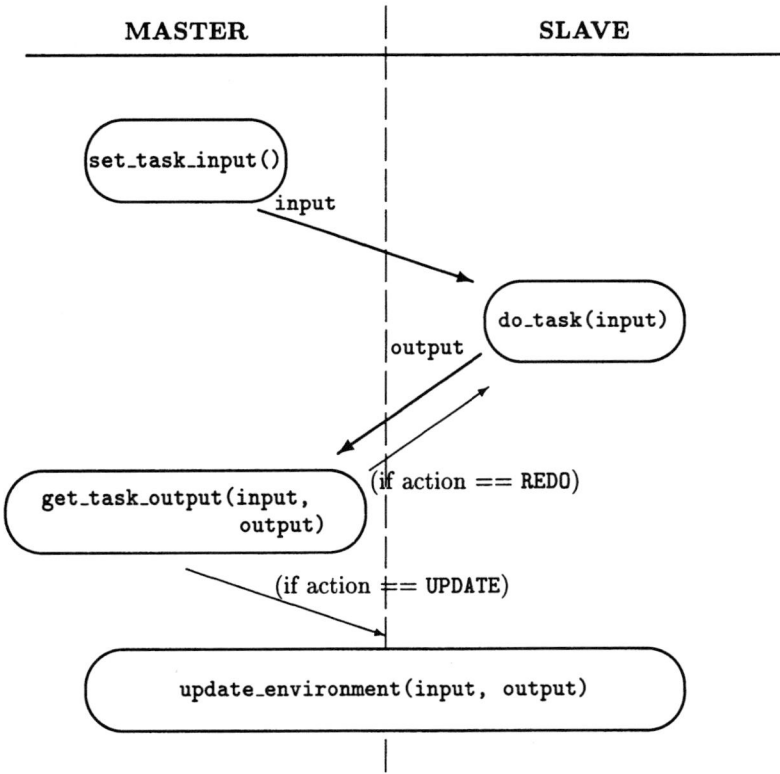

Figure 2.1. TOP-C Programmer's Model

A process always completes its current operation, before reading a pending message for the next operation. A message from the master to a slave requesting an update to the slave's copy of the environment always takes precedence over a message specifying a new task. A `REDO` action results in the original task input being sent back to the same slave, typically after a message to update the environment.

In addition to the task, the second key to the TOP-C model is the *environment* (global state). The environment, like the task, is not explicitly declared by the application. Rather, it is implicitly defined by the application routines. Each of the four application routines may read the most recent local environment. However, only update_environment() may modify the data in the environment. The environment is read and written only by the application routines, and not by any TOP-C system routine.

The most important issue for TOP-C is to allow tasks to concurrently read and make a request to modify the environment. As seen in figure 2., a decision to modify the environment can only happen if get_task_output() returns an UPDATE action. This action both allows TOP-C to record at what "time" the environment was last modified, and to then call update_environment(). In the case of distributed memory, update_environment() is called on each process, including the master. In the case of shared memory or sequential code, update_environment() is called only on the master.

3. Concurrency Issues for Shared Memory

Note that for any shared memory system (not just TOP-C), there is an inherent reader-writer problem when one thread (in this case the master) writes to a region of memory while another thread is reading the same region of memory. The TOP-C methodology reduces this to a single writer-multiple reader problem. The TOP-C solution is to allow both memory operations to proceed, but to later detect the memory collision and account for it. The method is analogous to the method of "optimistic concurrency" in distributed databases.

Concurrency is maintained in TOP-C in an application-specific manner. The system provides a utility, is_up_to_date(), callable from within the application routine, update_environment(). This routine will determine whether the environment was modified on the master after the task input under consideration was generated on the master, and before the task output was received by the master. Any memory collisions are a special case of this more general situation, and so will also be detected.

If the environment was not modified, then the application trivially attains perfect concurrency. If the environment was modified, then the application routine, get_task_output(), may either return a REDO action, or employ an application-specific technique to "patch" the task output to take account of the modified environment. The get_task_output() routine receives the task input, in addition to the task output, precisely to make it easier to patch the output.

The effect of this concurrency strategy is that the environment acts as a single large "page" of memory. If any task causes the page to be "touched", then all processes may have to read an update to the page. The page update is handled in a lazy manner, providing a type of latency hiding. However, the presence of only a single, atomic environment effectively means that false sharing of data is widespread within the system. This is the current state of TOP-C.

The issue of false sharing of a single monolithic environment tends to especially hurt TOP-C applications that require a shared memory model. This occurs because of a natural dichotomy in TOP-C applications. Applications that require only a smaller amount of memory for the environment tend to run comfortably in the distributed memory model, in which the environment is replicated among many processes. However, applications requiring a large amount of memory for the environment will prefer a shared memory environment. Otherwise, the cost of physical memory often makes it uneconomic to find a site with sufficient memory on each processor to allow the replication of a large environment within each process.

Thus, large environments favor a shared memory model. This software view of memory can be achieved either by an SMP architecture or by a DSM architecture on top of many workstations. The next section discusses an experimental version of TOP-C that better accommodates a shared view of memory by providing multiple pages, or segments, within the environment.

4. Multiple Segments within an Environment

In the experimental TOP-C model, the environment is replaced by multiple segments. The use of multiple segments forces us to change one command and one action in the TOP-C model: is_up_to_date() and UPDATE. All other aspects of the TOP-C retain the same simplicity.

Recall that the TOP-C environment is never explicitly declared. Rather it is implicitly defined by the application programmer as those portions of memory within a slave process that are read by do_task() and that are read or written to by update_environment(). (In addition, the master routines set_task_input() and get_task_output() may also read the environment.)

In our implementation of segments, we retain this idea that segments are implicit referenced, but never explicitly declared. Since the environment is replaced by segments, the utility is_up_to_date() must be extended to include a single parameter, specifying for which segments the query is being made. Currently, this parameter is specified as a string representing a set of numbers. For example, "1,3,5-7" represents segments 1, 3, and 5 through 7.

Second, the command update_environment() is now used to update one or more segments. It would be possible to add an additional requirement for the application programmer to have update_environment() return a string, such as "4-8", indicating which segments are being updates. This would allow TOP-C to maintain an internal table that updates a timestamp

for each segment, and then answer any application queries of the form
`is_up_to_date("1,3,5-7")`. However, it was felt to be a simpler syntax to
instead extend the `UPDATE` action returned by `get_task_output()`. Since the
application programmer already must return the action `UPDATE` (implemented
as a C constant), we now require the application programmer to instead
return a parametrized action such as `UPDATE("4-8")` (implemented as a
C function macro).

It is clear that the internal table of timestamps for each segment
can be maintained only on the master process, since queries of the form
`is_up_to_date()` and updates of the form `UPDATE()` both originate on the
master process. As each new task originates on the master, a new task ID
is issued as a monotonically increasing sequence. The timestamps for each
segment are then implemented as task ID's.

So an `is_up_to_date()` query can be answered by TOP-C simply
by determining the task ID of the current task being processed by
`get_task_output()`. That current task ID is compared with the maximum of
the timestamps for each segment being queried by `is_up_to_date()`. Those
timestamps are maintained by TOP-C in its internal table, and are task ID's
corresponding to the last `update_environment()` for each queried segment.
If the current task ID is "newer" (larger), then TOP-C returns true. Otherwise, it returns false.

Thus, the extensions to `is_up_to_date()` and `UPDATE()` impose a minimal additional burden on the TOP-C application programmer, while providing strong benefits in the form of higher concurrency. The partition of the
environment memory into segments by the application will often be a natural
extension of the application. For example, large application tables or other
arrays can be subdivided by partitioning the index set into equal subintervals. Object-oriented applications will often partition their environment by
associating an object ID with each object, and associating a TOP-C segment
with the memory used by an object. The object ID can then also be used as
a segment number.

5. TOP-C over Distributed Shared Memory

Existing DSM systems primarily provide physical memory management and
memory consistency. TOP-C provides memory management in the form
of implicitly specified TOP-C segments, where the user is responsible for
the memory organization, and the TOP-C framework provides consistency
management for this memory. Therefore the functionality of TOP-C and
a DSM system intersect in the area of memory management. This section
discusses the possible benefits and design of a combined system. There is not
yet an implementation of the ideas in this section.

The introduction of shared memory to TOP-C introduces a new problem
that was not present in the distributed memory of TOP-C. When the master

calls `update_environment()`, writes on the master take effect immediately on the slave, due to the shared memory. This is handled in SMP through a standard single-writer–multiple-reader solution by which readers may later re-read any modified segment through a REDO action. Nevertheless, this strategy also imposes a burden on the application writer in that `do_task()` may return a wrong answer after reading inconsistent data, but it must be guaranteed never to hang due to inconsistent data. DSM systems can emulate the lazy updates of TOP-C under distributed memory by implementing lazy release consistency.

Many DSM systems, such as TreadMarks [1], Quarks [16] and the earlier Munin [2] system, support release consistency. *Release consistency* allows for a weaker memory model in which an *acquire* operation is required before reading or writing a shared variable, and a *release* operation is required before another processor can acquire a shared variable. Release consistency allows initiation of a new acquire operation without waiting for pending reads to complete, and it allows a new write without waiting for pending release operations to complete.

A typical implementation of release consistency is to implement two library routines, acquire and release, (`Tmk_lock_acquire(lock_handle)` and `Tmk_lock_release(lock_handle)` in the case of TreadMarks), which operate on a lock handle (an integer in the case of TreadMarks). After an acquire operation, all writes by the application are noted by the DSM system until a corresponding release operation. (Interception of writes can be implemented by the UNIX system call, `mprotect()`.) If a second process acquires the same lock, then all of the modified pages will be replicated on the second process.

Release consistency is typically implemented in one of two variations. These two variations differ in how to handle write updates. The first variation is lazy release consistency. In lazy release consistency, a write update occurs only *after* the call to `release()` by the writing process, and when a second process then calls `acquire()` in an attempt to access the same page of memory. The second variation is eager release consistency. In this variation, modified pages are updated for all processes holding a copy of the page at the time of the call to `release()`. This update can be "batched" for efficiency, but the original call to `release()` may not be seen to complete by a second process until the second process has received the "eager" write updates.

The preferred DSM policy for TOP-C is one of *lazy release consistency* in which there are no page updates seen by other processes and no page invalidations until after the call to `release()` and at the time of a second call to `acquire()`. This mimics the TOP-C memory model of lazy, incremental updates. This fits well with the TOP-C methodology, in which writes to any one TOP-C segment are likely to be infrequent.

If TOP-C were implemented on top of a DSM system, this would require appropriate calls of `acquire()` and `release()` by TOP-C to the underlying DSM system. One would call `acquire()` before a call to `update_environment()` and `release()` after the call. Before a call to `do_task()` (on a slave), one would call `acquire()` immediately followed by `release()` in order to receive the modified pages.

If one has implemented multiple segments of the environment in TOP-C, one would invoke a different lock handle for each segment. It might become necessary for `update_environment()` to take an additional argument, specifying which segment to update. TOP-C would then guarantee to call `update_environment()` repeatedly, once for each segment that needs to be updated.

Plans are underway to test TOP-C on top of a DSM system. The experimental version of TOP-C (using shared memory) will be tested. This will provide important feedback about merging the TOP-C shared memory model with the shared memory model used by DSM.

1. C. Amza, A.L. Cox, S. Dwarkadas, P. Keleher, H. Lu, R. Rajamony, W. Yu, and W. Zwaenepoel, "TreadMarks: Shared Memory Computing on Networks of Workstations", IEEE Computer, Vol. 29, No. 2, pp. 18-28, February 1996.
2. J. Carter, J. Bennett, and W. Zwanpoel, Implementation and Performance of Munin, *Proc. 13th ACM Symp. Operating System Principles*, 1991, pp. 152–164.
3. R. Chow and T. Johnson, *Distributed Operating Systems and Algorithms*, Addison Wesley Longman, 1997.
4. G. Cooperman, "Practical Task-Oriented Parallelism for Gaussian Elimination in Distributed Memory", *Linear Algebra and its Applications* **275-276**, 1998, pp. 107–120.
5. G. Cooperman, GAP/MPI: Facilitating Parallelism, *Proc. of DIMACS Workshop on Groups and Computation II* **28**, *DIMACS Series in Discrete Mathematics and Theoretical Computer Science*, L. Finkelstein and W.M. Kantor (eds.), AMS, Providence, RI, 1997, 69–84.
6. G. Cooperman, STAR/MPI: Binding a Parallel Library to Interactive Symbolic Algebra Systems, *Proc. of International Symposium on Symbolic and Algebraic Computation (ISSAC '95)*, ACM Press, 126–132.
7. G. Cooperman, TOP-C: A Task-Oriented Parallel C Interface, 5th *International Symposium on High Performance Distributed Computing* (HPDC-5), 1996, IEEE Press, 141–150 (software at `ftp://ftp.ccs.neu.edu/pub/people/gene/top-c/`).
8. G. Cooperman, L.Finkelstein, M.Tselman and B.York, Constructing Permutation Representations for Matrix Groups, *J. Symbolic Computation* **24**, 1997, pp. 1–18.
9. G. Cooperman and V. Grinberg, "TOP-WEB: Task-Oriented Metacomputing on the WEB", *International Journal of Parallel and Distributed Systems and Networks* **1**, 1998, pp. 184–192; a shorter version appears as: "TOP-WEB: Task-Oriented Metacomputing on the Web", G. Cooperman and V. Grinberg, *Proceedings of Ninth IASTED International Conference on Parallel and Distributed Computing and Systems* (PDCS-97), IASTED/Acta Press, Anaheim, 1997, pp. 279–286.

10. G. Cooperman and G. Havas, Practical parallel coset enumeration, *Proc. of Workshop on High Performance Computation and Gigabit Local Area Networks*, G. Cooperman, G. Michler and H. Vinck (eds.), Lecture notes in control and information sciences **226**, Springer Verlag, pp. 15–27.

11. G. Cooperman, G. Hiss, K. Lux, and Jürgen Müller, The Brauer tree of the principal 19-block of the sporadic simple Thompson group, *J. of Experimental Mathematics* **6**(4), 1997, pp. 293–300.

12. G. Cooperman and M. Tselman, New Sequential and Parallel Algorithms for Generating High Dimension Hecke Algebras using the Condensation Technique, *Proc. of International Symposium on Symbolic and Algebraic Computation (ISSAC '96)*, ACM Press, 155–160.

13. G.C. Fox, W. Furmanski, M. Chen, C. Rebbi and J. Cowie, WebWork: Integrated Programming Environment Tools for National and Grand Challenges, *Proc, of Supercomputing '95*.

14. W. Gropp, E. Lusk and A. Skjellum, *Using MPI*, MIT Press, 1994.

15. J. Protić, M. Tomašević, V. Milutinović, Distributed Shared Memory: Concepts and Systems, IEEE Computer Society Press, 1998.

16. M. Swanson, L. Stoller, J. Carter, "Making Distributed Shared Memory Simple, Yet Efficient", *Proc. of the 3rd Int'l Workshop on High-Level Parallel Programming Models and Supportive Environments (HIPS'98)*, pages 2–13, March, 1998.

17. M. Tselman, Computing permutation representations for matrix groups in a distributed environment, *Proc. of DIMACS Workshop on Groups and Computation II* **28**, DIMACS Series in Discrete Mathematics and Theoretical Computer Science, L. Finkelstein and W.M. Kantor (eds.), AMS, Providence, RI, 1997, 371–382.

Metacomputing in the Gigabit Testbed West

Thomas Eickermann[1] and Ferdinand Hommes[2]

[1] Forschungszentrum Jülich, Germany
[2] GMD – Forschungszentrum Informationstechnik, Sankt Augustin, Germany

Abstract. The 'Gigabit Testbed West' is one of two Testbeds for the upgrade of the German scientific network to Gigabit capacity, that is planned for the year 2000. It currently uses a 2.4 Gigabit/second ATM link to connect the Research Centre Jülich and the GMD – National Research Center for Information Technology in Sankt Augustin. The testbed is the basis for several application projects ranging from metacomputing to multimedia. This contribution gives an overview of the infrastructure of the testbed and the applications.

1 Introduction

A common definition of metacomputing – the shared use of distributed supercomputing resources – contains different topics like unified access to the batch systems of different computing centers [1],[2] or the simultaneous use of several supercomputers by a single application. The first approach aims to simplify the access to supercomputers, the second should allow the solution of problems that could not be treated so far or solve problems more efficiently. The coupling of supercomputers offers a way to increase the peak CPU-performance and main memory accessible by a single application. This allows e.g. particle simulations with large numbers of particles, where main memory is often the limiting resource. Even more appealing is the so called 'heterogeneous metacomputing', which combines computers of different architecture, massively parallel computers, vector-computers or special-purpose machines like visualization servers [3].

A serious drawback is that the bandwidth and latency which are achievable over an external network — no matter if local or wide–area — can usually not compete with the performance of the internal network of a massively parallel computer. Because of that, only certain classes of applications can benefit from metacomputing. One such class is represented by so-called 'coupled fields' applications. Here, two or more space- and time–dependent fields interact with each other. An implementation of such applications can make explicit use of the performance hierarchy of the networks in the following way. The fields are distributed over the machines of the metacomputer and for each field, a parallelization via domain composition can be performed. Typically, the fields have to be exchanged over the network once per simulation timestep, while the calculation of each field often requires several iterations per timestep, and commmunication within each iteration. This means that although the requirements for the external network can be quite high, they are usually small compared to the internal communication needs. A second class of applications benefits from being distributed over supercomputers of different architecture, because they contain partial problems that can best be solved on massively parallel or vector-supercomputers. For other applications, real-time requirements are the reason to connect several machines.

2 The Gigabit Testbed West

In Germany, the network that connects research, science and educational institutions with each other and the rest of the internet is operated by the DFN–Verein, an association of these institutions founded in 1984. Since 1996 this network is based on ATM–technology and allows for access capacities up to 155 Mbit/s. An extension into the Gbit/s range on a national basis is planned for the year 2000. To prepare this transition, two testbeds have been set up in the western and southern parts of Germany. They will serve to evaluate new network technology as well as to gain experience with applications requiring bandwidths beyond the currently available 155 Mbit/s. In the area of scientific computation, such applications can e.g. be found in multimedia, distributed access to huge amounts of data and of course in metacomputing, which is the subject of this article.

The Gigabit Testbed West started as the first of the two German testbeds in August 1997. It is a joint project of the Research Centre Jülich and the GMD – National Research Center for Information Technology in Sankt Augustin close to Bonn. In the first year of operation the two locations — which are approximately 100 km apart — were connected by an OC–12 ATM link (622 Mbit/s) based upon Synchronous Digital Hierarchy (SDH/STM4) technology. In August 1998 this link has been upgraded to OC–48 (2.4 Gbit/s). The connection is provided by o.tel.o Service GmbH and uses the optical fiber infrastructure inside the power lines of the German power supplier RWE AG. In the framework of a beta-test Fore Systems ATM switches (ASX–4000) were used to connect the local networks of the research centers to the OC–48 line. Initial stability problems that were observed during the test turned out to be related to signal attenuation and timing. Those problems have been solved and both the SDH link and the switches are in stable operation now.

The application projects that use the testbed can rely on a solid base of installed supercomputer capacity. Jülich is equipped with 512–node Cray T3E–600 and 256–node T3E–900 massively parallel computers and a 16–processor Cray T90 vector-computer. An IBM SP2 and a 12 processor SGI Onyx 2 visualization server are installed in the GMD. Besides several institutes in the research centers in Jülich and Sankt Augustin other institutions participate in the testbed with their applications. These are the Alfred Wegener Institute for Polar and Marine Research (AWI), the German Climate Computing Center (DKRZ), the Universities of Cologne and Bonn, the National German Aerospace Research Center (DLR) in Cologne, the Academy of Media Arts in Cologne, and the industrial partners Pallas GmbH and echtzeit GmbH.

3   Supercomputer connectivity

A key factor for the success of metacomputing activities are communication networks that provide high–bandwidth and low–latency connections between the components of the metacomputer. Compared to the networking equipment for WAN-backbones, ATM-connectivity for supercomputers has evolved quite slowly. While 622 Mbit/s-Interfaces are now available for all common workstation platforms, solutions are still outstanding for the major supercomputers used in the Testbed West. For the Cray T3E, the Cray T90, and the IBM SP2 only 155 Mbit/s are available (and will be in the foreseeable future). For the SGI Onyx 2 a 622 Mbit/s ATM-interface is expected to be available in early 1999. Therefore a different solution had to be found to connect the supercomputers in Jülich and Sankt Augustin to the testbed. The best performing networking connection of the Cray supercomputers is the 'High Performance Parallel Interface' (HiPPI) which offers a peak performance of 800 Mbit/s when a low-level protocol and large transfer blocks (1 MByte or more) are used. Even with TCP/IP communication, transfer rates of more than 400 Mbit/s can be achieved within the local Cray complex in Jülich. This is mainly due to the fact, that HiPPI networks allow IP-packets of up to 64 KByte size (MTU size). One way to interconnect IP networks based on HiPPI and ATM technology is to use the ATM/HiPPI–gateway by Ascend Communications. A serious limitation of this solution is that on the HiPPI side only MTU sizes up to 9182 Byte are supported. Therefore we followed a different approach. A workstation is equipped with a Fore Systems 622 Mbit/s ATM interface and a HiPPI interface and acts as an IP-router between the HiPPI and the ATM network. Since the Fore ATM adapter supports large MTU sizes, IP packet sizes of 64 KByte are possible on each part of the network. We are currently using an SGI O200 and a SUN Ultra 30 as dedicated routers for the Cray systems in Jülich.

A similar solution was chosen to connect the IBM SP2 in Sankt Augustin to the testbed. 8 SP-nodes are equipped with 155 Mbit/s ATM adapters and one with a HiPPI interface. The ATM adapters are connected to the testbed via a FORE ASX 1000. The HiPPI network is routed by a SUN E5000 which has also a FORE 622 Mbit/s ATM adapter. Preliminary measurements show a throuphput of more than 370 Mbit/s between the Cray T3E in Jülich and the IBM SP2 in Sankt Augustin. The layout of the network as of September 1998 is depicted in figure 1. Throughput values that were measured in that network with various hardware are shown in table 1. The delay that is introduced by the 100 km SDH/ATM line is about 0.9 msec. This value is still below the delay introduced by the operating systems of the T3E ($\sim$3 msec)

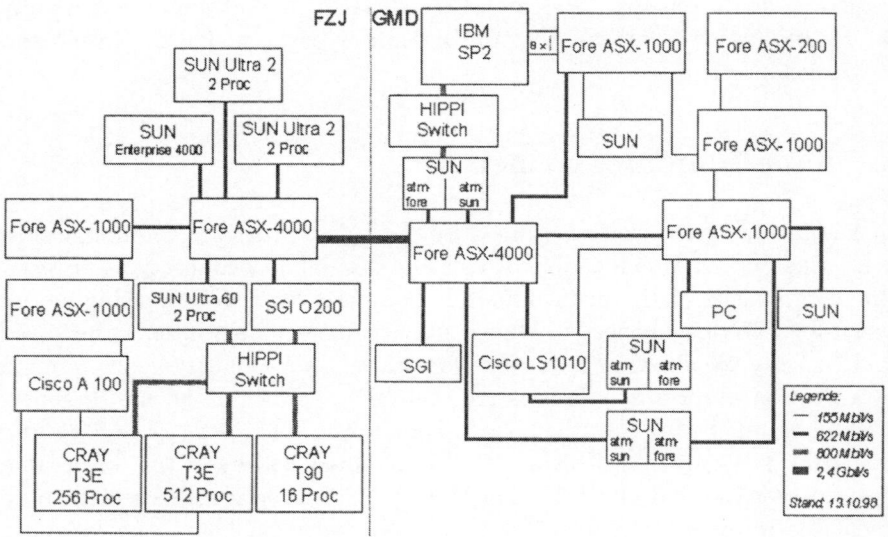

Fig. 1. Configuration of the Gigabit Testbed West in summer 1998. Jülich and Sankt Augustin are connected via a 2.4 Gbit/s ATM-link. The supercomputers are attached to the testbed via HiPPI-ATM gateways, several workstations via 622 or 155 Mbit/s ATM interfaces.

and the SP2 (~2 msec) which were measured with ping-pong tests in local networks.

Table 1. TCP throughput in ATM classical IP networks

	adapter [Mbit/s]	throughput [Mbit/s]
Sun Ultra 60, Solaris 2.6	622	530
Sun E5000, Solaris 2.6	622	501
SP2, Thin node, AIX 4.1.5	155	118
T3E-900	155	115
Onyx 2, IRIX 6.4	155	126

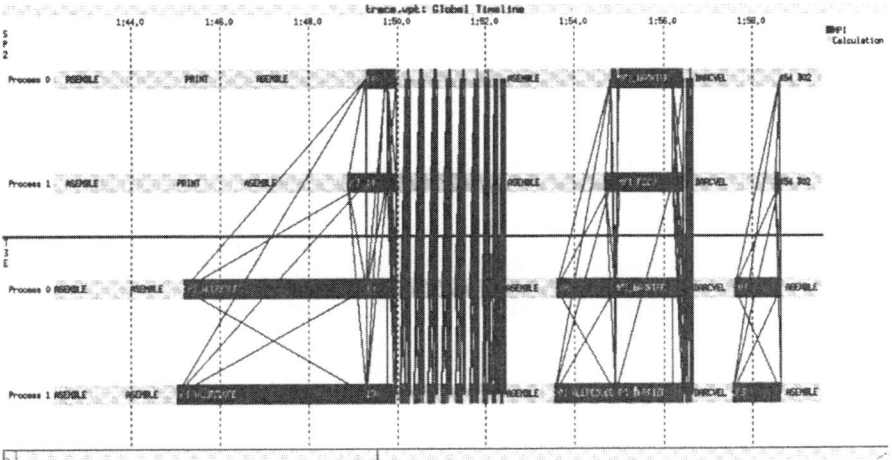

Fig. 2. VAMPIR timeline display of a metacomputing application running on two SP2 and two T3E nodes. The horizontal axis is the execution time, each horizontal bar represents a processor. Light parts of the bars depict calculations, dark parts MPI communication. The black lines represent MPI messages.

4 Tools

To make metacomputing usable for a broader range of users, the availability of at least a minimum set of tools is mandatory. Most important is a metacomputing-aware communication library. In the Gigabit Testbed West, it was decided to rely mainly on MPI [4] which has become the de–facto standard in distributed memory parallel computers. A couple of features that are useful for metacomputing applications are part of the MPI-2 [5] definition. Dynamic process creation and attachment e.g. can be used for realtime-visualization or computational steering; language-interoperability between C and FORTRAN is needed to couple applications that are implemented in different programming languages. When the project started no metacomputing-aware MPI-2 implementation was available (this is still true today, except for the LAM implementation, which implements the dynamic features of MPI-2 on workstation clusters [6]). Therefore such a development was assigned to Pallas GmbH. A first protoype was finished in September 1998. Until then, the PACX/MPI-library developed by the University of Stuttgart was used [7]. It supports a subset of MPI-1 and allows to couple Cray T3Es. This library has been ported to the IBM SP2 and optimized for high-speed networks by the project partners in Jülich and Sankt Augustin. For MPI point-to-point communication, throughput values of 73 Mbit/s with a latency of 6 msec have been observed between the Cray T3E in Jülich and the IBM SP2 in Sankt Augustin. For those measurements the 155 Mbit/s ATM interfaces have been used. First experiments with the HiPPI/ATM gateway show significant improvements compared to that value.

Also important are tools for performance evaluation and tuning. For message passing applications VAMPIR [8] is a well known product. It was developed in the Research Centre Jülich and is now distributed by Pallas GmbH. For the use in this project VAMPIR has been extended by some metacomputing features. Tracefiles, that have been created on the different machines of the metacomputer can be synchronized and merged and visualized in the timeline display. Figure 2 shows an example. A wrapper library for the instrumentation of PACX/MPI applications for the use with VAMPIR was also developed. No attempt has been made to develop a meta-debugger. With PACX/MPI messages that are exchanged between the machines can be traced. For other problems, parallel debuggers like Totalview have to be used separately on each machine.

5 Applications

A couple of application subprojects that touch different aspects of metacomputing have been defined within the Gigabit Testbed West. In the following the aims and the status in summer 1998 of each application is described briefly. More details will be presented in separate publications.

5.1 Solute transport in ground water

A typical 'coupled fields' szenario is the transport of solutants in ground water. The interacting fields are the velocity of the ground water flow and the concentrations of the solutants. Two independent programs that perform such kind of 3–D simulations have been developed in the Institute for Petroleum and Organic Geochemistry at the Research Centre Jülich. The program TRACE (Transport of Contaminants in Environmental Systems) simulates the flow of water in variably saturated, porous, heterogeneous media. It uses a finite–element discretization of the model equations and has been parallelized at the Central Institute for Applied Mathematics in Jülich based on a domain decomposition [9]. It is coded in FORTRAN 90 and uses MPI. The C++ program PARTRACE (PARticle TRACE) performs the simulation of the solutants using a Monte–Carlo method. In their original versions, the programs could only 'communicate' via files. TRACE simulates the water flow until a stationary flow evolves and writes the resulting fields into a file which is then used as input for the particle simulation that is done by PARTRACE.

It was considered a serious restriction of this approach that the simulation of particle transport is limited to stationary flows. To resolve this limitation, the applications were coupled using PACX/MPI. Each of them now runs in its own MPI-communicator and the water flow fields are exchanged via message-passing.

Currently, TRACE is run on the T3E in Jülich and PARTRACE on the SP2 in Sankt Augustin. In a typical run, 10 MBytes are transferred over the testbed at the beginning of each timestep. With one timestep taking approximately 2 seconds, this results in moderate average network load. Nevertheless, the peak rates are much higher, since all data are transferred in a single burst. Currently, work is underway to improve performance and scalability of both applications. This will also result in increasing network requirements. Futhermore it is planned to implement an online visualization of the computation.

5.2   MEG analysis

An application that can benefit from heterogeneous metacomputing emerges from the analysis of magnetoenzephalography data. The magnetic field around a human head is measured with an array of superconducting quantum interference devices (SQUIDs). From these data, the distribution of electric currents in the brain can be reconstructed by solving an inverse problem. In Jülich, this is done with the 'Multiple Signal Classification' (MUSIC) algorithm [10]. With MUSIC, parameters of a finite number of current dipoles are obtained in three phases [11].

- The number of dipoles is estimated using statistical methods that separate signal from noise.
- The positions of the dipoles are calculated. This is done by finding the extrema of a function, that measures how well a dipole placed at a given location is able to reproduce the signal estimated in the former step.
- In the last step, the time evolution of dipole strength and orientation are calculated.

The second phase is most time consuming but can be implemented on a massively parallel computer very efficiently. The first phase is better suited for a vector-computer. The reason for that is that it involves operations on matrices that are too small to be efficiently parallelized (typically 360x360). Separate measurements of a parallel program that implements the MUSIC algorithm on the T3E and the T90 confirm this. As soon as our MPI-2 implementation will be able to couple those machines, a distributed version of the program should be able to achieve an overall execution-time that is below the time needed on either the T3E or the T90.

5.3   Realtime fMRI

Another experiment in Jülich that deals with brain activity is based on functional Magnetic Resonance Tomography (fMRI) [12]. Here a test person is exposed to e.g. periodic visual or acoustic stimulations. The areas of brain

126

activity are identified by fitting the parameters of a model of the expected response of the brain with the MRI data. This not only improves the sensitivity of the measurement compared to simpler correlation methods but also allows to check those models. Head movements of the test person tend to produce artefacts in the detected activity. Therefore it is essential to correct for those movements. In order to allow interactive response of the experimentalist, all this should be done and visualized in realtime.

It is planned to implement this with the following setup. The raw data is transferred from the MRI scanner to the T3E, where it is processed. The resulting functional data is handed over to an SGI Onyx 2 at the GMD in Sankt Augustin. This machine creates an interactive 3–D representation of the brain on a Responsive Workbench that is again located in Jülich. For that purpose, two stereo-images have to be transferred over the gigabit testbed. In order to allow for interactive movement and slicing by a person operating the workbench, these images have to be updated several times a second. Currently only a simple 2–D visualization of the processed data is implemented. This setup is sketched in figure 3. It should be noted that a similar application has recently been demonstrated by the Pittsburgh Supercomputing Center [13].

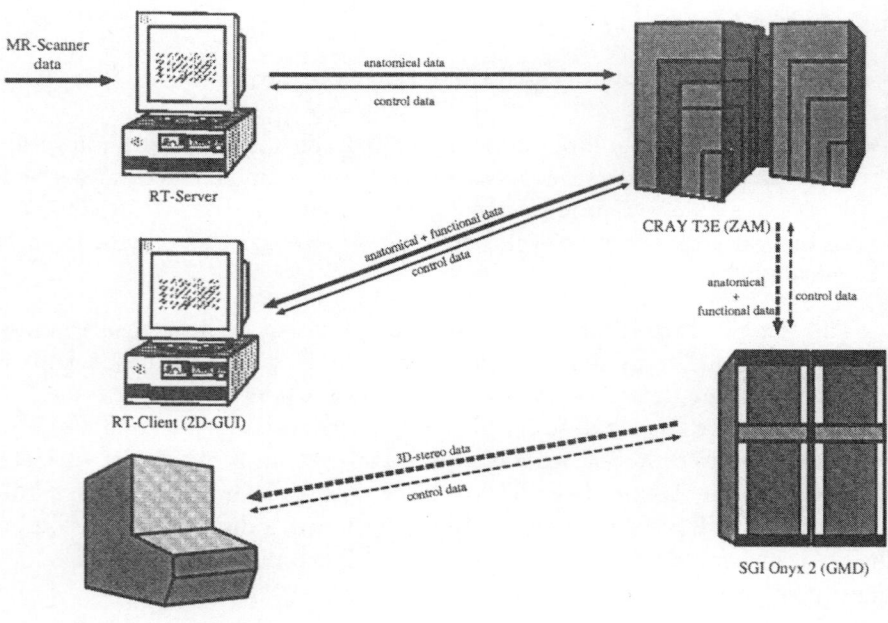

Fig. 3. Setup of the fMRI experiment. The raw scanner data are transferred through a front-end workstation to the T3E where they are processed. From there, anatomical and functional brain-images are transferred to either a workstation with a simple 2-D display or over the testbed to an Onyx 2 in the GMD. The rendered images are sent back over the testbed to a Responsive Workbench in Jülich.

5.4 Distributed climate and weather models

A second 'coupled fields' application in the gigabit testbed deals with the
distributed calculation of climate and weather models. Here, the Alfred–
Wegener–Institute (AWI), the German Climate Computing Center (DKRZ)
and the GMD will use the supercomputers in Jülich and Sankt Augustin for a
coupled simulation of atmospheric processes and the ocean–ice system. There
are two main differences to the ground water szenario. One is that here the
fields interact only at a 2–D interface, the ocean surface, whereas water and
solutants interact in the full 3–D simulation domain. This further reduces the
amount of data to be exchanged. Nevertheless, shorter simulation times for a
single timestep and higher model resolution lead to similar total bandwidth
requirements.

The second difference is that in the ground water case data flows in one
direction only — there is no feedback from the solutants to the ground water
flow. In contrast to that, both the ocean and the atmosphere models need the
fields from the other model as boundary conditions. Because of that, peak
bandwidth and latency of the network is much more critical here than in the
gound water problem.

5.5 Distributed fluid-structure interaction

A more general approach to 'coupled fields' type problems is pursued in the
EC funded project CISPAR. The idea there is to use well–established commercial computational fluid dynamics (STAR-CD) and structural mechanics
codes (PAM-SOLID, PERMAS) for problems that involve the interaction of a
fluid with flexible structures. Examples for such problems are artificial heart
valves, torque converters or ships. A standard interface for those codes as well
as a coupling library (COCOLIB) have been developed by the GMD and industrial project partners. Within the Gigabit Testbed West, the COCOLIB
will be ported to the metacomputer after the end of the CISPAR project in
1999.

5.6 New networks and applications

The testbed is currently extended by connecting new sites to the original link
between Jülich and Sankt Augustin and by defining new applications that
use those extensions. A dark fibre that links the national German Aerospace
Research Center (DLR) and the University of Cologne to the GMD has just
been set up. This line will be used for projects that range from distributed
traffic simulation and visualization to distributed virtual TV-production (in
cooperation between GMD, DLR, Academy of Media Arts in Cologne, and
echtzeit GmbH). The latter relies on the results of a multimedia project that
evaluates components for studio quality digital video transmission over ATM
in the testbed. A new 622 Mbit/s ATM-link between the University of Bonn

and the GMD will be the basis for metacomputing projects that deal with multiscale molecular dynamics and lithospheric fluids. Here the PARNASS-cluster [14] of the Institute for Applied Mathematics of the University of Bonn is connected to the IBM SP2 and the Cray T3E.

6 Conclusion

This contribution gave an overview over the metacomputing activities in the Gigabit Testbed West. The underlying 2.4 Gbit/s SDH and ATM technology for the wide area backbone seems to be mature, a neccessary condition for the upgrade of the German scientific network that is planned for the year 2000. In contrast to that, the networking capabilities of the supercomputers that are attached to the testbed have to be improved. The concept of a HiPPI/ATM gateway seems to be promising. A couple of applications that deal with various aspects of metacomputing are using the infrastructure of the testbed. Their results should enhance our understanding about the conditions under which distributed high–performance computing is feasible.

7 Acknowledgements

Most of the activities that are reported in this contribution are not the work of the authors but of several persons in the institutions that participate in the Gigabit Testbed West project. The authors wish to thank D. Conrads, W. Frings, D. Gembris, T. Graf, R. Niederberger, S. Posse, M. Sczimarowski, and H. Vereecken from the Research Centre Jülich, U. Eisenblätter, H. Grund, W. Joppich, G. Göbbels, M. Göbel, M. Kaul, E. Pless, R. Völpel, K. Wolf, P. Wunderling, and L. Zier at the GMD, W. Hiller and T.Störtkuhl at the AWI, V. Gülzow at the DKRZ and J. Henrichs and K. Solchenbach at Pallas GmbH, to mention but a few. We also wish to thank the BMBF for partially funding the Gigabit Testbed West and the DFN for its support. Special thanks to the University of Stuttgart for the PACX/MPI-library.

References

1. Erwin, D., The UNICORE Architecture and Project Plan, Workshop on Seamless Computing, ECMWF, Reading, September 16–17, 1997.
2. Sander, V., High Performance Computer Management, Workshop Hypercomputing, Rostock, September 8–11, 1997.
3. Eickermann, Th., Henrichs, J., Resch, M., Stoy, R., and Völpel, R., Metacomputing in gigabit environments: Networks, tools, and applications, Parallel Computing **24**, p. 1847-1872, 1998.
4. Message Passing Interface Forum, MPI: A Message-Passing Interface Standard, University of Tennessee, http://www.mcs.anl.gov/mpi/index.html, 1995.

5. Message Passing Interface Forum, MPI-2: Extensions to the Message-Passing Interface, University of Tennessee, http://www.mcs.anl.gov/mpi/index.html, 1997.

6. Burns, G.D., Daoud, R.B., Vaigl, J.R., LAM: An Open Cluster Environment for MPI, Supercomputing Symposium '94, Toronto, Canada, June 1994.

7. Beisel, T., Gabriel, E., Resch, M., An Extension to MPI for Distributed Computing on MPPs, in Marian Bubak, Jack Dongarra, Jerzy Wasniewski, Eds., Recent Advances in Parallel Virtual Machine and Message Passing Interface, p. 75-83, Springer-Verlag Berlin Heidelberg, 1997.

8. Nagel, W.E., Arnold, A., Weber, M., Hoppe, H.C., Solchenbach, K., VAMPIR: Visualization and analysis of MPI resources, Supercomp. **63**, Vol. XII, no. 1, p. 69–80, 1996.

9. Wimmershoff, R., Entwicklung und Implementierung einer dreidimensionalen Partitionierungsstrategie für das Programm TRACE auf einem massiv parallelen Rechner. Technical Report Forschungszentrum Jülich, Jül–3157, 1995, in German.

10. Mosher, J.C., Lewis, P.S., and Leahy, R.M., Multiple Dipole Modeling and Localization from Spatio–Temporal MEG DATA. IEEE Trans. Biomed. Eng. **39**, p. 541–557, 1992.

11. Beucker, R. and Schlitt, H.A., Objective Signal Subspace Determination for MEG, Forschungszentrum Jülich, ZAM, FZJ–ZAM–IB–9715, 1997.

12. Ogawa, S., Lee, T.M., Kay, A.R., Tank, D.W., Brain magnetic resonance imaging with contrast depending on blood oxygenation. Proc. Natl. Acad. Sci. USA **87**, p. 9868–9872, 1990.

13. Goddard, N.H., Hood, G., Cohen, J.D., Eddy, W.F., Genovese, C.R., Noll, D.C., and Nystrom, L.E., Online Analysis of Functional MRI Datasets on Parallel Platforms. Journal of Supercomputing, **11**, p. 295-318, 1997.

14. Griebel, M., Zumbusch, G., Parnass: Porting gigabit-LAN components to a workstation cluster, in W. Rehm ed., Proceedings of the 1st Workshop Cluster-Computing, held November 6-7, 1997, in Chemnitz, Chemnitzer Informatik Berichte, CSR-97-05 , p. 101-124, 1997.

High Performance Metacomputing in a Transatlantic Wide Area Application Testbed

Edgar Gabriel, Michael Resch, Paul Christ, Alfred Geiger, and Ulrich Lang[1]

High Performance Computing Center Stuttgart
Allmandring 30,
D-70550 Stuttgart,
Germany
{gabriel, resch}@hlrs.de

Abstract. During the last couple of years, a wide variety of tools and libraries have been developed to enable distributed computing and visualisation. This paper presents the technical background and the results of such a project meaned to couple different computational resources. A metacomputing implementation of MPI called PACX-MPI was used to make the applications run on such a cluster. Three applications were used for demonstration purposes. These applications had to be adapted for Metacomputing, to make them more latency tolerant.

1 Introduction

In 1997 the HLRS was involved in two transatlantic projects in the frame of the G7 Global Information Society Initiative "Global Interoperability of Broadband Networks" (GIBN). One was from PSC and HLRS and was focusing on the application aspect of metacomputing. The other one was from SNL and HLRS and was concentrating on distributed visualization in a virtual laboratory. During the first project phase it became clear that the projects should be merged into a Global Wide Area Application Test-bed (G-WAAT). This would allow to couple simulation and visualization in a metacomputing scenario.

The main targets of the merger were:

- To set up a production test-bed for metacomputing applications and distributed visualization
- To combine supercomputing forces in order to solve much larger problems than any of the partners could solve on his own resources.
- To integrate software components in order to establish a collaborative simulation and visualization environment

In a first step this meant to set up a network connection fast enough to allow distributed simulation and visualization. Second, it was necessary to find a communication software that enables metacomputing for one single application. Third applications had to be adapted to be able to fully exploit the provided metacomputer. Fourth, distributed visualization software had to be adapted and extended. In response to these needs a transatlantic network connection was set up. The communication issue was resolved by implementing a completely new communication library based on the MPI standard. An existing Collaborative Visualisation software was extended and improved [11].

The concept of the paper is as follows. The technical details of the testbed are described in section 2. Section 3 presents a library, which enables message-passing even between different Massively Parallel Processing Systems (MPP's) or Parallel Vector Processors (PVP's). The results achieved during the Supercomputing '97 in San Jose and Supercomputing '98 in Orlando using several applications are presented in section 4. A brief overview about the future work in this field is described in section 6.

2   A Transatlantic Network Connection

For a sufficient network throughput for metacomputing applications and collaborative working the most relevant network Quality of Service (QoS) requirements are small and constant delays and nearly no packet losses.

Measurements taken by HLRS on the standard internet connection which is provided and shared by the German DFN community, including a transatlantic link of 2*45 Mbps shared bandwidth showed, that the available QoS between HLRS and the US had a strong variance. During the working hours in Europe and the eastern part of USA the packet losses varied between 10% and 40% resulting in varying TCP throughputs, which were not sufficient for effective metacomputing and cooperative work.

Therefore a dedicated transatlantic test-bed was established connecting the two CRAY T3Es at HLRS and PSC based on a dedicated 2 Mbps ATM channel. For the Supercomputing '97 event, this network was extended to Sandia National Laboratories, Albuquerque New Mexico, and to San Jose. For the Supercomputing '98 event, the dedicated transatlantic ATM was rebuilt again based on a dedicated 10 Mbps ATM channel. Figure 1 shows the geographic extension and the participating network providers of the transatlantic metacomputing environment during SC' 98 (see also http://www.hlrs.de/-news/events/1998/sc98/).

2.1   Network Performance Measurements

With respect to latency, comparing the cross atlantic standard path provided by DFN and the dedicated ATM-Link, it was interesting to note the effect due to the number of routers involved and the translation of packet losses into additional delays. The results achieved on the network connection between a test workstation at HLRS and the CRAY T3E in Pittsburgh over the standard path and the dedicated link are depicted in the following Table. As already mentioned the network performance of the standard

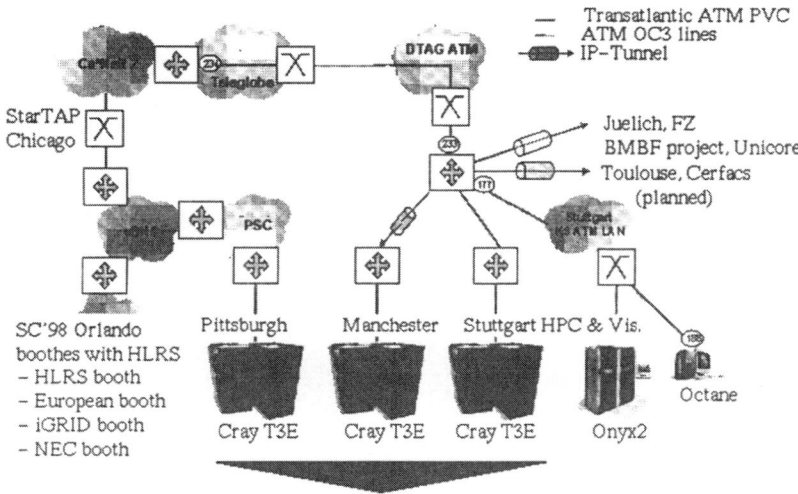

Fig. 1. Network for transatlantic metacomputing demonstrations during SC'98

Connection	Bandwidth [Mb/s]	no. of routers	tcp-throughput [kB/s]^3		packet losses [%]		delay [ms]	
			day	night	day	night	day	night
DFN	2*45	15	50	300	30	3	180^1	160
ATM-Link	2	5	200	-^{4}	0	0	150^2	-^{4}

Table 1. TCP throughput and network QoS between HLRS and PSC on standard Internet and dedicated ATM link, Summer 1997. ^{1}average value (variation between 160 and 300 ms). ^{2}variation between 150 and 155 ms. ^{3}the socket buffer used was 64 kB. ^{4}no tests done.

path is strongly influenced by the European and US working hours. During a small time-window in the European early morning hours, the packet loss and packet round-trip time were acceptable and a TCP throughput of approx. 300 kByte/s was achievable. However, during the daytime, the IP packet losses (measured with packetsize of 1 kByte) downgraded the TCP throughput to less than 50 kByte/s. The mean packet round-trip time on the standard path ranged from 160 to 300 ms.

134

On the dedicated ATM-link, there were practically no packet losses (during SC'97 a small number of packet losses appeared during the change over from CANARIES ATM-network to CA*Net II) with a nearly constant round-trip time of 150 ms. This good link performance resulted in a constant TCP throughput of 200 kByte/s, which is the maximum throughput available on an ATM-link with 2 Mbit/s bandwidth.

The higher number of routers on the standard path introduced a relatively small latency, so in the case of a small load as seen during European night-time hours the round-trip time on the standard path is comparable to that of the direct ATM link.

Figure 2 shows a comparison of the network delay and packet losses during a 24 hour period over the standard path and the dedicated ATM-link. The data on the dedicated ATM-link was captured during SC'97, the data on the standard path some time after SC'97.

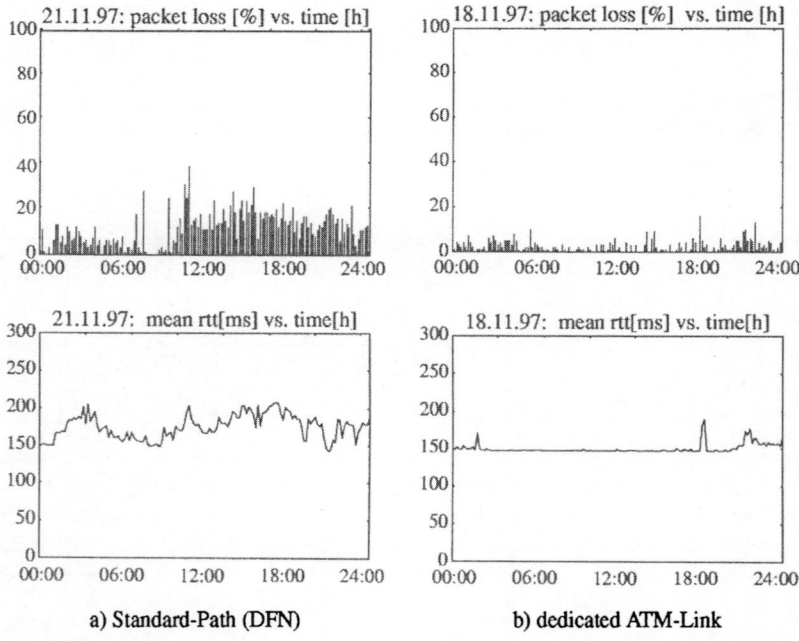

a) Standard-Path (DFN)   b) dedicated ATM-Link

Fig. 2. Round trip time and packet loss on the standard path and the direct ATM link during a 24 hour period.

As is well known, the TCP performance on links with large bandwidth times delay products is strongly dependant upon the TCP Window size, which is configured on the end-systems through the TCP socket buffer sizes. On the 2 Mbit/s ATM-link a socket buffer size of 64 kByte was sufficient for maximum TCP throughput.

3 Interoperable MPI

For metacomputing the question of communication is a crucial one. The library should be able to fully exploit the fast network of each single machine in the metacomputing scenario. At the same time it should be able to support the full communication functionality between different machines that an application requires. PACX-MPI (PArallel Computer eXtension MPI) was designed to enable message passing inside and over the bounderies of an MPP, too.

To realize this goal PACX-MPI has to distinguish messages which remain inside a machine, in this context called internal communication, and messages which have to be transferred to another MPP. The latter one will be called external communication. For the internal communication, PACX-MPI is using the vendor implemented MPI-library, since this is nowadays the only optimized and portable protocol, which is available on every system and which can fully exploit the capabilities of the underlying network. For the external communication PACX-MPI should use a standard protocol, and the decision was to implement as a first protocol TCP/IP.

To avoid, that each application node has to open a socket-connection to another node on a different machine when communicating, two so called daemon-nodes have been introduced. These two nodes take care of outgoing respectively incoming messages and are therefore transparent for the application.

Since PACX-MPI has the goal to support the whole MPI 1.2 standard, problems like the configuration of a global communicator had to be solved. Figure 3 is explaining the global configuration of MPI_COMM_WORLD on a metacomputer consisting of two machines.

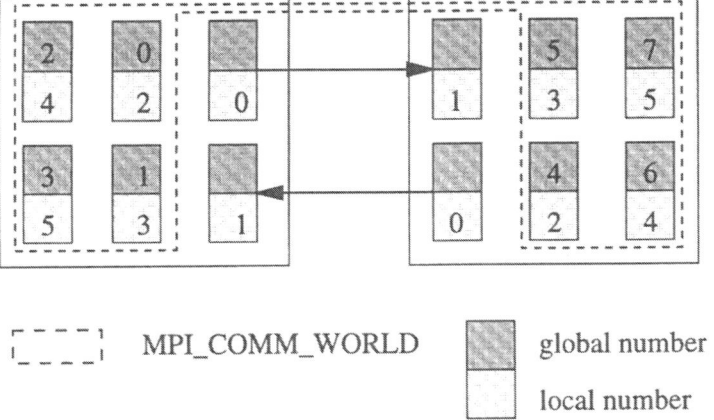

Fig. 3. MPI_COMM_WORLD on a metacomputer consisting of two MPP's

On the left machine, which shall be the machine with the number one, the first two nodes with ranks 0 and 1 are not part of MPI_COMM_WORLD, since these are the daemon nodes. The next node with the rank 2 is therefore the first node in our global communicator and gets the global rank number 0. All other application nodes get a global number according to their local ranks minus two, the last node on this machine has the rank 3. On the next machine, the daemon nodes again are not considered in the global MPI_COMM_WORLD. The node with the local rank 3 is number 4 in the global communicator, since the numbering on this machine starts with the last global rank on the previous machine plus. Introducing this renumbering and mapping of local pids to global ones, one gets a global MPI_COMM_WORLD without loosing the local information.

3.1 Point-to-point operations in PACX-MPI

A point-to-point operation in PACX-MPI can be briefly described as follows. The sender has to check first, whether the receiving node is on the same machine or not. If it is on the same machine, it can directly send the message to the receiving node using native MPI-commands. If it is on a different machine, like in the example of Figure 4, it has to create a header first, which contains all informations to identify a message, and then a data-package. Both packages are sent to a daemon node. The daemon node transfers both packages to the destination machine, where another daemon node receives the message and hands it over to the destination node.

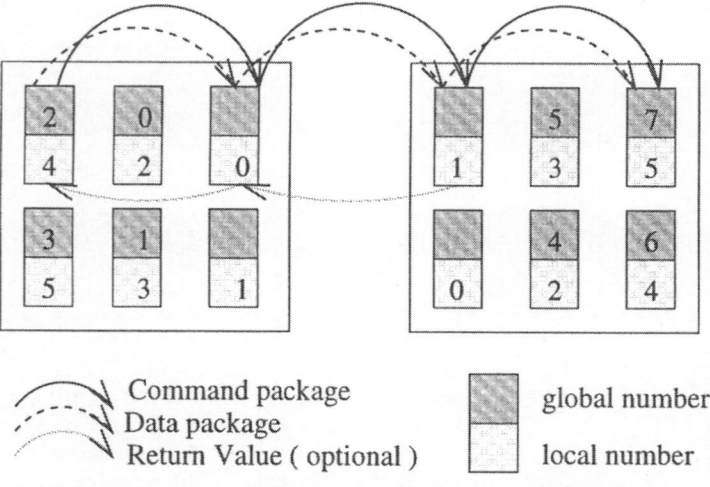

Fig. 4. Point-to-point operation from global node 2 to global node 7

The receiver has also to check whether the communication is internal or external. For an internal communication it can directly execute an MPI_Recv command. The only additional work which has to be done in this case is, that the MPI_Status has to be adapted, since global and local numbers are not identical (see Figure 3).

If the communication is external, the receiving node checks first, whether the expected message is already in the buffer. If not, it has to receive the header and the data-packets from a daemon node.

3.2   Global operations

Roughly speaking global operations in MPI can be split in two groups. The first group of operations has a root-node, which has to distribute (e.g. broadcast) or to receive (e.g. reduce) some global data. The second group has no such root-node, all nodes have the same status (e.g. barrier) or data (e.g. allreduce, all-to-all) after the global operation.

The first group of global operations, which have such a root-node, are splitted in two parts in PACX-MPI. One part is to distribute/collect the data between the machines, and a second part is a local operation inside the machine. The sequence of these two parts is depending on the the operation. For a broadcast data will be first distributed to all machines and afterwards the local broadcast will be performed. For a reduce-operation PACX-MPI has to perform first the local operation and only in the second step the global collecting of data will be performed.

For the second class of global operations whithout a root-node, there are several possibilities. The main difference between the algorithms is whether we are executing an all-to-all exchange of data between the machines, or whether we are collecting the global result on a dedicated node and distribute the global results in a second step.

Let's regard this situation using MPI_Allreduce. In the first algorithm each machine would execute a local MPI_Reduce, using a local root-node. In a second step, each machine would send its result to all other machines, calculate locally the global result and distribute this to all nodes on its machine. In this case we will have

$$N \cdot (N - 1)$$

number of messages which have to be exchanged between all machines, with N being the number of coupled machines. The advantage of this algorithm is, that all external communication steps can be performed theoretically in parallel.

In the second algorithm each machine is executing a local reduce again, but in the second step they all send their local result to a dedicated node, which calculates the global result and distributes than this result to all other machines. In this case we will have

$$2 \cdot N$$

external communication steps, but only N communications can be performed in parallel at the same time. Which of these two algorithms is performing faster, is an issue of actual investigations and strongly dependent of the network-configuration between the machines.

3.3 Related works

Their are several works related to this theme, each having a somewhat different approach. The well known Globus-project [7] tries to build up a whole bunch of metacomputing services, including also distributed computing for MPI-applications based on MPICH [6] and the NEXUS communication library. A disadvantage of this attempt is, that every external communcation step is done by direct node-to-node connection. This can lead for really big configurations to problems because of too many open ports/sockets. Additionally the underlying NEXUS-library has no support for global operations. Therefore the execution time for a broadcast-operation for example is stronlgy dependent of the distribution of nodes on the different MPP's.

The MagPIe-project [9] was setup to solve this problem. This project implemented global-operations for clusters of machines for MPICH, but on the other hand they still do not solve the problem of the direct point-to-point operations.

PVMPI [4] makes MPI applications run on a cluster of machines by using PVM for the communication between the different machines. Unfortunately the user can use only point-to-point operations and has to add some non MPI congruent calls. The subsequent project, MPI_Connect uses the same ideas but replaced PVM by a library called SNIPE [5], and supports now global operations too, in contrary to PVMPI.

A similar approach has been done by PLUS [1]. This library additionally supports communication between different message-passing libraries, like e.g. PARMACS, PVM and MPI. But again the user has to add some calls to his application.

Another project called Stampi [8] has been recently presented. This project already uses the MPI2 process model, but focuses mainly on local area computing. On the other hand they distinguish between one/two/three hop communication, and therefore a metacomputer need not perform direct node-to-node communication but can use some kind of daemon for the external communication.

4 Applications and Results

During the Supercomputing '97 in San Jose and Supercomputing '98 in Orlando a couple of demonstrations were done using PACX-MPI. In this section we will briefly describe the applications used in the metacomputing environment and we will also present some results achieved.

4.1 URANUS

The first application is a CFD-code called URANUS (Upwind Relaxation Algorithm for Nonequilibrium Flows of the University of Stuttgart) [2]. This program has been developed for simulating the reentry of a space vehicle in a wide altitude velocity range. The reason why URANUS was tested in such an environment is that soon two additional components of URANUS will have a great demand on memory: the nonequilibrium part has been finished in the sequential code and will be parallelized soon. Furthermore we will simulate the Crew-Rescue-Vehicle (X-38) of the new international space-station with more than 3 Million cells. Both components together require memory in the range of hundreds of Gigabytes, that cannot be provided by a single machine today. During the SC98 we simulated the European space-vehicle HERMES with 1.7 Million cells using 992 CPU's on two Cray T3E's.

The code is based on a regular grid decomposition, which leads to a very good load balancing and a simple communication pattern.

In the following we give the overall time it takes to simulate a medium size problem with 880.000 grid cells. For the tests we simulated 10 Iterations. We compared a single machine with 128 nodes and two machines with 2 times 64 nodes. Obviously the unchanged code is much slower on two machines. How-

Method	128 nodes using MPI	2*64 nodes using PACX-MPI
URANUS unchanged	102.4	156.7
URANUS modified	91.2	150.5
URANUS pipelined	-	116.7

Table 2. Comparison of timing results (sec) in metacomputing for URANUS

ever, the overhead of 50% is relatively small with respect to the slow network. Modification of the pre-processing does not improve the situation much. A lot more can be gained by fully asynchronous message-passing. Using so called 'Message Pipelining' [2] messages are only received if available. The receiving node may continue the iteration process without having the most recent data in that case. This helped to reduce the computing time significantly. The implication of this method is, that for convergence the number of iterations has to be increased by about 10 percent. Additionally one has to take care, that the messages are not older than two iterations, since this may prevent convergence at all. Tests for one single machine were not run because results are no longer comparable with respect to numerical convergence.

4.2 P3T-DSMC

The second application is P3T-DSMC. This is an object-oriented Direct Simulation Monte Carlo Code which was developed at the Institute for Computer Applications (ICA I) of Stuttgart University for general particle tracking problems [10].

Since Monte Carlo Methods are well suited for metacomputing, this application gives a very good performance on the transatlantic connection. For

Particles/CPU	60 nodes using MPI	2*30 nodes using PACX-MPI
1935	0.05	0.28
3906	0.1	0.31
7812	0.2	0.31
15625	0.4	0.4
31250	0.81	0.81
125000	3.27	3.3
500000	13.04	13.4

Table 3. Comparison of timing results (sec) in metacomputing for P3T-DSMC

small number of particles the metacomputing shows some overhead. But up to 125.000 particles timings for one time step are identical on one machine and in the metacomputing-testbed. This excellent behaviour is due to two basic features of the code. First, the computation to communication ratio is becoming better if more particles are simulated per process. Second, latency can be hidden more easily if the number of particles increases.

4.3 P3T-MD

The third application is also based on the P3T-toolkit, but instead of a Monte-Carlo code this program solves the molecular-dynamic equations to simulate the interactions between the particles. Therefore the code is stronger coupled compared to P3T-DSMC. During the SC98 event, a lot of tests have been performed with both P3T applications using up to 1024 processors. The results could not yet be fully evaluated.

Often molecular-dynamics simulations generate large sized outputs. It makes no sense to store a complete configuration of such a simulation in a metacomputing environment, since this would require a lot of additional time. P3T-MD does for big simulations a distributed postprocessing of the data. For example to calculate the force-distribution between the particles, each node is doing its own statistical analysis and only the result of this analysis is stored instead of the raw data.

5 Conclusions

The last two chapters have pointed out, that one has to invest a lot of work until the applications performs on a cluster of MPP's. An MPI-implementation suitable for such a cluster of machines has completly different requirements than an MPI-library working on a single machine. Optimizing point-to-point operations by dealing with different protocols is required but still not enough. The global operations have to be adapted to algorithms dealing with latencies of different ranges.

Additionally apllications have to be adapted to become more latency tolerant and to use less bandwidth. A CFD-application like URANUS is much more difficult to adapt for such an environment, since it is strongly coupled and it is not simple to save communication whithout loosing numerical performance. The key question in any case is, whether one succeeds in overlapping communication and computation.

A Monte-Carlo Method like P3T-DSMC is communicating less then the application above, and therefore fits automatically better for Metacomputing. Problems may still arise of dealing with huge amounts of data, which have to be transfered to a single machine for the final output. Thus some distributed postprocessing operations are inalienable to transfer and save only really important data.

These points may be resumed regarding the costs for Metacomputing. Since costs for the networks are depending on the reserved bandwidth and the time for which we are using the network, algorithms should be developed that consider economical apsects as well. Nevertheless, there are some applications, for which Metacomputing may be nowadays the only method to get some results, and for which it is worth to do the whole work.

6 Outlook

The future metacomputing activities of the High Performance Computing Center Stuttgart will be focused in a project called METODIS (MEtacomuting TOols for DIstributed Systems). This project is supported by the European Community and has the major goal to create a set of tools for Metacomputing. This will include a MetaMPI, based on PACX-MPI, a general ATM interface, that will be used by PACX-MPI and a metacomputing version of the performance analysis tool VAMPIR, which will be coupled to PACX-MPI.

To achieve a full support for MPI1.2, PACX-MPI has to be extended to support more functions. Up to now we've implemented mainly the MPI-calls according to our applications needs. Additionaly PACX-MPI will be extended to support not only TCP/IP for the external commmunication, but also to support other protocols, like ATM or HiPPI.

Acknowledgement

The authors would like to thank for the helpful support by networking organizations and groups, especially German Telekom, Teleglobe, CANARIE, STAR TAP, vBNS, ESNet. In addition we would like to thank Pittsburgh Supercomputing Center and the High Performance Computing Center Stuttgart for providing their machines for our tests and demonstrations.

References

1. Matthias Brune, Jörn Gehring and Alexander Reinefeld (1997), Heterogeneous Message Passing and a Link to Resource Management, Journal of Supercomputing, Vol. 1, 1-17
2. Thomas Bönisch and Roland Rühle (1998) Adapting a CFD Code for Metacomputing, 10th International Conference on Parallel CFD, Hsinchu/Taiwan, May 11-14.
3. Th. Eickermann, J. Heinrichs, M. Resch, R. Stoy, R. Völpel (1998) Metacomputing in gigabit environments: Networks, tools and applic ations, Parallel Computing 24, 1847-1872.
4. Graham E. Fagg, Jack J. Dongarra and Al Geist (1997) Heterogeneous MPI Application Interoperation and Process management under PVMPI, in Marian Bubak, Jack Dongarra, Jerzy Wasniewski (Eds.) 'Recent Advances in Parallel Virtual Machine and Message Passing Interface', 91-98, Springer.
5. Graham E. Fagg, Keith Moore, Jack J. Dongarra, Al Geist (1997) Scalable Networked Information Processing Environment (SNIPE), Technical Paper, Supercomputing 1997.
6. Ian Foster, Jonathan Geisler, William Gropp, Nicholas Karonis, Ewing Lusk, Georg e Thiruvathukal, Steven Tuecke (1998) Wide-Area Implementation of the Message Passing Standard, Parallel Computing 24.
7. Ian Foster, Carl Kesselman (1998) The Globus Project: A Status Report, Proc. IPPS/SPDP '98 Heterogeneous Computing Workshop, pg. 4-18, 1998.
8. Toshiya Kimura, Hiroshi Takemiya (1998) Local Area Metacomputing for Multidisciplinary Problems: A Case study for Fluid/Structure Coupled Simulation, 12th ACM International Conference on Supercomputing, Melbourne, July 13-17.
9. Thilo Kielmann, Rutger F.H. Hofman, Henri E. Bal, Aske Plaat, Raoul A.F. Bhoedjang (1998), MagPIe: MPI's Collective Communication Operations for Clustered Wide Area Systems, to appear at PPoPP'99, online version available at http://www.cs.vu.nl/albatross
10. Matthias Müller and Hans J. Herrmann (1998) DSMC - a stochastic algorithm for granular matter, in Hans J. Herrmann and J.-P. Hovi and Stefan Luding (Eds.) 'Physics of dry granular media', Kluwer Academic Publisher.
11. Andreas Wierse (1995) Performance of the COVISE visualization system under different conditions, in Visual Data Exploration and Analysis II, in Georges G. Grinstein, Robert F. Erbacher eds., *Proc. SPIE 2410*, pages 218-229, San Jose.

MILESS – A Learning and Teaching Server for Multi–Media Documents

Holger Gollan, Frank Lützenkirchen, and Dieter Nastoll

Computer Center, Essen University, Schützenbahn 70, 45117 Essen, Germany

Abstract. MILESS [7] is a joint project between the Computer Center and the Central Library of Essen University, together with the two pilot departments of linguistics and physics. The main purpose is to provide students and faculty of Essen University with a library server that supports several different functions that are needed within a digital library. Based on the IBM DB2 Digital Library product [4], MILESS can store and retrieve digital documents in any given format; moreover, searching is possible in a very elaborate way, and access control is supported as well.

In this article, we will first discuss why there is a growing need for digital library servers, followed by a description on how MILESS is build on top of the IBM DB2 Digital Library. We will describe the software techniques that are used to build the system, and we give a test case for the use of MILESS when referencing different articles within a mathematical journal.

1 The Need for a Digital Library

With the evolving web technologies, the amount of digital data that is accessible via the internet is enlarging in a dramatic way. Usually, these data will appear on certain websites, either personal or institutional. In addition, there might be commercial places on the web that hold lots of information in different digital formats. But this huge set of information leads to several problems.

- It is sometimes hard to find.
- It has no systematic order.
- It might vanish without further notice.

On the other hand, classical libraries have to find new ways to enable their customers to work with this new material in addition to the wellknown books and journals on the shelfs. Moreover, university students and faculty want to use digital and esp. multimedia material in learning, teaching and research. To face these problems and to meet these needs, a digital version of the classical library services is needed. It should support the use of such material by providing a reliable, permanent, and systematically ordered access to it.

2 MILESS and the IBM DB2 Digital Library

In late 1997, the Computer Center and the Central Library of Essen University started the MILESS project, which was funded by the local state ministry and the university. The idea was to install a digital library server that could solve the problems mentioned in the previous section. While the Computer Center brought in its knowhow in information technology and software development, the Central Library started to redefine the classical library techniques and services for the new types of digital and multimedia objects. In addition, two pilot departments (linguistics and physics) started to fill the digital library server with appropriate material.

To store and archive the digital documents, MILESS uses the IBM DB2 Digital Library product [4]. Its main parts consist of one library server and several object servers, where the object servers are responsible for the actual storage of the documents, but access is only possible via the library server that controls and manages the documents that are put into the digital library. Using this control mechanism, it is impossible to delete any documents in the object servers without notification of the library server, hence there can be no dead links within the system. The library server itself is running on top of a DB2 database, and the object servers can be connected to an ADSTAR Distributed Storage Manager (ADSM) [1] that handles the storage and archiving problems. E.g., documents can be archived when they haven't been used for a longer time.

The IBM DB2 Digital Library product offers a lot of features including several services that are useful for a digital library server. It handles storage and management of the documents via the object servers, and it enables access control via a rights management. Moreover, it has sophisticated search techniques, e.g. text mining and Query By Image Content (QBIC) to enable the user to find what he (she) is looking for within the stored digital documents.

While the IBM DB2 Digital Library product does an excellent job when it comes to the storage and retrieving problem, it is of no great help for the implementation of the MILESS data model. Since we wanted to have great flexibility in this respect, we adopted the Dublin Core [2] standard for the description of electronic resources, adding some additional features like contact information for the creators and contributors of the documents. Thus a document within MILESS can e.g. have several titles, several creators and contributors, several types and formats, etc. In particular, documents in MILESS can have several derivates in different formats; no standard format is required. Moreover, MILESS can handle hierarchical classifications that are widely used in science to help capturing the subjects of a document in a standardized way. This very complex and yet flexible data model for the metadata of electronic documents enables the librarians in the project to extend their classical library services to the digital material within the MILESS system.

Since the IBM DB2 Digital Library product can not handle such complex data models, additional software had to be written to enable MILESS to work with the Dublin Core standard. We will take a closer look on the new software in the next section. The following figure illustrates the different parts of the IBM DB2 Digital Library product.

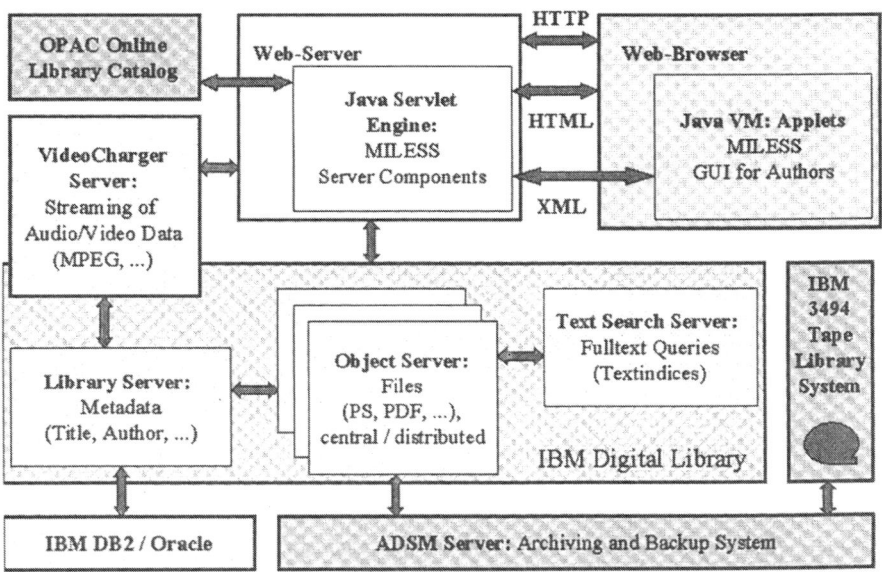

3 MILESS – A Closer Look

Besides the need for special software because of the complex data model, the additional MILESS software is divided into different layers, dependent on their functionality. To be platform independent, a design decision was made to use JAVA as the programming language, made possible by a JAVA API for the IBM DB2 Digital Library that can be used to connect JAVA code with the Digital Library product. Thus the bottom layer of the system is given by this programming interface that connects the Digital Library product with the outside world. This API is used by a so–called Persistency Layer that is responsible for the storing and retrieving of documents.

On top of that there is a collection of JAVA classes that implement the funtionalities for documents, legal entities (creators and contributors), etc.

Another part of the inner system is using JAVA servlets. MILESS is running inside a web server that is capable of using servlets, and these servlets are used for the connection and communication between the user and the system, e.g.

- A *DocumentServlet* is used to present the metadata of a document on a page within the web server.
- The *DerivateServlet* is needed to access a certain format of a specified document.
- The *SearchServlet* takes the user queries, connects to the Digital Library product to do the search, and presents the results on a page within the web server.

The following figure illustrates the different software layers of the MILESS system.

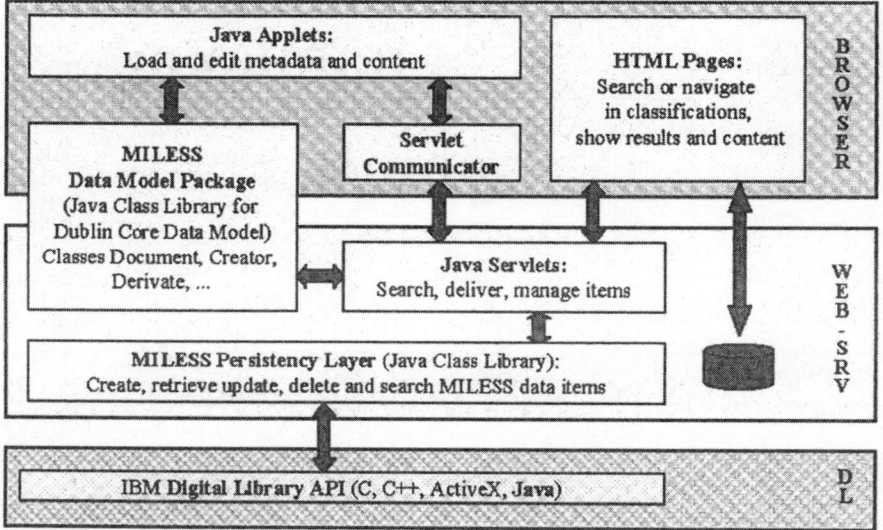

Besides the inner parts of the system that run on the server side, there are other parts that run on the side of the user client. Basically, the only thing the user needs is a web browser that connects him to the MILESS homepage at http://miless.uni-essen.de. From here, the user has access to the full functionality of the system; e.g. the search facilities can be reached just via normal HTML–pages. In addition, there is the possibility for an author to create or change documents inside the MILESS system. To do this, he (she) can call a graphical user interface (GUI) that runs as a JAVA applet inside the web browser. This GUI can be used to create new documents for the system, or to change already existing documents or personal data of creators and contributors. Moreover it helps navigating through hierarchical classifications to find the correct subjects for the documents. The communication between this GUI on the client side and the inner part of the system on the server side is handled via a Servlet Communicator, the data exchange is done via XML [3]. In the near future contributors can use XML directly to put material into the system. Thus the application on the client side doesn't have to know anything about the internal representations on the server side; especially the IBM DB2 Digital Library product and its internal structure is totally invisible to the outside world.

4 MILESS – From the user point of view

There are a lot of possible scenarios for a user of the MILESS system. Most people will use it just the way they use a classical library, but with enhanced features. One of these is the search facility. By making use of the search techniques of the underlying IBM DB2 Digital Library product, the user can not only search for certain authors or words and phrases within the title or the keywords, but it is also possible to search for words and phrases within the texts of the documents. Moreover, because of the way hierarchical classifications are implemented inside MILESS, he (she) can navigate through the hierarchy of such classifications, looking e.g. at all documents at a specific level.

Because of the capability of the system to handle documents in any given format, the user might have a problem with the actual format of a retrieved document. To overcome this difficulty, MILESS has a plug–in collection that could help the user and his browser to understand a strange format. Moreover, we are collecting different converters that could be used to create new formats from existing ones when putting new material into the system.

Creating new material is another scenario for the use of the MILESS system. With the help of the user GUI, anybody can create new documents in the system by providing the necessary metadata within the GUI and uploading the data files of the document into the system. This can be used e.g. by lecturers to put their lectures and exercises into the system, enabling students to work with this material online whenever they want.

Another scenario sees a lecturer preparing his next talk and searching the system for certain multimedia material he can use in the class. Such material can be JAVA applets or simulations/animations, audio-/video material, etc. To access video material, a video server [5] is included in the system that uses streaming techniques to deliver multimedia material in real time to multiple users.

Yet another use of MILESS is the linking between different articles of a mathematical journal; a first test case for this will be presented in the next section.

5 MILESS and the "Archiv der Mathematik" – A Test Case

This final section presents a collaboration with the Institute for Experimental Mathematics at Essen University, where six old volumes of the mathemat-

148

ical journal "Archiv der Mathematik" will be retrodigitized to make them available on the web in digital form (see [6]).

One small piece in this project is the automatic extraction of the bibliographic data like title, authors, author address, journal name, volume number, etc. These data, stored in an XML–format [3], can be used to fill the metadata fields of the Dublin Core standard automatically and to put the retrodigitized articles into the MILESS system, with restricted and controlled access because of copyright issues.

Another piece in this project is the automatic recognition of the cited references at the end of any article. Using Optical Character Recognition (OCR) and heuristics, an HTML–page is produced that contains the references and tries to link, where possible, to an online copy of the cited article. To do this, some standardization is needed to produce a correct link. To install a prototype for such a referencing functionality, we have put two articles of the "Archiv der Mathematik" into the MILESS system, namely

- William Crawley–Boevey, *Tameness of biserial algebras*, Arch. Math. **65**, 399–407.
- Christof Geiss, *On degenerations of tame and wild algebras*, Arch. Math. **64**, 11–16.

where the second one is a cited reference in the first article. After the automatic recognition of the references of the first article, the link to the second article will be produced automatically as

`http://miless.uni-essen.de/iem/Archiv_der_Mathematik/64/11`

where the volume number and the first page are used to uniquely identify the cited article. Upon this request the MILESS system starts an internal search to retrieve the referenced article. With this feature the reader can view the first article, realizing the citation, and following the link by just clicking on it. This can easily be extended to referenced articles being published in other journals, once these journals are available online and a unique way to reach their articles is realizable just from the bibliographic data as in the example above. An extension of the prototype in this direction is planned for the near future. Such an extension can lead to a distributed library for scientific journals, not necessarily restricted to mathematics, adding new features and functionalities for the user, but creating some demands on the underlying networks as well.

6 Acknowledgement

The MILESS project is financially supported by the local state government of Northrhine–Westfalia, Germany, and Essen University. Many people have been involved in the design and implementation of the system, including, but not restricted to D. Azkan, A. Bilo, E. Coelfen, B. Lix, V. Nordmeier, B. Schlesiona, A. Sprick.

References

1. ADSTAR Distributed Storage Management,
   http://www.storage.ibm.com/software/adsm/
2. The Dublin Core Standard, http://purl.oclc.org/dc/
3. Extensible Markup Language, http://www.w3.org/TR/REC-xml
4. IBM DB2 Digital Library, http://www.software.ibm.com/is/dig-lib/
5. IBM DB2 Digital Library Video Charger,
   http://www.software.ibm.com/data/videocharger/
6. G. O. Michler, "A Prototype of a Combined Digital and Retrodigitized Searchable Mathematical Journal", Preprint.
7. MILESS – Multimedialer Lehr- und Lernserver Essen, http://miless.uni-essen.de

Rural Educational System Network (RESNET): Design and Deployment

Salim Hariri and Wang Wei, Sung-Yong Park, Harvey Janelli

Center for Advanced TeleSysMatics (CAT)
University of Arizona
Tucson, AZ 85721
{hariri, wang}@ece.arizona.edu, www.ece.arizona.edu/~hpdc

Department of Computer Science
Sogang University
Seoul, Korea
sypark@ieee.org

Interactive Media Group, Inc.
14817 Sopras Circle
Addison, TX 75224
janelli_img@worldnet.att.net

ABSTRACT: The main objective of this project is to design and deploy the initial infrastructure of the Rural Education System Network (RESNET) in eastern Texas. We have selected the Asynchronous Transfer Mode (ATM) and single-mode fiber to build the RESNET infrastructure. The RESNET network operated initially at a backbone speed of OC-3c (155 Mbit/s), with the goal of upgrading to OC-12c (622 Mbit/s). The RESNET backbone connected the following sites: Tyler County Courthouse, Woodville ISD High School, Tyler County Hospital in the city of Woodville, Alabama & Coushatta Indian Reservation, Big Sandy ISD, Livingston High School, the Polk County Courthouse, the Polk County Hospital, main campus of Sam Houston State University in Huntsville, TX. The intended applications for RESNET are classified into three types: 1) Telecommunication Services, 2) Interactive Multimedia Services, and 3) Mutlimedia Services. The telecommunication services include: Switched data, voice, and video ATM service at 25 Mbps and 155 Mbps. In addition, it is the goal of the Tribes to establish a Call Center and Network Control Center. The interactive multimedia services include: Virtual Classroom, Virtual courtroom, Virtual County, Virtual Clinic, Teacher Network, Parent Network. The multimedia services include: Video-On-Demand (MPEG I & II), Education-On-Demand, Training-On-Demand, Multimedia Publishing, Electronic Publishing, and Intra/Internet Broadcasting. In this paper, we give an overview of the RESNET history, goals, design and technology adopted for RESNET, and conclude with future RESNET activities.

1. Overview of RESNET – Historical Perspective

The Rural Education System Network or RESNET was founded in 1992. It is an educational and public service partnership of private industry, local government, universities, hospitals, rural independent school districts and the Alabama-Coushatta Tribes. It was incorporated in the State of Texas on November 4, 1993. In 1994 the Tribes entered into the RESNET partnership to seek funding for the *Tribal Technology* project. In 1996 RESNET was successful in securing initial funding of $750,000 from the Houston Endowment (a private foundation). Money was provided for the construction of an 80-mile fiber optic network with an OC-3c (155Mbps) Asynchronous Transfer Mode (ATM) backbone. This network backbone connects three independent school districts (ISD) of Livingston, Big Sandy, and Woodville; two hospitals, the Indian Health Service (IHS) clinic and Tyler County hospital with UTMB Galveston; Sam Houston State University (a regional university); and the Polk, and Tyler county governments. In addition, the Department of Agriculture provided $340,000 of funding for the base line ATM electronics that connects these locations with the Alabama-Coushatta Tribes.

It was the consortium's intent that RESNET becomes a low cost community resource, operated and supported by tribal members. It provides a high-speed interface to the Internet, and serves as an Intranet for the Tribes, hospitals, local governments, and educational components.

RESNET has assembled a solid partnership of local business, international technology providers, and world class academics. All of the partnerships have been consummated with the use of a formal teaming agreement. This binding agreement clearly defines the commitment of all the parties. The majority of the partners have been involved for over three years and continue to get more and more involved. Current business and technology partners include: Lucent Technologies (4 years); Entergy Corp. (3 years); Sam Houston Electric Cooperative (SHECO) (2.5 years); the CASE Center at Syracuse University (2 years); and the newest partner, the IBM Corporation, Network Hardware Division. All of the current RESNET ISD's mentioned have been committed for over four years and have played an active role in the definition of end-user needs. Sam Houston State University, Rice University, Texas A&M University, University of Texas, and the University of Houston are committed to providing distance learning programs and higher education courseware for the network. This has been a "grass-roots" effort, primarily in support of the Tribes and their children. Unlike other programs, the Alabama-Coushatta Tribes will continue to share resources with the surrounding community.

RESNET recently entered into a special partnership with the SHECO. Both parties have a joint sheath agreement whereby SHECO in addition to giving RESNET pro-bono pole contacts, matches mile for mile additional fiber installation. For every mile of fiber RESNET installs, SHECO installs a mile and both parties exchange fibers to form a common network. In addition, SHECO maintains the entire fiber optic system upon completion of installation by RESNET.

2. RESNET Goals and Strategies

The main goal of RESNET is the development of a community networking environment to cooperatively develop applications that utilize advanced communication technologies. The RESNET will provide high-speed network connectivity to all the RESNET sites (e.g., schools, court houses, hospitals, and Indian reservations) and state of the network-based applications (e.g., switched telecommunication services, interactive multimedia services and multimedia services).

The RESNET also provides a high-speed connection to the Internet, including NGI (Next Generation Internet), and the NII (National Information Infrastructure). This federally mandated program was established by the High Speed Computing Act of 1991 (a.k.a. "NREN" - the National Research and Education Network). Its original goal was to establish a multimedia broadband fiber-optic network, connecting over 1200 national universities and research facilities in the U.S. and eventually overseas. The NII now has evolved into the Internet. It provides access to a myriad of information sources. It is in effect, an extension of thousands of electronic assets worldwide, with very high-bandwidth requirements. Very large databases, multimedia database with very large files, high-bandwidth applications (e.g., the ability to log on to the Hubbell telescope and concurrently view on-going experiments) are all part of this "mega-network".

It is RESNETs' goal to see that the children in rural east-Texas are not left behind while this Information Superhighway (NII) comes to fruition. The population base of rural America is highly economically disadvantaged and highly populated by minorities. It is, in effect, a mirror image of the inner city, with a more favorable environment and lower population density. The RESNET technology goal is to install Fiber-to-the-Schools (FTTS) using single-mode fiber backbone, with ATM transport at OC-12. There will be ATM switches at each ISD campus and ATM network interface cards (25 Mbps) installed in each student workstation.

The technology involved in our migration toward NII/NGI compatibility is fairly simple. We must work with what is presently installed, wherever possible, and upgrade to the compatible technology whenever possible. Our goal in the pilot is to extend the fiber installed at the University of Houston and their full services, into the three ISD campuses of the RESNET pilot. RESNET takes care of the solutions to insure the security and privacy of the data and network between the two sites.

3. RESNET Design and Technology

The RESNET is a broadband fiber-optic based, private wide-area ATM network, whose scope, by the year 2002, will include ten counties in eastern Texas area.

154

The first segment of this consortium-owned, not-for-profit managed network was started around the Alabama-Coushatta Indian Reservation in Polk County, Texas, and it was designed in a hierarchical way such that each town has a backbone switch, several small-to-medium sized ATM switches, and customer premise equipment. Therefore, the initial design of RESNET (Phase I), as we can see from Figure 1, has been focused on building a high-speed ATM backbone (OC-3) by connecting Indian Reservation to Livingston, Woodville and Dallardsville (Big Sandy ISD), and on connecting the backbone switch to the community facilities such as schools and hospitals in each town. The next phases of RESNET design (Phase II, III, and IV) will be extended to cover other ISDs located in west, north, and south regions as shown in Figure 1.

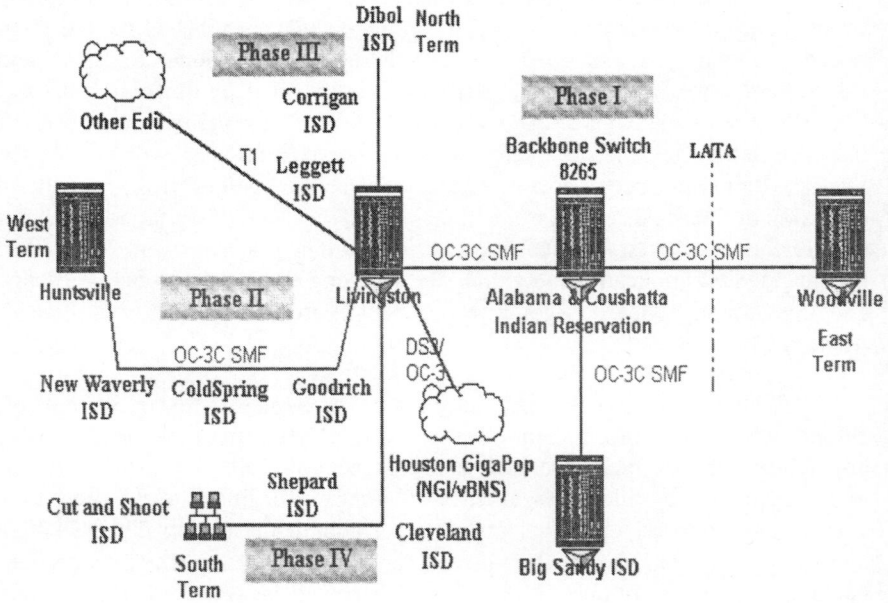

Figure 1: Overview of RESNET Design

The RESNET design is implemented mainly by using ATM products from IBM. For example, in each RESNET site, we have installed a backbone switch (IBM 8265 or IBM 8260 Nways ATM switch), and a combination of IBM 8260 Nways Multiprotocol Switching Hubs and IBM 8285 workgroup ATM switches to build ATM clusters operating at 155 and 25 Mbps speed. Each ATM switch runs PNNI version 1.0 and UNI 3.0/3.1 to connect to other ATM switches and workstations. Each 8265 backbone switch has been equipped with a Multiprotocol Switched Services (MSS) module that provides various routing and switching services. The MSS is a key component of IBM's Switched Virtual Network architecture and supports various features such as Classical IP over

ATM (CLIP) [1], LAN Emulation (LANE) [2] services and Next Hop Resolution Protocol (NHRP) [3]. It also supports various routing (e.g., RIP, OSPF, etc.) and bridging protocols.

For the administrative purpose, each town is configured with different IP subnet and each subnet runs either CLIP or LANE services based on its requirements. In this case, the MSS module installed in each backbone switch provides the necessary functions such as Address Resolution Protocol (ARP) server and LANE servers (LECS, LEC, and BUS). Since any cut-through routing (e.g., NHRP) capabilities has not been configured at the time of this writing, the communication between any two towns still passes through two routers, which may be the performance bottleneck as the inter-town traffic grows. However, we envision that the problem of this performance bottleneck can be resolved when standard-based short-cut routing schemes such as Multiprotocol over ATM (MPOA) [4] or Multiprotocol Label Switching (MPLS) [5] are implemented within the MSS module.

The current RESNET backbone operates at the speed of 155 Mbps (OC-3) and ATM is the main transport mechanism over the RESNET. However, with the advances in high-speed interfaces (e.g., OC-12, OC-48, and OC-192) and optical technologies (e.g., Dense Wavelength Division Multiplexing (DWDM)), the future RESNET backbone is expected to be upgraded to support gigabits or terabits applications.

RESNET also allows the sharing of networked resources within a school district, town, county or region. A high-speed connection (DS-3 or OC-3) to the Houston *GigaPop* is being sought by RESNET in order to access Rice University, Texas A&M, University of Texas, and the University of Houston, most of which have a connection to the vBNS (very-high-speed Backbone Network Service) national ATM network. Sharing of the computational resources and the cooperative educational programs (both K-12 and adult education) permits RESNET participants to access to the resources that they could not afford on an individual basis.

Figures 2 through 5 show the actual designs for RESNET sites, Livingston, Alabama Coushatta, Woodville, and Big Sandy ISD, respectively.

156

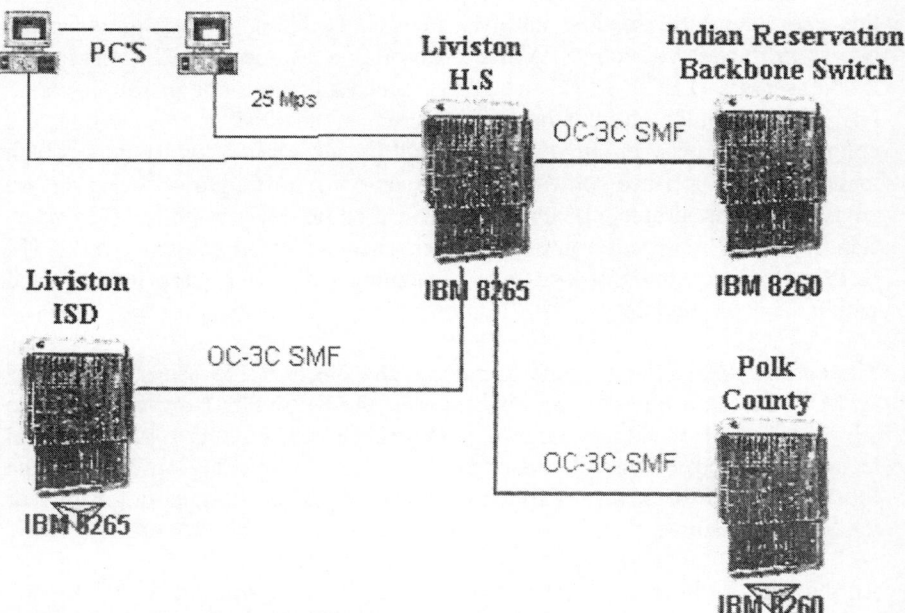

Figure 2: Detailed View of Livingston

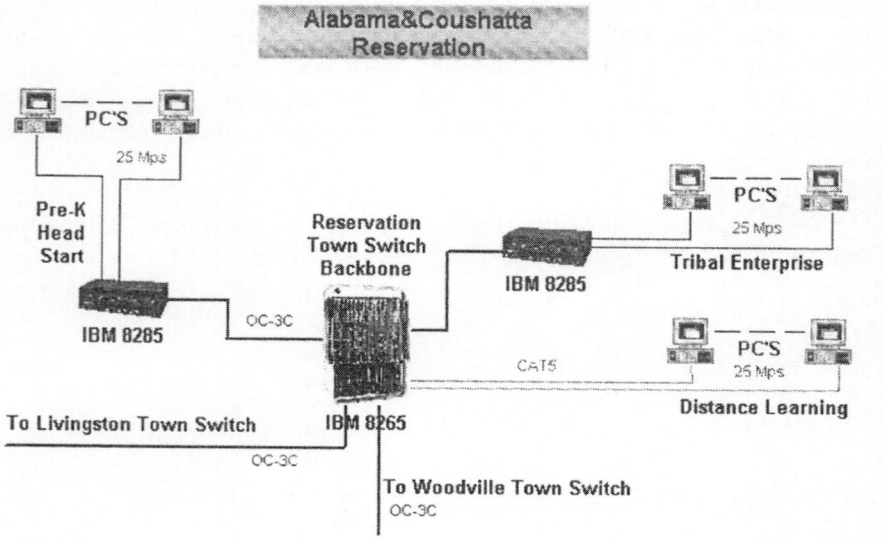

Figure 3: Detailed View of Alabama-Coushatta Indian Reservation

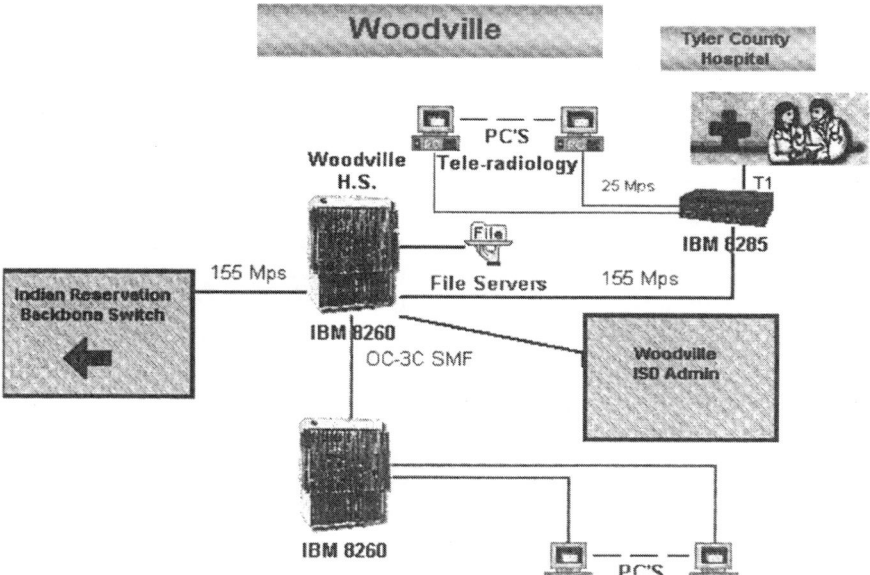

Figure 4: Detailed View of Woodville

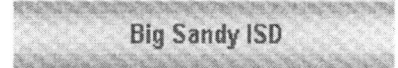

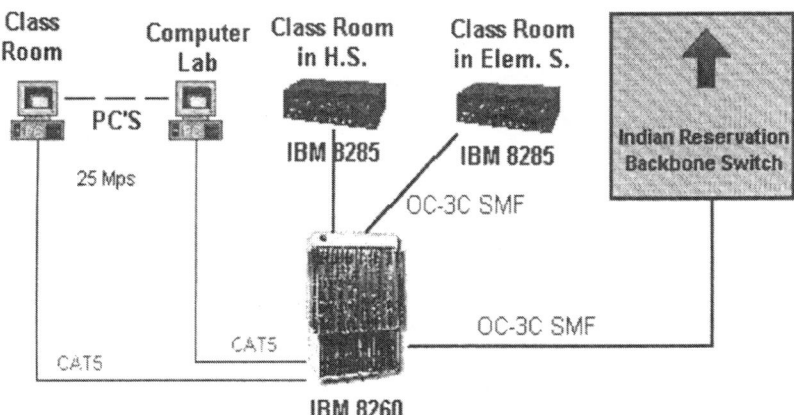

Figure 5: Detailed View of Big Sandy ISD

4. RESNET Demonstration

One of the design goals of RESNET is to develop and deploy network-based multimedia applications that can fully utilize the high-speed ATM network connectivity across RESNET sites. For this purpose, we have created three proof-of-concept scenarios and demonstrated them at the RESNET opening ceremony in November, 1997. The demonstrations were presented at the Alabama Coushatta Indian Reservation and three scenarios were demonstrated: 1) Video-conferencing demonstration, 2) Video on demand demonstration, and 3) Voice over IP demonstration. These three applications were selected based on the current needs of different RESNET sites and on the future business plans using the RESNET infrastructure. In what follows, we briefly review the demonstrations and some of the experience from the demonstrations.

4.1 Video-conferencing Demonstration

Video-conferencing is one of the most important applications that provide users with advanced video collaboration solutions both for education and business applications. In this demonstration, the First Virtual's ATM-based video-conferencing solutions [6] have been installed both at Alabama Coushatta Indian Reservation and at Big Sandy ISD. The First Virtual's video-conferencing solution consists of a PC equipped with a plug-and-play 25 Mbps ATM NIC card from First Virtual (VC-NIC) and an MVIP-capable video-conferencing equipment from PictureTel. Unlike other ATM NIC cards, the VC-NIC card is specially designed to support video-networking applications and includes an industry standard MVIP interface on the board. This onboard MVIP interface provides direct connection to MVIP-capable multi-vendor video-conferencing equipment and allows the video traffic to bypass the system bus. The VC-NIC card is fully compliant with standard UNI 3.x signaling protocol and LANE 1.0 protocol. The video data is transmitted at 384 Kbps speed. Although the 128 Kbps video-conferencing based on a single ISDN line meets the needs of face-to-face conferencing, the ATM-based 384 Kbps conferencing provides higher quality video enough to support most business applications and distance learning applications.

4.2 Video on Demand Demonstration

In the business and educational applications, it is common to record video presentations, classes, and movies, and store them in a centralized storage system in order to eliminate the need for the replication and distribution of tape. Retrieving this information from the remote PCs on demand is an important application since most of the ISDs in RESNET can share a wealth of course material and educational movies.

In order to implement this scenario, a PC server (IBM 330) with a 155 Mbps ATM interface card (LANE interface) has been installed and connected to the backbone ATM switch at Alabama-Coushatta Indian Reservation. This PC server runs Windows NT Server 4.0 operating system and is equipped with RAID-5 disk array to provide a high-level of fault tolerance and better performance. Several DVD movies were downloaded into the disk array so that client PCs in remote sites can access the movies simultaneously over the network. On the other hand, we have installed IBM Turboways ATM 25 Mbps NICs into several client PCs located at different RESNET sites (e.g., Livingston, Woodville, Big Sandy) and connected them to their local IBM 8285 workgroup ATM switches. The client PCs were also equipped with MPEG-2 decoder board and software DVD player to play the DVD movies stored in the PC server. The public domain Network File System (NFS) software (NFS server and NFS client) was used to implement the communication between the client and server. For example, the PC server exports its file system and the client PCs mount the remote file system into the network drive. Once the mount operations are properly executed, each client PC opens the network drive and plays the DVD movies stored in the PC server over the ATM network.

4.3 Voice over IP Demonstration

One of the main advantages of using ATM is that any types of data (voice, video, data) can be mixed and transferred over the same network infrastructure. With the explosive growth of Internet and the increasing interests in building Next Generation Network (NGN) (NGN is a future communication infrastructure that integrates voice, data, and video traffic into a single common packet network), Voice over IP has been gaining increasing popularity among researchers and creating a lot of opportunities for business and educational applications.

In our demonstration, we have installed two Tempest Data Voice Gateway (DVG) from Franklin Telecom [7] at Alabama-Coushatta Indian Reservation and Woodville High School. The two places are 15 miles apart and belong to different LATAs. The Tempest DVG is self-contain, PC-based standalone box with three interface cards (DSP board, Telephone interface board, LAN interface board). This box runs Linux operating system and contains system software from Franklin Telecom. One of the problems we have met was that the data interface provided by Tempest DVG was Ethernet only, thereby the direct connection to the RESNET ATM backbone was not an option. In order to solve this problem, two local Ethernet subnets were created both at Alabama-Coushatta Indian Reservation and at Woodville High School. In each subnet, we have also installed a Windows NT-based router (a PC with dual data interfaces (Ethernet ad ATM)) so that the Ethernet traffic generated from the Tempest DVG is routed and trasmitted across the RESET ATM backbone to the Woodville High School. The router at the Woodville High School in turn routes the data to the local Ethernet subnet. Although each voice packet has to pass through two routers, the quality of the voice was quite impressive. As we increase more simultaneous voice sessions, the quality of voice might be degraded due to the nature of Ethernet and the two intermediate routers. Installing a T1 board from Franklin Telecom and connecting it directly to the backbone switch (IBM 8265/8260 has an interface module for T1/E1) is another option to improve the throughput and guarantee the quality of simultaneous voice sessions. Also, in a real environment, we can create an ATM PVC between two Ethernet subnets and bridge the voice traffic (or tunelling) to improve the performance.

5. Summary and Concluding Remarks

In this paper, we presented the design and deployment of the Rural Educational System Network (RESNET) in eastern Texas. We reviewed how this project started, funded and the steps involved in implementing the RESNET backbone network. We also reviewed in further detail the technology adopted to design each RESNET site. We are currently working with Texas A & M university to take the responsibility of managing all the RESNET services. In addition, we are currently pursuing aggressively initiatives to provide high-speed connectivity to the national high-speed backbone (vBNS). Once this connection is established, we will work with the Researchers at the Center for Advanced TeleSysMatics (CAT) at the University of Arizona and Texas A&M to establish an Adaptive Distributed Virtual Computing Environment (ADViCE) [8] on RESNET.

References

[1] M. Laubach, "Classical IP and ARP over ATM", RFC 1577, January 1994.
[2] ATM Forum, "LAN Emulation over ATM Specification – ver 1.0", February 1994.
[3] D. Katz, D. Piscitello and B. Cole, "NBMA Next Hop Resolution Protocol", Internet Draft, December 1995.
[4] ATM Forum, "Multiprotocol over ATM – ver 1.0", July 1997.
[5] R. Callon, P. Doolan, N. Felman, A. Fredette, G. Swallow and A. Viswanathan, "A Framework for Multiprotocol Label Switching", Internet Draft, November 1997.
[6] http://www.fvc.com
[7] http://www.ftel.com
[8] Salim Hariri et al., "The design and evaluation of a virtual distributed computing environment", Cluster Computing, Vol. 1, May 1998, pp. 81-93.

Some Performance Studies in Exact Linear Algebra

George Havas[1] and Clemens Wagner[2]

[1] Centre for Discrete Mathematics and Computing
   Department of Computer Science and Electrical Engineering
   The University of Queensland, Queensland 4072, Australia
   havas@csee.uq.edu.au
   http://www.it.uq.edu.au/~havas/
[2] Fachgruppe Praktische Informatik
   Fachbereich Elektrotechnik und Informatik
   Universität-GHS Siegen
   D-57078 Siegen, Germany
   wagner@informatik.uni-siegen.de
   http://pi.informatik.uni-siegen.de/clemens/

Abstract. We consider parallel algorithms for computing the Hermite normal form of matrices over Euclidean rings. We use standard types of reduction methods which are the basis of many algorithms for determining canonical forms of matrices over various computational domains. Our implementations take advantage of well-performing sequential code and give very good performance.

1 Introduction

Algorithms for exact linear algebra have been much studied. Many different strategies for calculation of canonical forms of matrices have been proposed. A comprehensive bibliography and a number of earlier methods for integer matrices are examined in [8], including references to various polynomial time algorithms. Some parallel and some more recent methods are described in [16,9,6,22,20,11,12,21]. Reduction methods for general Euclidean rings are studied in detail in [23].

We concentrate on algorithms which use reduction as their underlying principle. In spite of the fact that the worst case performance of reduction methods can be exponentially bad (see [4] and [13]), such techniques provide the basis for many sequential implementations. We do not consider modular methods in this paper. Rather we study other recent implementations which focus on finding well-performing algorithms and good heuristics for reduction methods. We start by outlining the mathematical background. We then show how to extend sequential algorithms to a parallel environment. We finish by presenting some sample performance figures and outlining recommendations for choosing appropriate parallel algorithms.

2 Mathematical background

A commutative ring $\mathbf{R}$ with identity 1 is Euclidean if there is a value function $\varphi : \mathbf{R}^* \to \mathbb{N}_0$ (where $\mathbf{R}^* = \mathbf{R} \setminus \{0\}$ and $\mathbb{N}_0$ is the set of nonnegative integers) such that the following properties hold for $a \in \mathbf{R}$ and $b \in \mathbf{R}^*$.

1. For $a \neq 0$, $\varphi(ab) \geq \varphi(a)$.
2. There exist $q, r \in \mathbf{R}$ with $a = qb + r$, such that either $r = 0$ or $\varphi(b) > \varphi(r)$.

Paradigm examples of Euclidean rings are $\mathbb{Z}$ (the ring of integers, with absolute value as value function) and $\mathbb{F}[x]$ (the ring of univariate polynomials with coefficients in a field $\mathbb{F}$, with degree as value function).

An element $a \in \mathbf{R}$ is a unit if it has an inverse $a^{-1} \in \mathbf{R}$ such that $aa^{-1} = a^{-1}a = 1$. The set $U(\mathbf{R})$ of all units of $\mathbf{R}$ is a multiplicative group. Elements $a, b \in \mathbf{R}$ are associates if there exists a unit $c \in \mathbf{R}$ such that $a = bc$ and we write $a \sim b$. Association is an equivalence relation with equivalence classes $[a] := \{b \in \mathbf{R} \mid a \sim b\}$. A subset $R \subseteq \mathbf{R}$ is a representative set for $\mathbf{R}$ if: $\{[a] \mid a \in R\} = \mathbf{R}/\sim$; and $\forall a, b \in R, a \not\sim b$.

For a Euclidean ring $\mathcal{R}$ a function $\rho : \mathbf{R} \times \mathbf{R}^* \to \mathbf{R}$ is called a residue class system if for all $a, a' \in \mathbf{R}$ and $b \in \mathbf{R}^*$

$$\rho(a, b) \quad \in \quad \{a - qb \mid q \in \mathbf{R}\},$$
$$\varphi(\rho(a, b)) \quad < \quad \varphi(b), \qquad \text{and}$$
$$\rho(a, b) = \rho(a', b) \iff \exists t \in \mathbf{R} : a = a' + tb.$$

Let $M \subset \mathbf{R}$ with $M \neq \{0\}$ be a finite, nonempty subset of $\mathbf{R}$. The greatest common divisor of M ($\gcd(M)$) is the equivalence class $[g]$ such that: $\forall a \in [g]$, $a \mid M$; and $\forall b \in \mathbf{R}$ with $b \mid M$, $b \mid [g]$. If we have a representative set R for $\mathbf{R}$ and $d \in \gcd(M) \cap R$ then d is uniquely determined. Further background material on Euclidean rings is given in [18,7,5].

Matrices A and B with entries in a Euclidean ring $\mathcal{R}$ are column equivalent if there exists a unimodular matrix V such that $A = BV$. Matrix V corresponds to a sequence of elementary column operations: multiplying a column by a unit of $\mathcal{R}$; adding any multiple by a ring element of one column to another; or interchanging two columns.

For any matrix B over a Euclidean ring $\mathcal{R}$ with representative system R there exists a unique lower triangular matrix H which is column equivalent to B and which satisfies the following conditions. Let r be the rank of B.

1. The first r columns of H are nonzero and the remaining columns are zero.
2. For $1 \leq j \leq r$ let $H_{i_j, j}$ be the first nonzero entry in column j. Then $i_1 < i_2 < \ldots < i_r$.
3. $H_{i_j, j} \in R \setminus \{0\}$, for $1 \leq j \leq r$.
4. For $1 \leq k < j \leq r$, $H_{i_j, k} = \rho(H_{i_j, k}, H_{i_j, j})$.

This matrix is called the column Hermite normal form (HNF) of the given matrix B and has many important applications. As already mentioned, there are many algorithms based on reduction methods for computing the HNF. Descriptions of such methods for canonical form computation in Euclidean rings (sometimes specialized to the integers) in the literature include [18,7,5,19].

3 Sequential algorithms

Deterministic polynomial-time HNF algorithms (non-modular) include those of Kannan and Bachem [14], Chou and Collins [3], and Havas, Majewski and Matthews [12] for the integers; of Kannan [15] for $\mathbb{Q}[x]$; and of Wagner [23] for $\mathbb{F}_q[x]$. Heuristic algorithms (often faster and/or "better") include those of Havas and Majewski [9] for the integers and of Wagner [23] for $\mathbb{F}_q[x]$. All in all, even the sequential algorithm situation is a quite complicated story, which is addressed in much more detail in [23]. We do not go into this further here, but rather build parallel algorithms based upon effective sequential ones.

4 Parallel implementations

The problem we consider is: Given $A \in \mathrm{Mat}_{m \times n}(\mathbf{R})$, compute in parallel H the HNF of A together with a unimodular matrix V such that $H = AV$. To unify the computation we actually compute the HNF of a working matrix $W = \left(\begin{smallmatrix} A \\ I_n \end{smallmatrix} \right)$. Let K be the Hermite normal form of W. The first m rows of K are the Hermite normal form of A and the last n rows of K give a unimodular transformation matrix V.

A *parallel computer* $\mathcal{P} := \{\pi_0, \dots, \pi_{N-1}\}$ consists of N processors with distinct memory and a communication network. Let

$$\ell(l, \tau) := \left\lfloor \frac{l}{N} \right\rfloor + \begin{cases} 1 & \text{if } \tau < (l \bmod N) \\ 0 & \text{otherwise} \end{cases}$$

for each $0 \leq \tau \leq N - 1$.

We use the matrix distribution model from [24]. Each processor π_τ on the parallel computer stores part of the working matrix W corresponding to a $(\ell(m, \tau) + 1) \times n$ matrix $A^{(\tau)}$ and a $\ell(n, \tau) \times n$ matrix $V^{(\tau)}$ which are submatrices of the working versions of A and the multiplier $V \in \mathrm{GL}_n(\mathbb{F}[x])$, respectively. An extra row, row $\ell(m, \tau) + 1$ of $A^{(\tau)}$ is used to control the computations. We call this the *computation row*.

The matrix is distributed rowwise to processors, but in stripes **not** in blocks. For each operational row we also do a broadcast. Thus, we distribute the input matrix A by storing the ith row of A in row i' of matrix $W^{(\tau)}$ on processor π_τ where $\tau := (i-1) \bmod N$ and $i' := \lfloor (i-1)/N \rfloor + 1$ for $1 \leq i \leq m$. The parallel, hybrid HNF algorithm PARALLEL-HNF is given by pseudocode in Figure 1. It uses the standard DIV operator for the appropriate Euclidean ring and calls two subprocedures: COMPUTE-GCD and PARALLEL-ROD.

PARALLEL-HNF$(W^{(\tau)}, \alpha, \beta)$

input τ: processor index

$W^{(\tau)}$: π_τ-part of a full column-rank, distributed matrix W

α, β: non-negative integers

$k \leftarrow 1$

$(i_1, \ldots, i_n) \leftarrow 0$

while $k \leq n$ **do**

    $r \leftarrow 0$

    $i \leftarrow 1$

    $s \leftarrow \min\{k + \alpha - 1, n\}$

    $f \leftarrow k$

    while $r < s$ **do**

        $f \leftarrow \max\{r + 2, f\}$

        $\mu \leftarrow (i - 1) \bmod N$

        if $\mu = \tau$ **then**

            $j \leftarrow \lfloor (i - 1)/N \rfloor + 1$

            broadcast $W^{(\tau)}_{j,r+1}, W^{(\tau)}_{j,f}, \ldots, W^{(\tau)}_{j,s}$ **to all other**

        else

            $j \leftarrow \ell(m, \tau) + 1$ ▷ *Index of computation row*

            receive $W^{(\tau)}_{j,r+1}, W^{(\tau)}_{j,f}, \ldots, W^{(\tau)}_{j,s}$ **from** μ

        if $i = i_{r+1}$ **then**

            for $l \leftarrow f$ **to** s **do**

                $q \leftarrow \text{DIV}(W^{(\tau)}_{j,l}, W^{(\tau)}_{j,r+1})$

                $W^{(\tau)}_{*,l} \leftarrow W^{(\tau)}_{*,l} - q W^{(\tau)}_{*,r+1}$

        $W^{(\tau)} \leftarrow \text{COMPUTE-GCD}(W^{(\tau)}, j, r + 1, f, s)$

        if $W^{(\tau)}_{j,r+1} \neq 0$ **then**

            $(i_1, \ldots, i_n) \leftarrow (i_1, \ldots i_r, i, i_{r+1}, \ldots, i_{n-1})$

            $r \leftarrow r + 1$

    $W^{(\tau)} \leftarrow \text{PARALLEL-ROD}(W^{(\tau)}, r, \beta, i_1, \ldots, i_r)$

    $k \leftarrow k + \alpha$

return $W^{(\tau)}$

Fig. 1. Parallel hybrid Hermite normal form algorithm

A call COMPUTE-GCD(B, i, j_0, j_1, j_2) takes a matrix B with n columns, an integer $i \geq 1$ and $1 \leq j_0 < j_1 \leq j_2 \leq n$ as input, where i is a valid row index of B. It produces a right equivalent matrix B' as output where

$$B'_{i,j_0} = \gcd(B_{i,j_0}, B_{i,j_1}, \ldots, B_{i,j_2}) \text{ and } B'_{i,j} = 0$$

for $j_1 \leq j \leq j_2$. This algorithm computes B' from B by *unimodular column operations* (*swapping* of two columns, *multiplying* a column with an unit, and *adding a multiple of one column to another column*). There are various different methods to obtain B' from B (e. g. [1,2,10,23]). By using gcd algorithms whose execution depends only on the entries in the operational row we need no additional communication for this purpose.

The function PARALLEL-ROD reduces the off-diagonal entries. It is a hybrid algorithm which is controlled by parameter $\beta \in \mathbb{N}_0$. For $\beta + 1$ greater than or equal to the rank of the input matrix it is a parallel variant of the standard reduction algorithm. For $\beta = 1$ this algorithm is a parallel version of Chou-Collins' reduction method. If we choose β equal to zero the algorithm does not change the input matrix. A more detailed description of Parallel-ROD algorithm can be found in [24].

Theorem 1. *Let $A \in \mathrm{Mat}_{m \times n}(\mathbf{R})$ with rank r be in echelon form. Let $1 \leq \beta \leq r$ and $q := \lfloor \frac{r-2}{\beta} \rfloor$. Then the PARALLEL-ROD algorithm uses $\frac{q(q+1)}{2}\beta + r$ broadcasts to transfer at most $\frac{r^2 + r + q^2\beta + q\beta}{2}$ ring elements.*

A proof is given in [24].

The PARALLEL-HNF algorithm divides the input matrix into vertical blocks of width α. For $k := \alpha, 2\alpha, \ldots, (p-1)\alpha, n$ (where $p := \lceil \frac{n}{\alpha} \rceil$) the HNF of the leading k-column submatrix is computed. This is shown in Figure 2.

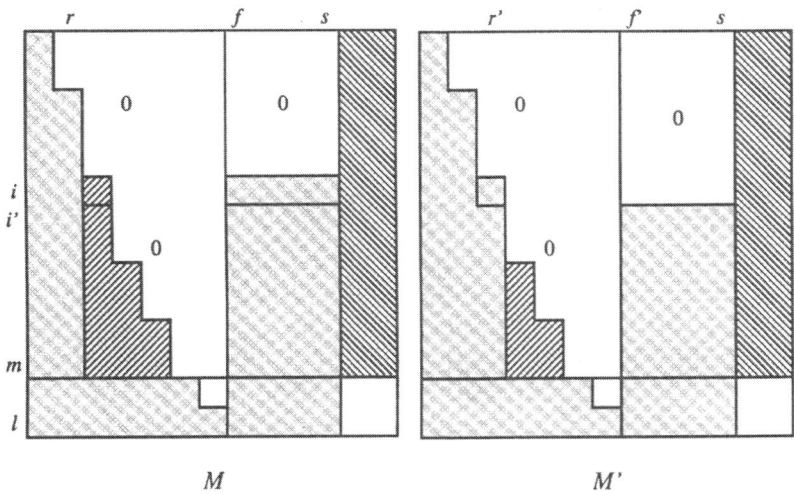

Fig. 2. Computing the HNF of the leading k columns, for $k = \alpha, 2\alpha, \ldots$

Theorem 2. *Let $A \in \mathrm{Mat}_{m \times n}(\mathbf{R})$ with rank r and $1 \leq \alpha, \beta \leq n$. Then the PARALLEL-HNF algorithm uses $O(\frac{n^3}{\alpha\beta} + \frac{nm}{\alpha})$ broadcasts with distributed $W := \binom{A}{I_n}, \alpha$ and β as input. It transfers $O(\frac{n^3}{\alpha})$ ring elements.*

For $\mathbf{R} = \mathbb{F}[x]$ the procedure uses $O((m + n - r)n^5\beta^2 4^\alpha \|A\|^2)$ field operations. At most $O(\frac{n^4\beta 2^\alpha \|A\|}{\alpha})$ field elements are transferred via broadcasts.

Proof. Transforming a $(m + n) \times s$ principal submatrix of W with $s \in \{\alpha, 2\alpha, \ldots, (p-1)\alpha, n\}$ into echelon form requires at most $m + n - r$ broadcasts. Transforming the echelon form into HNF requires (by Theorem 1) $\frac{q(q+1)}{2}\beta$ with $q := \left\lfloor \frac{s-2}{\beta} \right\rfloor$ broadcasts. In total this leads to the broadcast bound

$$p(m+n-r) + \sum_{s=\alpha,\ldots,(p-1)\alpha,n} \left(\frac{1}{2} \cdot \left\lfloor \frac{s-2}{\beta} \right\rfloor \left(\left\lfloor \frac{s-2}{\beta} \right\rfloor + 1 \right) \beta + s \right)$$

$$\leq p(m+n-r) + \sum_{i=1}^{p} \left(i\alpha + \left\lfloor \frac{i\alpha - 2}{2} \left(\frac{i\alpha - 2}{\beta} + 1 \right) \right\rfloor \right)$$

$$\leq p(m+n-r) + \alpha \frac{p(p+1)}{2} + \sum_{i=1}^{p} \left\lfloor \frac{(i\alpha)^2}{2\beta} + \frac{i\alpha}{2} \right\rfloor$$

$$= \underbrace{p(m+n-r)}_{\in O(n(n+m)/\alpha)} + \underbrace{\alpha \frac{p(p+1)}{2}}_{\in O(n^2/\alpha)} + \left\lfloor \underbrace{\alpha^2 \frac{2p^3 + 3p^2 + p}{12\beta}}_{\in O(n^3/(\alpha\beta))} + \underbrace{\alpha \frac{p(p+1)}{4}}_{\in O(n^2/\alpha)} \right\rfloor$$

$$\in O\left(\frac{n^3}{\alpha\beta} + \frac{nm}{\alpha} \right)$$

To compute the echelon form we do not need to broadcast the linearly dependent rows. Thus, the number of broadcast ring elements can be majorized by

$$\sum_{i=1}^{s} (1 + (\alpha - 1)) = s\alpha$$

for a $(m + n) \times s$ principal submatrix of W. Transforming the echelon form to Hermite normal form requires broadcasting at most $O(r^2)$ ring elements. Computing the Hermite normal form of all $(m + n) \times s$ principal submatrices of W with $s \in \{\alpha, 2\alpha, \ldots, (p-1)\alpha, n\}$ we need

$$O\left(n\alpha + \sum_{i=1}^{p-1} i\alpha^2\right) + pO(r^2) \subseteq O\left(n\alpha + \frac{p(p-1)}{2}\alpha\right) + O(pr^2)$$

$$\subseteq O(n\alpha) + O(\frac{n^2}{\alpha}) + O(\frac{n^3}{\alpha}) \subseteq O(\frac{n^3}{\alpha})$$

ring elements to be broadcast.

The proofs of the other two estimates are quite lengthy. They can be found in [23].

5 Performance examples

We have implemented this and related algorithms in C/C++ on the IBM SP2 at the GMD in Sankt Augustin. We have used the xlC compiler and the message passing library MPL (both IBM products). We have used the Sorting-GCD algorithm, due to Majewski-Havas [17], for the implementation of the COMPUTE-GCD function, where we used a heap for determining the polynomial with the largest degree or the integer with the largest absolute value, respectively in a subvector.

We have done many practical studies with these algorithms. In this paper we give some details of the behavior of PARALLEL-HNF for some random matrices over $\mathbb{F}_2[x]$ and $\mathbb{Z}$. Thus we used an input matrix over $\mathbb{F}_2[x]$ which is a random 80×80 matrix, where the degree of each entry is less than or equal to 80. The rank of this matrix is $r = 80$. The input matrix over $\mathbb{Z}$ is a random 100×100 matrix. The absolute value of each entry is less than or equal to 64.

Table 1 and Figures 3 and 4 show the results of experiments in which we varied α and β. We used 16 nodes of the SP2. The first row of each measurement is the total running time (*minutes:seconds.hundredth*). The second row gives the maximum degree ($\mathbb{F}_2[x]$), or the number of bits in the largest absolute value ($\mathbb{Z}$) which arose during the computation. In Figure 3 the x-axis shows α while the y-axes show run times and maximum degrees. In Figure 4 the x-axis shows α while the y-axes show run times and maximum number of bits.

	$\mathbb{F}_2[x]$				$\mathbb{Z}$		
α	$\beta = 1$	$\beta = \alpha$	$\beta = n+ \frac{}{1-\alpha}$	α	$\beta = 1$	$\beta = \alpha$	$\beta = n+ \frac{}{1-\alpha}$
1	06:53.08	06:57.10	06:52.68	1	00:20.35	00:20.28	00:18.92
	12708	12708	12717		1401	1401	1412
3	05:34.19	07:14.12	10:28.21	3	00:23.25	00:22.91	00:21.99
	12708	15073	18646		1698	1698	2138
5	05:45.01	06:41.58	10:51.69	5	00:19.68	00:19.59	00:22.39
	13701	15016	23377		1598	1744	2639
8	04:43.10	05:06.79	10:02.18	8	00:20.57	00:20.35	00:21.78
	12232	19614	26823		1641	2374	3146
16	04:11.13	04:40.08	09:05.69	20	00:21.01	00:22.65	00:27.84
	12384	27387	33377		1263	5851	6506
20	04:28.23	05:13.12	08:40.16	40	00:31.24	00:40.86	00:46.43
	11270	29951	35593		1464	13977	14266
40	03:49.84	06:12.17	06:02.98	60	00:43.38	01:14.43	01:11.70
	9982	37468	37883		2043	20222	20222
60	03:32.68	08:11.38	05:00.42	80	01:37.72	03:09.35	01:49.23
	11270	35593	30719		4544	64587	51291
80	01:33.32	07:01.55	01:32.47	100	11:36.60	11:36.20	11:38.67
	9955	43486	9955		20987	20987	20987

Table 1. Effect of varying α and β

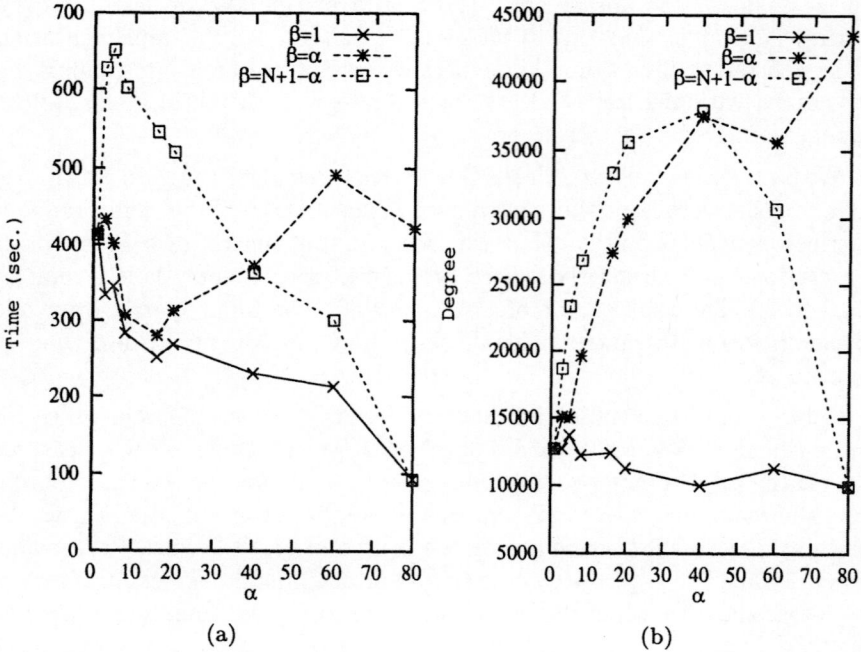

Fig. 3. Effect of varying α and β for a matrix over $\mathbb{F}_2[x]$: (a) running times, (b) maximum degree

6 Concluding Remarks

The reuse of sequential code for parallel implementations is well supported by the operation row concept. We have used earlier sequential GCD algorithms for our parallel implementations. This leads to parallel implementations for Hermite normal form computation with good speed-up. In the integer case the hybrid algorithm gives best results for small α and $\beta = n$. For $\mathbf{R} = \mathbb{F}_2[x]$ we get the best performance for $\alpha = n$ and small β.

For *integer* matrices with *large* rank, the new procedures are fastest. For distributed computation it is good to use the parallel, hybrid procedure with *small* $\alpha(< 10)$ and $\beta := n$. These procedures produce very good transformation matrices (i.e., entries with small absolute value) if the rank of the matrix is less than the number of columns. The transformation matrices are almost as good as ones obtained using LLL lattice basis reduction methods, which are orders of magnitude slower. For matrices over $\mathbb{F}_2[x]$ and for *integer* matrices with *small* rank, "Gaussian elimination" type procedures ($\alpha := n$) are fastest. For distributed computation use *small* $\beta(\in \{1, 2\})$.

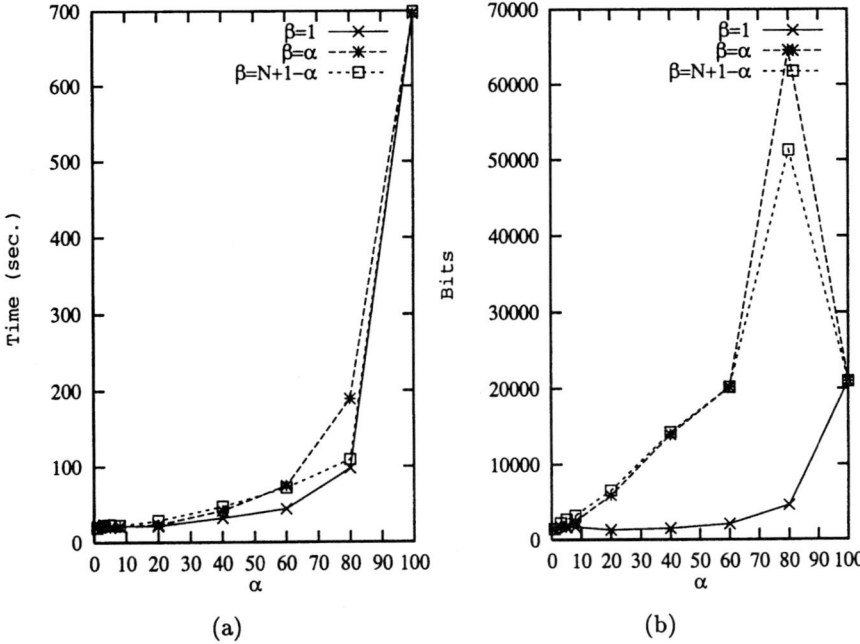

Fig. 4. Effect of varying α and β for an integer matrix: (a) running times, (b) maximum number of bits needed

Acknowledgements

The first author was partially supported by the Australian Research Council.

References

1. W. A. Blankinship. A new version of the Euclidian algorithm. *Amer. Math. Monthly* **70** (1963) 742–745.
2. G. H. Bradley. Algorithm and bound for the greatest common divisor of n integers. *Comm. ACM* **13** (1970) 433–436.
3. T-W.J. Chou and G.E. Collins. Algorithms for the solution of systems of linear Diophantine equations. *SIAM J. Comput.* **11** (1982) 687–708.
4. X.G. Fang and G. Havas. On the worst-case complexity of integer gaussian elimination. *ISSAC'97* (Proc. 1997 Internat. Sympos. Symbolic Algebraic Comput.), ACM Press (1997) 28–31.
5. K.O. Geddes, S.R. Czapor and G. Labahn. *Algorithms for Computer Algebra.* Kluwer Academic Publishers, 1992.
6. M. Giesbrecht. Fast computation of the Smith normal form of an integer matrix. *ISSAC'95* (Proc. 1995 Internat. Sympos. Symbolic Algebraic Comput.), ACM Press (1995) 110–118.
7. B. Hartley and T.O. Hawkes. *Rings, Modules and Linear Algebra.* Chapman and Hall, 1976.

8. G. Havas, D.F. Holt and S. Rees. Recognizing badly presented Z-modules. *Linear Algebra Appl.* **192** (1993) 137–163.

9. G. Havas and B.S. Majewski. Hermite normal form computation for integer matrices. *Congressus Numerantium* **105** (1994) 87–96.

10. G. Havas and B.S. Majewski. Extended gcd calculation. *Congressus Numerantium* **111** (1995) 104–114.

11. G. Havas and B.S. Majewski. Integer matrix diagonalization. *J. Symbolic Computation* **24** (1997) 399–408.

12. G. Havas, B.S. Majewski and K.R. Matthews. Extended gcd and Hermite normal form algorithms via lattice basis reduction. *Experimental Mathematics* **7** (1998) 125–135.

13. G. Havas and C. Wagner. Matrix reduction algorithms for Euclidean rings. *Proc. 1998 Asian Symposium on Computer Mathematics*, Lanzhou University Press (1998) 65–70.

14. R. Kannan and A. Bachem. Polynomial algorithms for computing Smith and Hermite normal forms of an integer matrix. *SIAM J. Comput.* **8** (1979) 499–507.

15. R. Kannan. Solving systems of linear equations over polynomials. *Theoretical Computer Science* **39** (1985) 69–88.

16. E. Kaltofen, M. S. Krishnamoorthy, and B. D. Saunders. Parallel algorithms for matrix normal forms. *Linear Algebra Appl.* **136** (1990) 189–208.

17. B. S. Majewski and G. Havas. A solution to the extended gcd problem. *ISSAC'95* (Proc. 1995 Internat. Sympos. Symbolic Algebraic Comput.), ACM Press (1995) 248–253.

18. M. Newman. *Integral Matrices.* Academic Press, 1972.

19. C.C. Sims. *Computation with finitely presented groups.* Cambridge University Press (1994).

20. A. Storjohann. Near optimal algorithms for computing Smith normal forms of integer matrices. *ISSAC'96* (Proc. 1996 Internat. Sympos. Symbolic Algebraic Comput.), ACM Press (1996) 267–274.

21. A. Storjohann. Computing Hermite and Smith Normal Forms of Triangular Integer Matrices. *Linear Algebra Appl.* **282** (1998) 25–45.

22. A. Storjohann and G. Labahn. Asymptotically fast computation of Hermite normal forms of integer matrices. *ISSAC'96* (Proc. 1996 Internat. Sympos. Symbolic Algebraic Comput.), ACM Press (1996) 259–266.

23. C. Wagner. *Normalformberechnung von Matrizen über euklidischen Ringen.* PhD thesis, Institut für Experimentelle Mathematik, Universität-GH Essen, 1997. Published by Shaker-Verlag, 52013 Aachen/Germany, 1998.

24. C. Wagner. Fast parallel Hermite normal form computation of matrices over $F[x]$. *Euro-Par'98 Parallel Processing*, Lecture Notes Comput. Sci. **1470** (1998) 821–830.

Performance Analysis of Wavefront Algorithms on Very-Large Scale Distributed Systems

Adolfy Hoisie, Olaf Lubeck and Harvey Wasserman
<hoisie, oml, hjw> @lanl.gov

Scientific Computing Group
Los Alamos National Laboratory
Los Alamos, NM 87545

Abstract. We present a model for the parallel performance of algorithms that consist of concurrent, two-dimensional wavefronts implemented in a message passing environment. The model combines the separate contributions of computation and communication wavefronts. We validate the model on three important supercomputer systems, on up to 500 processors. We use data from a deterministic particle transport application taken from the ASCI workload, although the model is general to any wavefront algorithm implemented on a 2-D processor domain. We also use the validated model to make estimates of performance and scalability of wavefront algorithms on 100-TFLOPS computer systems expected to be in existence within the next decade as part of the ASCI program and elsewhere. On such machines our analysis shows that, contrary to conventional wisdom, inter-processor communication performance is not the bottleneck. Single-node efficiency is the dominant factor.

1. Introduction

Wavefront techniques are used to enable parallelism in algorithms that have recurrences by breaking the computation into segments and pipelining the segments through multiple processors [1]. First described as "hyperplane" methods by Lamport [2], wavefront methods now find application in several important areas including particle physics simulations [3], parallel iterative solvers [4], and parallel solution of triangular systems of linear equations [5-7].

Wavefront computations present interesting implementation and performance modeling challenges on distributed memory machines because they exhibit a subtle balance between processor utilization and communication cost. Optimal task granularity is a function of machine parameters such as raw computational speed, and inter-processor communication latency and bandwidth. Although it is simple to model the computation-only portion of a single wavefront, it is considerably more complicated to model multiple wavefronts existing simultaneously, due to potential overlap of computation and communication and/or overlap of different communication or computation operations individually. Moreover, specific message passing synchronization methods impose constraints that can further limit the available parallelism in the algorithm. A realistic scalability analysis must take into consideration these constraints.

Much of the previous parallel performance modeling of software-pipelined applications has involved algorithms with one-dimensional recurrences and/or one-dimensional processor decompositions [5-7]. A key contribution of this paper is the development of an analytic performance model of wavefront algorithms that have recurrences in multiple dimensions and that have been partitioned and pipelined on multidimensional processor grids. We use a "compact application" called SWEEP3D, a time-independent, Cartesian-grid, single-group, "discrete ordinates" deterministic particle transport code taken from the DOE Accelerated Strategic Computing Initiative (ASCI) workload. Estimates are that deterministic particle transport accounts for 50-80% of the execution time of many realistic simulations on current DOE systems; this percentage may expand on future 100-TFLOPS systems. Thus, an equally-important contribution of this work is the use of our model to explore SWEEP3D scalability and to show the sensitivity of SWEEP3D to per-processor sustained speed, and MPI latency and bandwidth on future-generation systems.

Efforts devoted to improving performance of discrete ordinates particle transport codes date back many years and have extended recently to massively-parallel systems [8-12]. Research has included models of performance as a function of problem and machine size, as well as other characteristics of both the simulation and the computer system under study. For example, Koch, Baker, and Alcouffe [3] developed a parallel efficiency formula that considered computation only, while Baker and Alcouffe [9] developed a model specific to CRAY T3D put/get communication. However, these previous models had limiting assumptions about the computation and/or the target machines.

In this work, we model parallel discrete ordinates transport and account for both computation and communication. We validate the model on several architectures within the realistic limits of all parameters appearing in the model. Sections 2 and 3 of the paper briefly describe the algorithm and its implementation. Sections 4 and 5 derive the performance model and give validation results. In the final sections of the paper, the model is used to estimate SWEEP3D performance on future generation parallel systems, showing the sensitivity of this application to system computation and communication parameters.

Note that although we present results for three different parallel systems, no comparison of achieved system performance or scalability is intended. Rather, measurements from the three systems are presented in an effort to demonstrate generality of the performance model and sensitivity of application performance to machine parameters.

2. Description of Discrete Ordinates Transport

Although much more complete treatments of discrete ordinates neutron transport have appeared elsewhere [12-14], we include a brief explanation here to make clear the origin of the wavefront process in SWEEP3D. The basis for neutron transport simulation is the time-independent, multigroup, inhomogeneous Boltzmann transport equation, which is formulated as

$$\nabla \cdot \Omega \Psi(\mathbf{r},E,\Omega) + \iint \sigma(\mathbf{r},E)\psi(\mathbf{r},E,\Omega) =$$
$$\iint dE' d'(\mathbf{r},E' \to E,\Omega \cdot \Omega')\Psi(\mathbf{r},E',\Omega') \ +$$
$$(1/4\pi)\iint dE' d\Omega' \chi(\mathbf{r},E' \to E)v\sigma \ (\mathbf{r},E')\Psi(\mathbf{r},E',\Omega') \ + Q(\mathbf{r},E,\Omega).$$

The unknown quantity is Ψ, which represents the flux of particles at the spatial point $\mathbf{r}$ with energy E traveling in direction Ω.

Numerical solution involves complete discretization of the multi-dimensional phase space defined by $\mathbf{r}$, Ω, and E. Discretization of energy uses a "multigroup" treatment, in which the energy domain is partitioned into subintervals in which the depedence on energy is known. In the discrete ordinates approximation, the angular-direction Ω is discretized into a set a quadrature points. This is also referred to as the S_N method, where (in 1D) N represents the number of angular ordinates used. The discretization is completed by differencing the spatial domain of the problem on to a grid of cells.

The numerical solution to the transport equation involves an iterative procedure called a "source iteration" (see Ref. 13). The most time-consuming portion is the "source correction scheme," which involves a transport sweep through the entire grid-angle space in the direction of particle travel. A lower triangular matrix is obtained, as such one needs to go through the grid only once in inverting the iteration matrix. In Cartesian geometries, each octant of angles has a different sweep direction through the mesh, and all angles in a given octant sweep the same way.

For a given discrete angle, each grid cell has a spatially-exact particle "balance equation" with seven unknowns. The unknowns are the particle fluxes on the six cell faces and the flux within the cell. Boundary conditions and the spatial differencing approximation are used to provide closure to the system. Boundary conditions (typically vacuum or reflective) allow the sweep to be initiated at the object's exterior. Thereafter, for any given cell, the fluxes on the three incoming cell planes for particles traveling in a given discrete angle are known and are used to solve for the cell center and the three cell faces through which particles leave the cell. Thus, each interior cell requires in advance the solution of its three upstream neighboring cells – a three-dimensional recursion. This is illustrated in Figure 1 for a 1-D arrangement of cells and in Figure 2 for a 2-D grid.

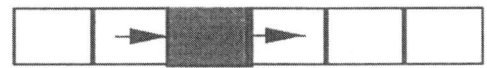

Figure 1. Dependences for a 1-D Transport Sweep.

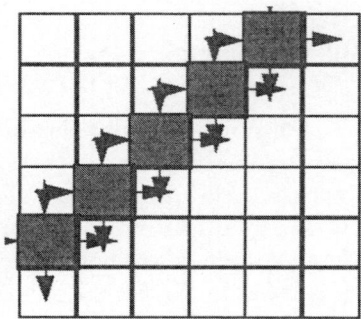

Figure 2. 2-D Transport Sweep along a Diagonal Wavefront.

3. Parallelism in Discrete Ordinates Transport

The only inherent parallelism is related to the discretization over angles. However, reflective boundary conditions limit this parallelism to, at most, angles within a single octant.

The two-dimensional recurrence may be partially eliminated because solutions for cells within a diagonal are independent of each other (as shown in Figure 2). The success of this "diagonal sweep" scheme on SIMD computers such as single-processor vector systems (using 2-D plane diagonals) and the Thinking Machines, Inc. Connection Machine (using 3-D body diagonals) has been demonstrated [3].

Diagonal concurrency can also be the basis for implementation of a transport sweep using a decomposition of the mesh into subdomains using message passing to communicate the boundaries between processors, as described in [12] and shown in Figure 3. The transport sweep is performed subdomain by subdomain in a given angular direction. Each processor's exterior surfaces are computed by, and received in a message from, "upstream" processors owning the subdomains sharing these surfaces.

However, as pointed out by Baker [9] and Koch [3], the dimensionality of the S_N parallelism is always one order lower than the spatial dimensionality because recursion in one spatial direction cannot be eliminated.

Because of this, parallelization of the 3-D S_N transport in SWEEP3D uses a 2-D processor decomposition of the spatial domain.

Parallel efficiency would be limited if each processor computed its entire local domain before communicating information to its neighbors. A strategy in which blocks of planes in one direction (k, in the current implementation) and angles are pipelined through this 2-D processor array improves the efficiency, as shown in Figure 3. Varying the k- and angle-block sizes changes the balance between parallel utilization and communication time.

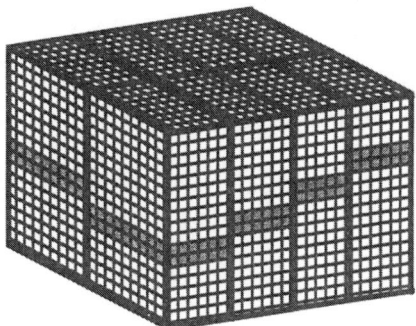

Figure 3. Illustration of the 2-D Domain decomposition on eight processors with 2 k-planes per block. The transport sweep has started at top of the processor in the foreground. Concurrently-computed cells are shaded.

4. A Performance Model for Parallel Wavefronts

This section describes a performance model of a message passing implementation of SWEEP3D. Our model uses a pipelined wavefront as the basic abstraction and predicts the execution time of the transport sweep as a function of primary computation and communication parameters. We use a two-parameter (latency/bandwidth) linear model for communication performance, which is equivalent to the LogGP model [15]. We use the term latency to mean the sum of L and o in the LogGP framework, and bandwidth to mean the inverse of G. Since different implementations of MPI use different buffering strategies as a function of message size, a single set of latency/bandwidth parameters describes a limited range of message sizes. Consequently, multiple sets are used to describe the entire range. Computation time is parameterized by problem size, the number of floating-point calculations per grid point, and a characteristic single-CPU floating-point speed.

4.1 Pipelined Wavefront Abstraction

An abstraction of the SWEEP3D algorithm partitioned for message passing on a 2-D processor domain (ij plane) is described in Figure 4. The inner-loop body of this algorithm describes a wavefront calculation with recurrences in two dimensions. Each processor must wait for boundary information from neighboring processors to the north and west before computing on its subdomain. For convenience, we assume that the implementation uses MPI with synchronous, blocking sends/receives. There is little loss of generality in this assumption since the subdomain computation must wait for message receipt. Multiple waves initiated by the octant, angle-block and k- block loops are pipelined one after another as shown in Figure 5, in which two inner loop bodies (or "sweeps") are executing on a *Px* by *Py* processor grid. Each diagonal line of processors is executing the same k-block loop iteration in parallel on a different subdomain; two such diagonals are highlighted in the figure.

Using this pipeline abstraction as the foundation, we can build a model of execution time for the transport sweep. The number of steps required to execute a computation of N_{sweep} wavefronts, each with a pipeline length of N_s stages and a repetition delay of d is given by equation (1).

$$Steps = N_s + d(N_{sweep} - 1),\qquad(1)$$

The first wavefront exits the pipeline after N_s stages and subsequent waves exit at the rate of $1/d$.

The pipeline consists of both computation and communication stages. The number of stages of each kind and the repetition delay per wavefront need to be determined as a function of the number of processors and shape of the processor grid. The cost of each individual computation/communication stage is dependent on problem size, processor speed and communication parameters.

```
FOR EACH OCTANT DO
  FOR EACH ANGLE-BLOCK IN OCTANT DO
    FOR EACH K-BLOCK DO
      IF (NEIGHBOR_ON_WEST) RECEIVE FROM WEST (BOUNDARY DATA)
      IF (NEIGHBOR_ON _NORTH) RECEIVE FROM  NORTH (BOUNDARY)
        COMPUTE_MESH (EVERY I,J DIAGONAL; EVERY K IN K-BLOCK;
                      EVERY ANGLE IN ANGLE-BLOCK)
      IF (NEIGHBOR_ON_EAST) SEND TO EAST(BOUNDARY DATA)
      IF (NEIGHBOR_ON_SOUTH) SEND TO SOUTH(BOUNDARY DATA)
    END FOR
  END FOR
END FOR
```

Figure 4. Pseudo Code for the wavefront Algorithm

4.2 Computation Stages

Figure 5 shows that the number of computation stages is simply the number of diagonals in the grid.

A different number of processors is employed at each stage but all stages take the same amount of time since processors on a diagonal are executing concurrently. The cost of one computational stage is thus the time to complete one COMPUTE_MESH function (see algorithm abstraction above) on a processor's subdomain. The discussion can be summarized with two equations. Equation (2) gives the number of computation steps in the pipeline,

$$N_s^{comp} = P_x + P_y - 1 \qquad(2)$$

and Equation 3 gives the cost of each step,

$$T_{cpu} = (\frac{N_x}{P_x} + \frac{N_y}{P_y} + \frac{N_z}{K_b} + \frac{N_a}{A_b})\frac{N_{flops}}{R_{flops}} \qquad (3)$$

where N_x, N_y, and N_z are the number of grid points in each direction; K_b is the size of the k-plane block; A_b is the size of the angular block; N_{flops} is the number of floating-point operations per gridpoint; and R_{flops} is a characteristic floating-point rate for the processor. The next sweep can begin as soon as the first processor completes its computation so the repetition delay, d^{comp}, is 1 computational step (i.e., the time for completing one diagonal in the sweep).

4.3 Communication Stages

The number and cost of communication stages are dependent on specific characteristics of the communication system. The effect of blocking synchronous communications is that messages initiated by the same processor occur sequentially in time and messages must be received in the same order that they are sent. As implemented, the order of receives is first from the west, then from the north, and the order of sends is first to the east and then to the south. These rules lead to the ordering (and concurrency) of the communications for a 4 x 4 processor grid as shown in Figure 6 for a sweep that starts in the upper-left quadrant.

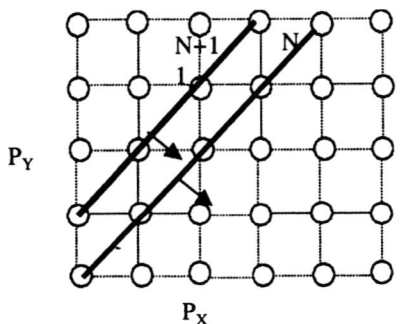

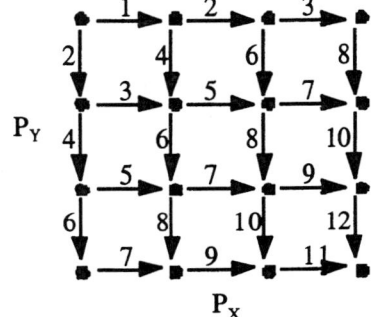

Figure 5. Multidimensional Pipelined Wavefronts

Figure 6. Communication Pipeline.

In Figure 6 edges labeled with the same number are executed simultaneously and the graph shows that it takes 12 steps to complete one communication sweep on a 4 x 4 processor grid. We assume that a logical processor mesh can be imbedded into the machine topology such that each mesh node maps to a unique processor and each mesh edge maps to a unique router link. One can generalize the number of stages to a grid of P_x by P_y processors by observing that communication for each row of processors is initiated by a message from a north neighbor in the first column of processors. South-going messages in the first column of processors occur on every other step since each processor in the column a) has no west neighbor, and b) must send east before sending south. Thus the last processor in the first column receives a message on step $2(P_y-1)$. This initiates a string of west-going messages along the last row that are also sent on every other step, and the number of stages in the communication pipeline is given by

$$N_s^{comm} = 2(P_y-1) + 2(P_x-1) \quad (4)$$

Analogous to the computational pipeline, different stages of the communication pipeline have different numbers of point-to-point communications. However, since these occur simultaneously, the cost of any single communication stage is the time of a one-way, nearest neighbor communication. This time is given by:

$$T_{msg} = t_0 + \frac{N_{msg}}{B} \quad (5)$$

where latency + overhead (t_0) and bandwidth (B), are defined in LogGP as noted above.

The repetition delay for the communication pipeline, d^{comm}, is 4 because a message sent from the top-left processor (processor 0) to its east neighbor (processor 1) on the second sweep cannot be initiated until processor 1 completes its communication with its south neighbor from the first sweep (Figure 6).

4.4 Combining Computation and Communication Stages

In the previous two sections, we derived formulas for the modeling of SWEEP3D that are general for any pipelined wavefront computation. We can summarize the discussion in two equations that give the separate contributions of computation and communication:

$$T^{comp} = [(P_x + P_y - 1) + (N_{sweep} - 1)] * T_{cpu} \quad (6)$$

$$T^{comm} = [2(P_x + P_y - 2) + 4(N_{sweep} - 1)]*T_{msg} \quad (7)$$

The major remaining question is whether the separate contributions, T^{comp} and T^{comm}, can be summed to derive the total time. They would not be additive if there were any additional overlap of communication with computation not already accounted for in each term. To see that this is not the case, consider the task graph for an execution consisting of two wavefronts on a 3 x 3 processor grid (Figure 7). This graph shows communication tasks (circles numbered with a send/receive processor pair) and computation tasks (squares numbered by a computing processor). The total number of stages in the combined communication/computation pipeline is equal to the number of nodes (of each type) in the longest path through the graph (the critical path) shown in dotted boxes in the figure. The critical path for the first sweep can be counted from Figure 7: 5 computational tasks and 8 communication tasks. This result is exactly the number given by eqns. (2) and (4). One can further verify that there is no further overlap between two pipelined sweeps other than the predicted sum of

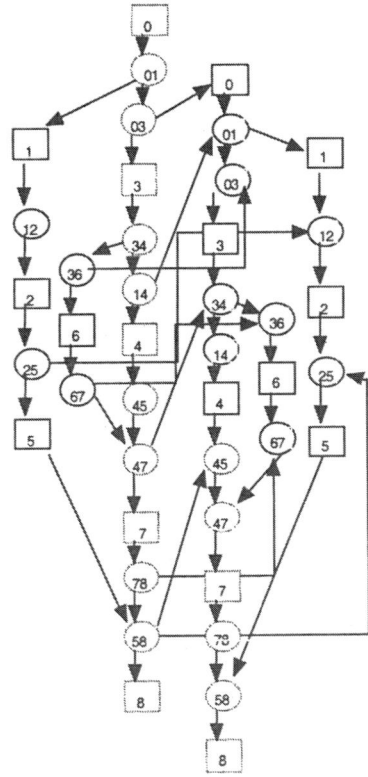

Figure 7. Pipelined Wavefront Task Graph.

eqns. (6) and (7). The second sweep completes exactly 1 computation and 4 communication steps after the first.

In summary, total time for the sweep algorithm is the sum of eqns. (6) and (7), where Tcpu is given by eqn. (3) and Tmsg is given by eqn. (5). The validation of the model against experiment involves the measurement and/or modeling of Tmsg and Tcpu. We take Tmsg to be the time needed for the completion of a send/receive pair of an appropriate size and Tcpu to be the computational work associated with the subgrid computation on each processor.

5. Validation of the Model

In this section, we present results that validate the model with performance data from SWEEP3D on three different machines, with up to 500 processors, over the entire range of the various model parameters. Inspection of eqns. (6) and (7) leads to identification of the following validation regimes:

$N_{sweep} = 1$: This case validates the number of pipeline stages in T^{comp} and T^{comm}, as functions of $(P_x + P_y)$, in the available range of processor configurations.

$N_{sweep} \sim (P_x + P_y)$: Validation of a case where the contributions of the $(P_x + P_y)$ and N_{sweep} terms are comparable.

$N_{sweep} \gg (P_x + P_y)$: This case validates the repetition rate of the pipeline.

For each of these three cases, we analyze problem sizes chosen in such a way as to make:

$T^{comp} \gg T^{comm}$; (validate eqn. (6) only)
$T^{comp} = 0$; (validate eqn. (7) only)
$T^{comp} \sim T^{comm}$; (validate the sum of eqns. (6) and (7)).

5.1 $N_{sweep} = 1$

For a single sweep, the coefficients of T_{msg} and T_{cpu} in equations 6 and 7 represent the number of communication and computation stages in the pipeline, respectively. Any overlap in communication or computation during the single sweep of the mesh is encapsulated in the respective coefficients. In hypothetical problems with $T_{msg} \sim T_{cpu}$, and in the limit of large processor configurations (large $P_x + P_y$), equations 6 and 7 show that the communication component of the elapsed time would be twice as large as the contribution of the computation time. In reality, for problem sizes and partitionings reasonably designed (small subgrid surface-to-volume ratio), T_{cpu} is considerably larger than T_{msg}. Computation is the dominant component of the elapsed time.

This is apparent in Figure 8, which presents the model-experiment comparison for a weak scalability analysis of a 16 x 16 x 1000 subgrid size sweeping only one octant. This size was chosen to reflect an estimate of the subgrid size for a 1-billion cell-problem running on a machine with about 4,000 processors; the former is a canonical goal of ASCI and the latter is simply an estimate of the machine size that might satisfy a 3-TFLOPS peak performance requirement. In a "weak scalability" analysis, the problem size scales with the processor configuration so that the computational load per processor stays constant. This experiment shows that the contribution of communication is small (in fact, the model shows that it is about 150 times smaller than computation), and the model is in very good agreement with the experiment.

We note that in the absence of communication our model reduces to the linear "parallel computational efficiency" models used by Baker [9] and Koch [3] for S_N performance, in which parallel computational efficiency is defined as the fraction of time a processor is doing useful work.

To validate the case with $N_{sweep} = 1$ and "comparable" contributions of communication and computation we had to use a subgrid size that is probably unrealistic for actual production simulation purposes (5 x 5 x 1). Even with this size computation outweighs communication by about a factor of 6. Figure 9 depicts a weak scalability analysis on the SGI Origin 2000 for this size. The model-experiment agreement is again very good.

Validation of cases where $T^{comp} = 0$ involved the development of a new code to

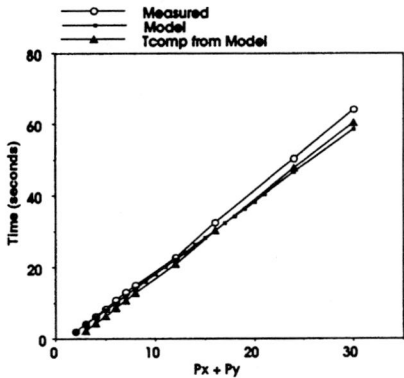

Figure 8. T^{comp} **dominant.**
$N_{sweep} = 1$. **IBM RS/6000.**

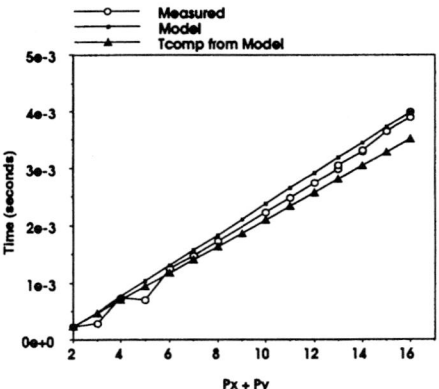

Figure 9. $T^{comp} \sim T^{comm}$. $N_{sweep} = 1$.
SGI Origin.

simulate the communication pattern in SWEEP3D in the absence of computation. The code developed for this purpose simply implements a receive-west, receive-north, send-south, send-east communication pattern enclosed in loops that initiate multiple waves. Figure 10 shows a very good agreement of the model with the measured data from this code.

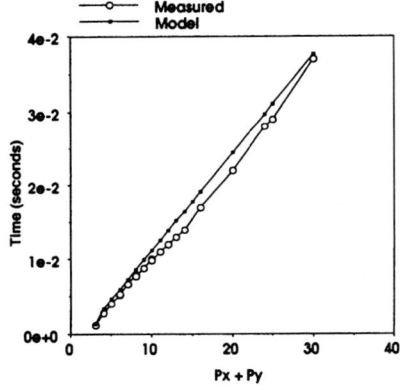

Figure 10. $T^{comp} = 0$. $N_{sweep} = 1$.
SGI Origin. .

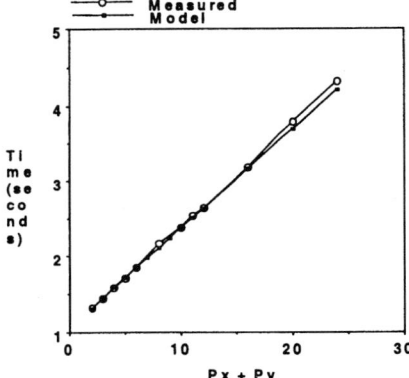

Figure 11. T^{comp} **dominant.**
$N_{sweep} = 10$. **SGI Origin.**

5.2 $N_{sweep} \sim (P_x + P_y)$

As described in Section 4, sweeps of the domain generated by successive octants, angle blocks, and k-plane blocks are pipelined, with the depth of the pipeline, N_{sweep}, given by the product of the number of octants, angle blocks, and k-plane blocks. We can select k-plane and angle block sizes so that $N_{sweep} = 10$, which, in turn, balances the contribution of N_{sweep} and (P_x+P_y) for processor configurations used in this work. In Figure 11 the comparison using a data size for which T^{comp} is dominant is presented, showing an excellent agreement with the measured elapsed time. The case with no computation is in fact a succession of 10 sweeps of the domain, with the communication overlap described by equation 6. Figure 12 shows a very good agreement with experimental data for this case.

An excellent model-experiment agreement is similarly shown in Figure 13, for a subgrid size 5 x 5 x 1, which leads to balanced contributions of the communication and computation terms to the total elapsed time of SWEEP3D.

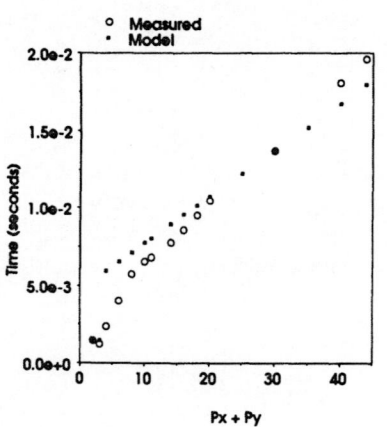

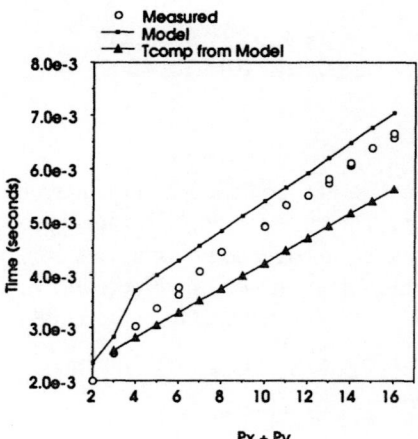

Figure 12. $T^{comp}=0$. $N_{sweep} = 10$.
CRAY T3E.

Figure 13 T^{comp} **dominant.**
$N_{sweep}=10$. **SGI Origin.**

5.3 Nsweep >> Px+Py

We present model-data comparisons using weak scalability experiments for cases in which N_{sweep} is large compared with $(Px+Py)$ in Figure 14 (6 x 6 x 360 subgrid; $T^{comp} \sim T^{comm}$) and in Figure 15 (16 x 16 x 1000 subgrid; T^{comp} dominant). The model is in good agreement with the measured execution times of SWEEP3D in both cases.

5.4 Strong Scalability

In a "strong scalability" analysis, the overall problem size remains constant as the processor configuration increases. Therefore, T_{msg} and T_{cpu} vary from run to run as the size of the problem size per processor decreases. In Figure 16 the comparison between measured and modeled time for the strong scalability analysis out to nearly 500 processors on the problem size 50 x 50 x 50 is shown. The agreement is excellent.

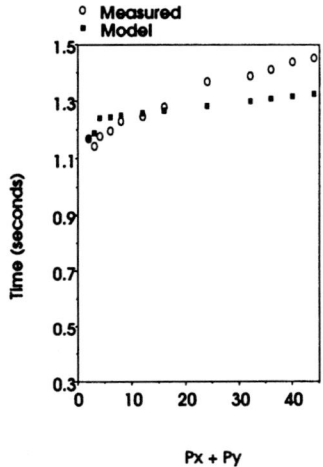

Figure 14. $T^{comp} \sim T^{comm}$.
6 x 6 x 360. N_{sweep} large.
CRAY T3E. $K_b = 10$.

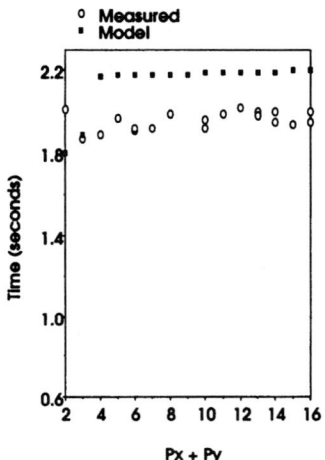

Figure 15. T^{comp} dominant.
16 x 16 x 1000. N_{sweep} large.
IBM RS/6000 SP.

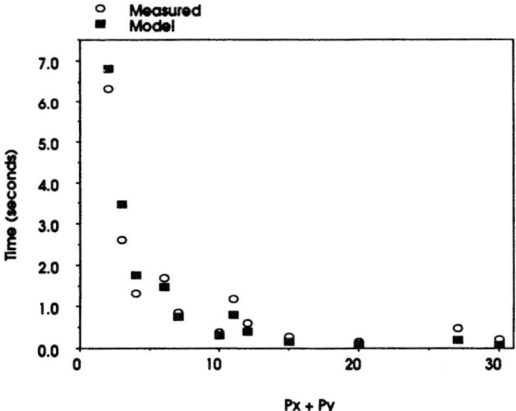

Figure 16. Strong Scalability. CRAY T3E.

6. Applications of the Model. Scalability Predictions

Performance models of applications are important to computer designers trying to achieve proper balance between performance of different system components.

ASCI is targeting a 100-TFLOPS system in the year 2004, with a workload defined by specific engineering needs. For particle transport, the ASCI target involves $O(10^9)$ mesh points, 30 energy groups, $O(10^4)$ time steps, and a runtime goal of about 30 hours. With 5,000 unknowns per grid point, this requires about 40 TBytes total memory. In this section we apply our model to understanding the conditions under which the runtime goal might be met.

Two sources of difficulty with such a prognosis are (1) making reasonable estimates of machine performance parameters for future systems, and (2) managing the SWEEP3D parameter space (i.e., block sizes). We handle the first by studying a range of values covering both conservative and optimistic changes in technology. We handle the second by reporting results that correspond to the shortest execution time (i.e., we use block sizes that minimize runtime).

We assume a 100-TFLOPS-peak system composed of about 20,000 processors (5 GFLOPS peak per processor, an extrapolation of Moore's law). With this processor configuration, given the proposed size of the global problem, the resulting subgrid size is approximately 6 x 6 x 1000.

Plots showing dependence of runtime with sustained processor speed and latency for MPI communications are shown in Figures 17 and 18 for several k-plane block sizes and using optimal values for the angle-block size. Table 1 collects some of the modeled runtime data for a few important points: sustained processor speeds of 10% and 50% of peak, and MPI latencies of 0.1, 1, and 10 microseconds. Our model shows that the dependence on bandwidth (1/G in LogGP) is small, and as such no sensitivity plot based on ranges for bandwidth is presented. Table 1 data assumes a bandwidth of 400 Mbytes/s.

One immediate observation is that the runtime under the most optimistic technological estimates in Table 1 is still larger than the 30-hour goal by a factor of two. The execution time goal could be met if, in addition to these values of processor speed and MPI latency (L+o in LogGP), we used what we believe to be an unrealistically high bandwidth value of 4 GBytes/s.

Assuming a more realistic sustained processor speed of 10% of peak (based on data from today's systems), Table 1 shows that we miss the goal by about a factor of six, even when using 0.1 µs MPI latency. With the same assumption for processor speed, but with a more conservative value for latency (1 µs), the model predicts that we are a factor of 6.6 off. In fact, our results show that the best way to decrease runtime is to achieve better sustained per-processor performance. Changing the sustained processor rate by a factor of five decreases the runtime by a factor of three, while decreasing the MPI latency by a factor of 100 reduces runtime by less than a factor of two. This is a result of the relatively low communication/computation ratio that our model predicts. For example, using values of 1 µs and 400 MB/sec for the communication latency and bandwidth, and a sustained processor speed of 0.5 GFLOPS, the communication time will only be 20% of the total runtime.

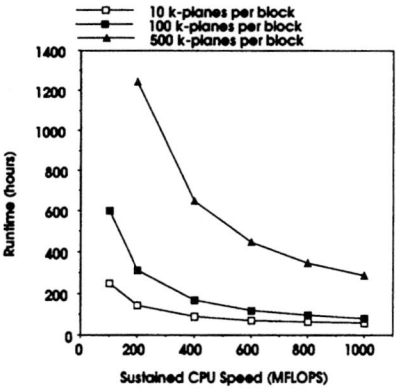

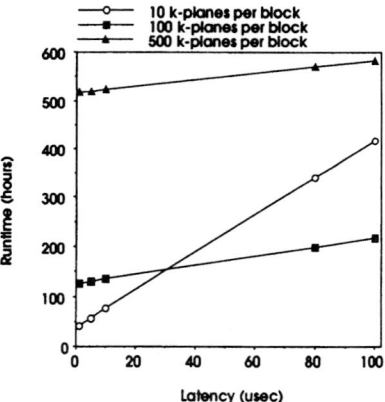

Figure 17. Sensitivity of the billion-point transport sweep time to sustainedper-processor CPU speed on a hypothetical 100-TFLOPS system as projected by the model for several k-plane block sizes and with MPI latency = 15 μs, and bandwidth = 400 Mbytes/s.

Figure 18. Sensitivity of the billion-point transport sweep time to MPI latency on a hypothetical 100-TFLOPS system as projected by the model for several k-plane block and with sustained per-processor CPU speed = 500 MFLOPS, bandwidth = 400 Mbytes/s.

Table 1. Estimates of SWEEP3D Performance on a Future-Generation System as a Function of MPI Latency and Sustained Per-Processor Computing Rate

MPI Latency	10% of Peak		50% of Peak	
	Runtime (hours)	*Amount of Communication*	*Runtime (hours)*	*Amount of Communication*
0.1 μs	180	16%	56	52%
1.0 μs	198	20%	74	54%
10 μs	291	20%	102	58%

7. Conclusions

A scalability model for parallel, multidimensional, wavefront calculations has been proposed with machine performance characterized using the LogGP framework. The model accounts for overlap in the communication and computation components. The agreement with experimental data is very good under a variety of model sizes, data partitionings, blocking strategies, and on three different parallel architectures. Using the proposed model, performance of deterministic transport codes on future generation parallel architectures of interest to ASCI has been analyzed. Our analysis showed that contrary to conventional wisdom, interprocessor communication performance is not the bottleneck. Single-node efficiency is the dominant factor.

8. Acknowledgements

We would like to thank Ken Koch and Randy Baker of LANL Groups X-CM and X-TM for many helpful discussions and for providing several versions of the SWEEP3D benchmark. We thank Vance Faber and Madhav Marathe of LANL Group CIC-3 for interesting discussions regarding mapping problem meshes to processor meshes. We acknowledge the use of computational resources at the Advanced Computing Laboratory, Los Alamos National Laboratory, and support from the U.S. Department of Energy under Contract No. W-7405-ENG-36. We also thank Pat Fay of Intel Corporation for help running SWEEP3D on the Sandia National Laboratory ASCI Red TFLOPS system, and SGI/CRAY for a generous grant of computer time on the CRAY T3E system. We also acknowledge the use of the IBM SP2 at the Lawrence Livermore National Laboratory.

References

1. G. F. Pfister, In Search of Clusters – The Coming Battle in Lowly Parallel Computing, Prentice Hall PTR, Upper Saddle River, NJ, 1995, pages 219-223.

2. L. Lamport, The Parallel Execution of DO Loops," Communications of the ACM, 17(2):83:93, ?., 19?.

3. K. R. Koch, R. S. Baker and R. E. Alcouffe, "Solution of the First-Order Form of the 3-D Discrete Ordinates Equation on a Massively Parallel Processor," Trans. of the Amer. Nuc. Soc., 65, 198, 1992.

4. W. D. Joubert, T. Oppe, R. Janardhan, and W. Dearholt, "Fully Parallel Global M/ILU Preconditioning for 3-D Structured Problems," to be submitted to SIAM J. Sci. Comp.

5. J. Qin and T. Chan, "Performance Analysis in Parallel Triangular Solve," In Proc. of the 1996 IEEE Second International Conference on Algorithms & Architectures for Parallel Processing, pages 405-412, June, 1996.

6. M. T. Heath and C. H. Romine, "Parallel Solution of Triangular Systems on Distributed Memory Multiprocessors," SIAM J. Sci. Statist. Comput. Vol. 9, No. 3, May 1988,

7. R. F. Van der Wijngaart, S. R. Sarukkai, and P. Mehra, "Analysis and Optimization of Software Pipeline Performance on MIMD Parallel Computers," Technical Report NAS-97-003, NASA Ames Research Center, Moffett Field, CA, February, 1997.

8. R. E. Alcouffe, ``Diffusion Acceleration Methods for the Diamond-Difference Discrete-Ordinates Equations," Nucl. Sci. Eng.{64}, 344 (1977).

9. R. S. Baker and R. E. Alcouffe, "Parallel 3-D S_N Performance for DANTSYS/MPI on the CRAY T3D, Proc. of the Joint Intl'l Conf. On Mathematical Methods and Supercomputing for Nuclear Applications, Vol 1. page 377, 1997.

10. M. R. Dorr and E. M. Salo, "Performance of a Neutron Transport Code with Full Phase Space Decomposition and the CRAY Research T3D," ???

11. R. S. Baker, C. Asano, and D. N. Shirley, "Implementation of the First-Order Form of the 3-D Discrete Ordinates Equations on a T3D, Technical Report LA-UR-95-1925, Los Alamos National Laboratory, Los Alamos, NM, 1995; 1995 American Nuclear Society Meeting, San Francisco, CA, 10/29-11/2/95.

12. M. R. Dorr and C. H. Still, "Concurrent Source Iteration in the Solution of Three-Dimensional Multigroup Discrete Ordinates Neutron Transport Equations," Technical Report UCRL-JC-116694, Rev 1, Lawrence Livermore National Laboratory, Livermore, CA, May, 1995.

13. E. E. Lewis and W. F. Miller, Computational Methods of Neutron Transport, American Nuclear Society, Inc., LaGrange Park, IL, 1993.

14. R. E. Alcouffe, R. Baker, F. W. Brinkley, Marr, D., R. D. O'Dell and W. Walters, "DANTSYS: A Diffusion Acclerated Neutral Particle Transport Code," Technical Report LA-12969-M, Los Alamos National Laboratory, Los Alamos, NM, 1995.

15. D. Culler, R. Karp, D. Patterson, A. Sahay, E. Santos, K. Schauser, R. Subramonian, and T. von Eiken, "LogP: A Practical Model of Parallel Computation," Communications of the ACM, 39(11):79:85, Nov., 1996.

16. H. J. Wasserman, O. M. Lubeck, Y. Luo and F. Bassetti, "Performance Evaluation of the SGI Origin2000: A Memory-Centric Characterization of LANL ASCI Applications," Proceedings of SC97, IEEE Computer Society, November, 1997.

17. C. Holt, M. Heinrich, J. P. Singh, E. Rothberg, and J. L. Hennessy, "The Effects of Latency, Occupancy, and Bandwidth in Distributed Shared Memory Multiprocessors," Stanford University Computer Science Report CSL-TR-95-660, January, 1995.

The DFN Gigabitwissenschaftsnetz G-WiN

Eike Jessen,

DFN/TU München

Abstract. The German national scientific networking association, DFN, will provide a gigabit network (G-WiN) in spring 2000. The paper analyzes history and trend of DFN network throughput, bandwidth, and cost, and the traditional and innovative load to be carried by the network is evaluated and forecast. Testbeds, which promote gigabit applications and pilot technology, are described. The current status of G-WiN specification is compared to US projects.

1 Deutsches Forschungsnetz

G-WiN is the abbreviation for Gigabit-Wissenschaftsnetz. It will go into operation in spring 2000. It is provided to research and education in Germany by the Verein zur Förderung eines Deutschen Forschungsnetzes (DFN), an association of 400 members, mainly universities and polytechnics, research institutes and other technical and scientific institutions. The purpose of DFN is to

- provide networking for research and education in Germany; DFN does so by specifying and procuring network services; it acts as a cooperative for the interest of its members
- promote efficiency and quality of research and education by innovative network usage

In 1998, DFN has a turnover of 174 Mio DM, of which 110 Mio DM are spent for data networking services, including the links from Germany to foreign countries. The cost of networking services is paid by the participating institutions, except transient funding by the Federal Ministry of Education and Science, Research and Technology (BMBF), for the introduction of new network generations.

2 Evolution of Network Throughput and Cost

The first DFN Wissenschaftsnetz, the X.25 WiN, began its operation in 1990. Since then, the original network has been upgraded to access rates of 2 Mb/s (1992) and been widely replaced by the Breitbandwissenschaftsnetz B-WiN (broad band science network), with access rates of up to 155 Mb/s. These networks have been specified by DFN and are operated by the Deutsche Telekom. The G-WiN will offer access rates of up to 2488 Mb/s.

Fig. 1 shows the evolution of the annual average throughput and the bandwidth of the DFN networks. The bandwidth is the maximum throughput that can be carried by the network, given the existing configuration and routing and a traffic distribution as observed in current operation. The bandwidth has to surpass the average throughput by the peak hour factor (1.8), and a burstiness surcharge factor (2..3 at least). In 1995/96 the X.25 WiN suffered by heavy congestion, as the average throughput approached network bandwidth.

Network average throughput has grown by factors of 2...5 per year, where there was only one year, that of the introduction of B-WiN, with the high factor. For future, one can expect a growth of 2..2.5 per year for the traditional network load. In Fig. 1, this leads to the forecast area marked by the two diverging straight lines, roughly a factor of 5 from summer 1998 to summer 2000.

Besides, the network will enable its users to go into new forms of communication, principally characterized by high bandwidth realtime applications. With large uncertainty, a further factor of 1.5 respects this innovative usage (i.e. 2.5 times the capacity of the B-WiN for today). So, summing it up, G-WiN should have approximately 8 times the throughput in 2000, compared to B-WiN in mid 1998, or 4.5 Gb/s. Our estimates are conservative, compared to estimates for future US nets.

The diagram of fig. 1 describes the cost of science nets in Germany (core network for connecting 600 institutions) by lines of constant cost. The lines rise over time by a factor of 1.4 per year, i.e. networks of this class have - at constant bandwidth - become cheaper by a factor of 1.4 (or 30 % less cost) per year. This trend will increase in the years to come, and an additional factor of 2 per year seems credible in the next 3 years, because of excess fiber capacities in Germany (investment of the competing line and network providers), multiplex usage of existing fibers (wave length division multiplexing, WDM), and an effective competition on the market, which in the high speed data communication area only begins to work. This specific forecast makes the iso-cost-lines bend in 1998.

The distance of the iso-cost lines describes the cost of bandwidth at constant time. It is slightly better (from the point of view of the network customer) than Grosch's law in the first two decades of the computer market. The customer pays only square root of n times the price for a n-fold increase in power. This is an „economy of scale"-effect. During their history, the DFN science networks have taken their way up the slope of cost; for the 400 fold increase in bandwidth, only 5 times the cost is paid (core net only!), as a consequence of the trend in time and the progress to higher bandwidth, using the economy of scale. Extrapolating the analysis to the year 2000 and to the G-WiN with 8 times the bandwidth of B-WiN, the cost of G-WiN should be well under 2 times that of B-WiN, maybe 1.4. By the way, this means that the price per bit x kilometer will be 6 times lower, recommending G-WiN not only as a high performance network for innovative usage, but also a cheap medium for mass data transport.

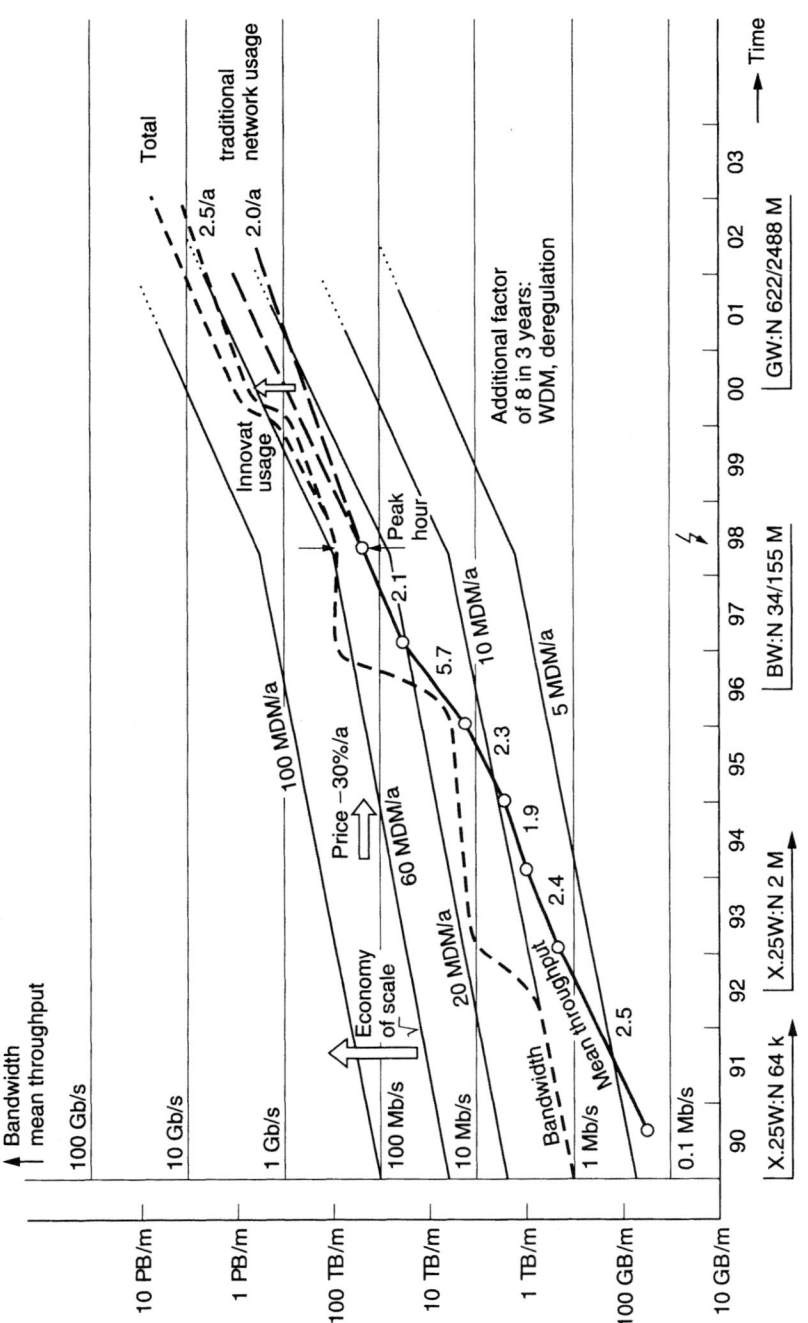

Fig. 1: Network performance and cost evolution

3 Network Load

Traditional network load will grow by a factor 4 to 6 between spring 1998 and spring 2000. This seems to be well established by the experiences of the past, see fig. 1. The network will, however, enable its users to put a far more demanding load on the net, in particular high bandwidth communication under realtime constraints. Fig.2 gives a survey on the types and limiting factors of innovative load and of the resulting traffic volume in the network and of the peak bandwidth per connection.

Among the real-time communication types, metacomputing (distributed supercomputing) has the highest requirements for peak bandwidth per connection. Data may be fed into the network at speeds of 0.8 and 1.6 Gb/s. The number of concurrent computations in the network, however, will be small, perhaps 5, and the repetition frequency and volume of the communication bursts will be small, resulting in a medium throughput. There is competition with local computing, realistic bandwidth cost and security arguments will limit the usage.

4 Innovative Load

Visualisation and control of remote computing processes, executed on special and high performance computers, will be widely used, but - under adequate compression techniques - not render high data throughput per connection. Even virtual reality can be reduced to multiple video and graphics channels with only moderate bandwidth (some Mb/s each) per connection.

Interpersonal communication by network phone/video, as well as interpersonal cooperation will be very frequent and therefore result in a high bandwidth demand, though the demand per connection is also moderate. Media servers, for instance for distribution of distant learning and teaching multimedia material, will be few in number, but with high throughput, though widely not in realtime, where it does not belong into our survey of realtime innovative load. There is, however, a class of realtime high bandwidth media communication visible, where studio media data are transported to remote processing systems and retransmitted in realtime, so that during the recording the processed signal can be checked and used for the control of the recording.

Type	Factors Limiting the usage	Network volume total	Peak bandwidth per connection
realtime			
Distributed computation	Algorithms, economy, security	Medium	high
Remote visualisation & control of computation	Competition with Workstations	Low	low ... high
Interpersonal communication & cooperation (phone, video, CSCW, virtual reality	Terminal devices	Medium	low ... medium
Media Servers	Server technology, portable media	Low	low ... medium
Signal processing	Competition with local processing	Medium	low ... high
non-realtime			
Update, distribution, backups, non-interactive visualisation, experiments	Competition with portable media	Low	medium

Fig. 2: Classes of innovative network load and probable volume (in total) and peak bandwidth (per connection)

As the network has much lower costs for the volume x distance product, non-realtime mass data transports may be attractive, compared to the shipment of physical media. This will bring software and data distribution onto the network, as backups, experimental data and non-interactive visualisation output of remote computations, to an extent far increased compared to today. The limiting factor will be the progress in the technology of portable media. This kind of load may be considerable in volume, it is not defining network peak performance, as it may be delayed.

5 Gigabit Testbeds (1997-2000)

DFN runs two gigabit testbeds (see Fig. 3) for application promotion and for technology piloting. The testbeds are to deliver experience for the specification of the Gigabitwissenschaftsnetz G-WiN.

The first testbed was established in 1997 and is situated in North Rhine-Westfalia. It connects institutions in Jülich, Bonn, Cologne and Essen. It has been upgraded to 2.5 Gb/s and studies applications in metacomputing, remote visualisation, media processing and simulation. The problems come from physics, material sciences, geosciences, traffic control, and studio data processing.

The second testbed was established in 1998 and connects Munich, Erlangen, and Berlin. It pilots wavelength division multiplexing (3x2488 Mb/s per fiber) and optical amplification. Its applications come from the same areas and from medicine and distant teaching.

6 G-WiN Specification

The general requirements for G-WiN are, as of late autumn 1998,

- data communication infrastructure for German science, as the successor of B-WiN, i.e. a full scale network, comprising high performance and commodity traffic
- built with components from the market (this seems achievable)
- a production network (not merely a research object), though probably beyond state of the market in 2000
- roughly 10 fold bandwidth, compared to B-WiN 1998
- more configuration flexibility, to use advantageous regional opportunities and to offer more flexible conditions of usage to the participating institutions

- efficient IP-based communication; it is not clear in 1998 whether the network will, in the long run, base IP on ATM and SDH, as B-WiN (and vBNS e.g.) has done with very good operational results, or base IP directly on SDH, as the Abilene network in USA will do, or base the IP protocol stack directly on the „black" fiber, multiplexed by WDM (which can be combined with the first two options as well).

The decision reflects mainly the tradeoff between bandwidth consumption, complexity, and bandwidth allocation granularity. In any way, the network is to cooperate with ATM based access systems, and it is likely that G-WiN should start with ATM and skip it when the competing architectures are mature.

- guaranteed quality of service; it is, however, by autumn 1998, by no means clear how this can be achieved; the solution is closely related to the ATM/SDH variants. B-WiN offers ATM permanent virtual channels as the only means of guaranteed service. ATM traffic classes still lack the embedding in application oriented quality of service requirements. Besides, ATM will not generally be available on an end-to-end basis between the host computers. So there is much debate on schemes based fully on IP, as RSVP and (more recently) MPLS (multi-protocol label switching). These techniques are, however, not yet well understood in actual operation. So, at least now, the way of providing adequate quality of service to the flows in the network remains open.
- specification for potential line and network providers until the end of 1998, start of G-WiN operation in spring 2000

7 Comparison with other Scientific Networks

Fig. 4 puts the G-WiN into perspective with similar projects of the preceding generation and that of tomorrow. vBNS is structurally very similar to B-WiN, but operates on a higher level of bandwidth and lower level of load; the distance between G-WiN and Abilene (one year ahead) will be smaller. Both US scientific nets are mainly (80%) funded by sponsors, the participating institutions see only a small percentage of the actual network costs. vBNS and Abilene are defined not to carry the lower bandwidth commodity traffic, which deliberately is left to commercial providers, though it probably could be carried by the broadband networks at lower cost. B-WiN connects a much larger number of institutions (even neglecting the 4000 users of the B-WiN dial-up service WiNShuttle)

8 Gigabit Testbeds (1997-2000)

* Technology piloting
* Application promotion

West	South
Jülich, Bonn, Cologne, Essen ... ?	Erlangen, München, Berlin, Suttgart (?)
Start: August 97	July 98
Lines: 0.6, 2.5 Gb/s	0.6, 2.5 Gb/s WDM
Provider: o.tel.o	DTAG
HiPPI, ATM/SDH	ATM/SDH (2 Gb/s ATM achieved!)
Applications	
Metacomputing + Visualisation * Molecular Dynamics * Earth Shell	 * Black Holes * Surface Effects
Media Processing * Distrib. virtual TV Prod.	 * Distrib. TV Prod./Server * Virtual Laboratory * Distance Learning
Simulation + Visualisation * Traffic	 * Medicine

Fig. 3: DFN Gigabit Testbeds (1997-2000)

Network	vBNS	B-WiN	Abilene	G-WiN
Operational since	1996	1996	1999?	2000?
Ordered by	NSF	DFN	UCAID	DFN
Provider selected	MCI	DTAG e.a.	Qwest, UCAID	to be decided
Paid by	MCI/NSF/instit	instit/BMBF	Qwest/instit	instit/BMBF
Protocols	IP/ATM/Sonet	IP/ATM/SDH	IP/Sonet	IP/?
Sites	71	600	120	600
Usage restriction	"projects"	(science)	(no commodity traffic)	(science)
Trunks Gb/s	622 Mb/s	41... 94 Mb/s	2.5 Gb/s	0.6/2.5 Gb/s
Access Rates	155/622 Mb/s	2...155 Mb/s	155/622 Mb/s	2...622 Mb/s
QoS	SVCs, PVCs	PVCs	MPLS	SVCs, PVCs MPLS?

Fig. 4: Comparison between G-WiN and other scientific networks.

Abbreviations:

ATM: Asynchronous Transfer Mode
B-WiN: Breitbandwissenschaftsnetz
DTAG: Deutsche Telekom AG
MPLS: Multiprotocol Label Switching
NSF: National Science Foundation
SDH: Synchronous Digital Hierarchy
S/PVC: Switched/Permanent virtual circuit
UCAID: University Cooperation for Advanced Internet Development
vBNS: very high-speed Backbone Network services

On Network Resource Management for End-to-End QoS*

Ibrahim Matta

Computer Science Department
Boston University
Boston, MA 02215, USA
matta@cs.bu.edu

Abstract. This article examines issues and challenges in building distributed Quality-of-Service (QoS) architectures. We consider architectures that facilitate cooperation between the applications and the network so as to achieve the following tasks. Applications are allowed to express in their own language their QoS (performance and cost) requirements. These application-specific requirements are communicated to the network in a language the network understands. Resources are appropriately allocated within the network so as to satisfy these requirements in an efficient, scalable, and reliable manner. Furthermore, the applications and the network have to cooperate "actively" (or continually) to be able to achieve these tasks under time-varying conditions.

1 Introduction

Advanced network-based applications have received a lot of attention recently. An example of such applications is depicted in Figure 1. Here, a number of scientists are collaborating in a videoconference to control a distributed simulation. This application requires distributing simulator output to the scientists, the exchange of audio and video signals among them, etc. Such application demands some high-quality services from the network. For example, the simulator data should be distributed to the scientists quickly and with no loss. The audio and video signals should be transmitted to other participants in a timely and regular fashion to ensure interactivity, etc. The network in turn should provide these services while utilizing the network resources efficiently (to maximize revenue). Also, the network should deliver these services in a scalable and reliable way.

In this article, we examine architectures designed to provide such advanced applications with varying QoS. We start by discussing in Section 2 traditional network services that do not provide QoS support as well as traditional applications that do not express their QoS or are not aware of the QoS that the network is providing them with. We will argue why we need

---

* This work was done while the author was with the College of Computer Science of Northeastern University. The work was supported in part by NSF grants CAREER ANIR-9701988 and MRI EIA-9871022.

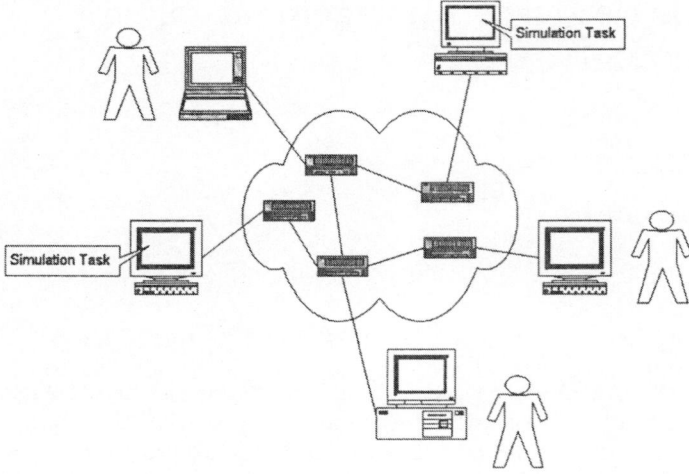

Fig. 1. Example of an advanced network-based application.

the network and applications to be aware of (and sensitive to) QoS, and we discuss in Section 3 how QoS support is achieved. Then, for these QoS-aware applications and network to cooperate, we present in Section 4 a generic integrated architecture and describe its components and discuss its main features. These features include the scalability of the architecture to large networks and large number of applications or users, and also its robustness to various dynamics. We conclude in Section 5.

2 QoS-oblivious Architectures

In this section, we discuss traditional network architectures and their shortcomings due to their lack of QoS support.

2.1 QoS-oblivious Network Services

Traditionally, a network, such as the current Internet, provides only a best-effort service, which means that the data which applications send can experience arbitrary amounts of loss or delays. On the positive side, the network could only implement simple mechanisms for traffic control. For example, any application can access (or send data over) the network at any time. Network devices (like switches or routers) can serve packets carrying application's data in a simple first-come-first-serve fashion. Packets can be routed to their destination over paths that are optimal with respect to a single metric, for example paths that have the minimum number of links (or that traverse the least number of routers). With these mechanisms, however, the application's QoS may not be met. For example, shortest-distance paths may be loaded and not provide the least delay and so may not satisfy the requirements of delay-critical applications. This kind of best-effort network has served quite well traditional data applications (e.g. Telnet, FTP, E-mail) that have relatively modest QoS requirements.

More advanced applications need additional support, for example, for multicasting a data packet to many destinations. The network should then be able to establish a delivery tree rooted at the source and whose branches lead to the various destinations. In the case of many senders, the network can build one tree for each sender. This is what current multicast protocols like DVMRP (Distance Vector Multicast Routing Protocol) [27] and MOSPF (Multicast Open Shortest Path First) [23] do. The tree usually consists of paths with minimum delay from the source to each destination. Figure 2 shows an example three-node network with two senders $S1$ and $S2$ and with a receiver at each node. A packet from $S1$ is replicated on both links to reach the two other nodes/receivers. Similarly, for $S2$. So this multicast uses 4 links or the total cost (in number of links) is 4. This cost may reflect the overhead of replication or the amount of bandwidth consumed by this communication group. Also, each packet experiences a maximum delay of going over 1 link.

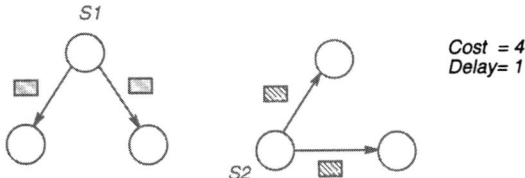

Fig. 2. Sender-based multicast trees.

Other multicast routing protocols would build a single shared tree that spans all members of the group. A major objective of such protocol is to minimize the sum of the link costs, or build a so-called "Steiner tree." Since it is well known that it is computationally expensive (or NP-complete) to find such minimum-cost tree [11], protocols usually implement heuristics that are less expensive and give close to optimal trees. With such trees, the goal is to minimize the cost of replication and bandwidth, possibly at the expense of higher delays from a source to a destination. Figure 3 shows a shared tree. Here, the data packet from $S2$ needs to be generated only once, as opposed to being replicated in Figure 2 when source-rooted trees are used. This shared tree uses 3 links only, as opposed to 4 links with source-routed trees. However, the packet from $S2$ experiences a maximum delay of traversing 2 links, as opposed to 1 link with source-rooted trees. Some protocols like PIM-sparse [8] and CBT (Core Based Tree) [3] try to achieve a balance between cost and delay by having a single shared tree with the root at a center node and minimum-delay paths built from the center to members of the multicast group.

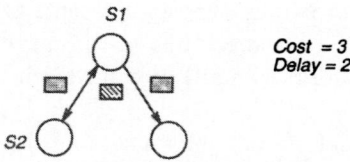

Fig. 3. A shared multicast tree.

One can easily see that regardless of what type of tree a *best-effort* multicast routing protocol builds, this tree may not be appropriate. This depends on several factors such as the location of the group members, the layout (topology) of the network, the requested QoS, etc. Figure 4(a) shows a case where a source-based tree should be built rather than a shared tree (with boldfaced links) as it costs less (using 2 links as opposed to 3). Figure 4(b) shows a case where a shared tree achieves lower cost (using 2 links as opposed to 3). This illustrates the need for adaptive network services that establish appropriate structures so as to efficiently utilize the network resources as well as satisfy the QoS requirements of applications. We later discuss such QoS-sensitive multicast routing protocols.

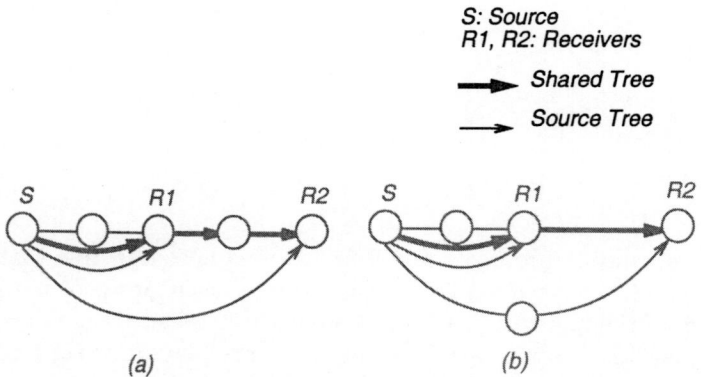

Fig. 4. (a) A source-based tree costs less, (b) a shared tree costs less.

2.2 QoS-oblivious Applications

As we traditionally had networks that are insensitive to various parameters for the state of the network and applications, we traditionally had applications that are insensitive to the state of the network and what kind of QoS they are getting from the network. A major consequence of this is that the application can experience "arbitrary" QoS. For example, under loaded network conditions, a video application can start losing its frames and suffering arbitrary degradation in quality. If the application had adapted its coding strategy to further compress its data and thus sent fewer frames, this may have increased the likelihood that the transmitted frames make it through the network. As a result, the application gets a consistent service (although of lesser quality due to compression). Another example of QoS-oblivious applications is one that "arbitrarily" assigns a distributed computation over the network.

Figure 5 shows a video example where the decoder (receiver) could decode either MPEG- or JPEG-coded streams. MPEG is more expensive to decode than JPEG. Thus, if there is enough computation cycles, and communication is expensive, we could just send MPEG-coded video and have a matching MPEG decoder, where video traffic is routed over the shortest (one-link) path. On the other hand, if there are not enough cycles and communication is cheap, we could use the decoder in JPEG mode to reduce computation cost and transparently insert an intermediate hardware-based (computationally inexpensive) transcoder that translates from MPEG format to JPEG format. Here, video traffic is routed through the transcoder over a longer (two-link) path, which is acceptable as communication is assumed to be cheap. This example illustrates the advantages of having applications that could adapt their operation mode based on the state of the system.

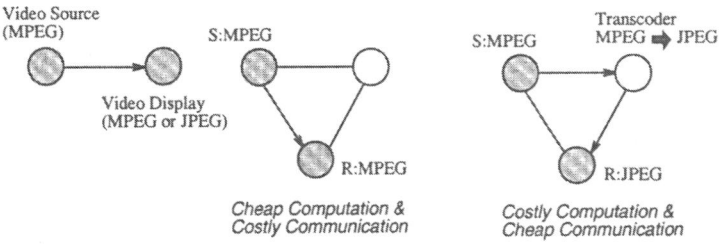

Fig. 5. Video transport example.

3 QoS-sensitive Architectures

Now that we have argued for the flexibility and potential benefits of QoS-sensitive network and applications over traditional ones, we next discuss various issues and challenges that must be addressed to implement them.

The network has to distinguish among traffic streams (or flows) requiring different QoS. A flow generally defines a stream of data packets that belong to the same application or a pre-defined aggregated traffic class. The works of several standardization groups, such as the IETF integrated-services [6] and differentiated-services [4] working groups and the ATM Forum Traffic Management working group [14], are based on allocating each traffic flow a given (absolute or relative) share of resources (bandwidth, buffers, etc.). This provides different flows with different services; better service (but typically more expensive) to higher priority flows at the expense of worse service (and usually cheaper) to lower priority flows.

Figure 6 shows a general architecture of a QoS network. A QoS manager is responsible for receiving requests for some QoS through some signaling protocol, such as the Internet RSVP protocol [7] or the ATM signaling protocol. The manager communicates with a routing component to find the outgoing link(s) or path that can likely satisfy the request and over which the flow would be routed. This path selection is typically based on an outdated view of the state of the network. The manager then communicates with an admission control component which decides whether indeed there are enough resources on the selected link(s) to satisfy the given QoS without violating the QoS already promised to existing flows. If the flow request is accepted (admitted), the QoS manager installs appropriate state in other components: A classifier that recognizes packets belonging to the flow. A route lookup component that forwards the flow's packets over the selected path. A shaper (or dropper) that shapes the flow (or drops excess traffic) according to the initial traffic specifications declared by the request and based on which resource allocations have been made. Finally, the manager has to set the parameters of the scheduler to allocate to the flow the necessary resources that are needed to satisfy the requested QoS.

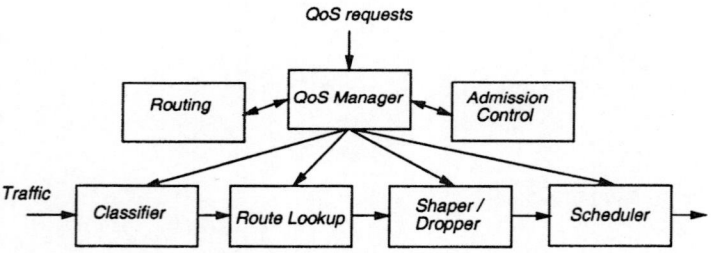

Fig. 6. A general architecture of a QoS network.

Figure 7 shows the architecture of a general scheduler and a traffic shaper. The scheduler isolates different traffic flows by allocating them fractions of the link's resources (bandwidth, buffer space). The traffic of a flow is shaped before entering the scheduler to compete for resources. The shaper shown is called "token bucket shaper." Tokens accumulate (fill the bucket) at some specified rate. A packet is allowed to enter the scheduler only if there is one or more tokens to drain. The depth of the bucket allows for the flow to burst (i.e. packets can enter back-to-back), but then the rate at which data enters so as to contend for resources is bounded by the rate at which tokens are generated. This bounded traffic specification makes it possible to test whether a new flow could be supported by considering its worst-case behavior.

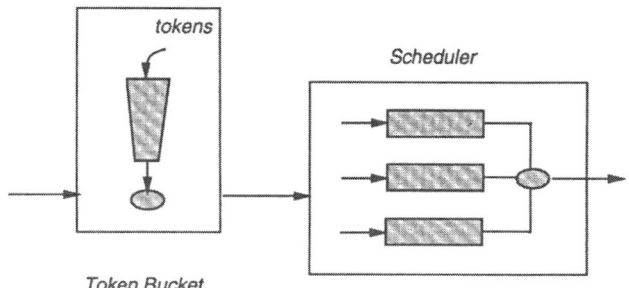

Fig. 7. A general scheduler with a traffic shaper.

As for routing (multicast) traffic, the routing protocol should choose a path that satisfies the QoS requirements of (all) the receiver(s). Thus, in multicasting, the type of multicast tree that should be constructed depends on the state of the network. Consider the example in Figure 8, where the numbers shown reflect the maximum link delays in the presence of a three-participant application, two of which send and receive at nodes A and C. Assume the application's QoS requirements are maximum end-to-end delay of 13 from the sender to any receiver, and jitter (or maximum difference between individual end-to-end delays) of 7. A shared tree would violate the application's delay and jitter requirements since the maximum end-to-end delay is 20 and the jitter is 10. Thus, under this network state, source-based trees should be constructed. Figure 9 shows the network in a different state, where a shared tree should be constructed.

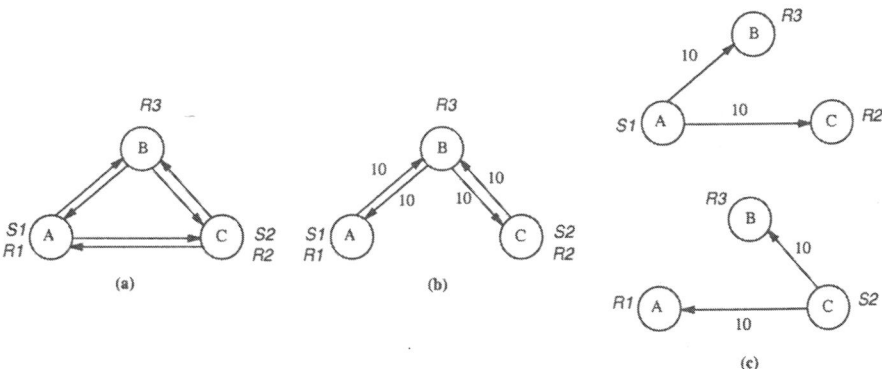

Fig. 8. (a) Original network, (b) shared tree, (c) source-based trees. Example where source-based trees should be constructed.

Clearly, to be QoS-sensitive, protocols have to account for the various dynamics at all levels: application, host, and network. Protocols have to be aware of the current participants in the application. Who communicates with

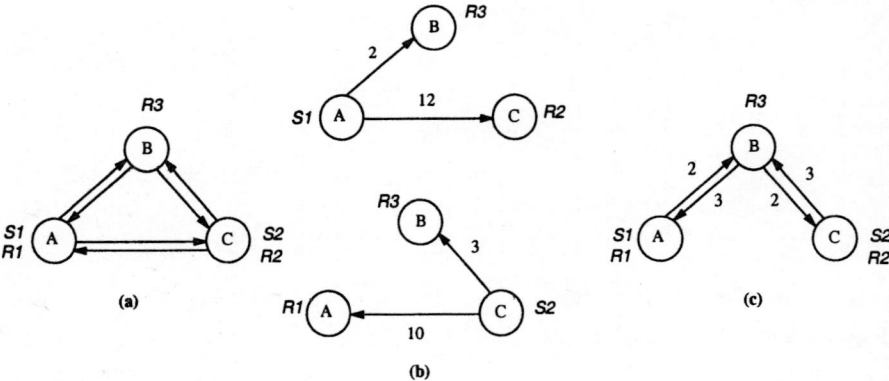

Fig. 9. (a) Original network, (b) source-based trees, (c) shared tree. Example where a shared tree should be constructed.

whom? Is a participant aware of the quality of her transmission so she can adapt to varying qualities? How is data generated during this transmission? What are the delivery requirements? How much resources are available at hosts (or end-systems) and at the network (switches/routers)? How does the layout (topology) of the network look like now? By accounting for all these dynamics, we would have a system that is capable of satisfying a wide variety of QoS requirements for diverse applications at all times, while operating efficiently.

Going back to multicast routing as an example of network control, the system may migrate from one multicast tree to another in response to changes at the application, host or network level. Figure 10 shows an example, where if the multicast protocol builds a single shared tree, then 5 links are used, and only $S2$'s packet needs to be replicated. Also, if, say, each source transmits at x bits/sec, then to guarantee rate for $R2$ to receive from both senders, we need to reserve $2x$ bits/sec on link $(S2, R2)$. If the available bandwidth on $(S2, R2)$ were less than $2x$, then a QoS network would not admit this application. However, if the multicast routing protocol adapts its tree construction mechanism and switches to source-based trees, this application could be admitted since the traffic is now better distributed over the network.

Thus, it is sometimes beneficial to switch to building a new type of tree. Building source-based trees in the previous configuration had allowed for admitting the application, although at the expense of more replication and more routing state information as we have to maintain 2 trees (one for each source) as opposed to 1 tree. This previous example also contradicts the common view that a minimum-cost shared tree reduces the amount of bandwidth consumed compared to sender-based trees, especially when we have multiple senders as the path from some sender to a receiver may turn out to be very long. Thus, a QoS system must employ a more intelligent tree construction strategy that adapts the shape of the tree dynamically, where we can trade-

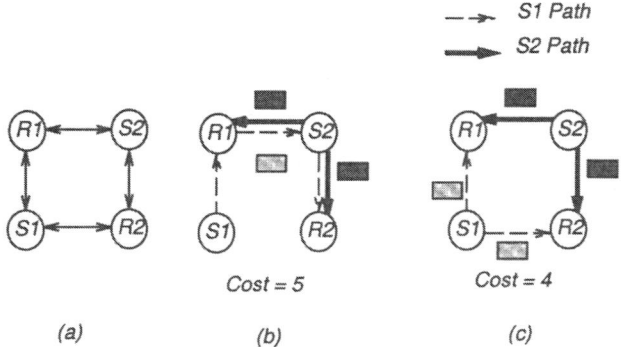

Fig. 10. (a) Original network, (b) shared tree, (c) source-based trees. Example illustrates benefits of adaptive multicast tree construction.

off between revenue from QoS support and cost of overhead to provide this support. We present an example of such strategy later.

4 Integrated QoS Architecture

The objective is to build an architecture that allows for protocols to adapt their behavior so as to account for various parameters and dynamics. The architecture should facilitate the exchange of information between the applications and network; the application should express an acceptable QoS region and be able to adapt its behavior to one that matches the level of QoS that the network currently delivers. The network should of course communicate that QoS level to the application. The goal is to efficiently utilize the network (or maximize its revenue) under the QoS constraints imposed by existing applications.

Figure 11 shows a generic integrated QoS architecture. In this architecture, applications express their QoS requirements in application-specific terms to an application-specific QoS manager. This manager understands the various attributes of a specific type of applications, and maps the application-dependent requirements into application-independent and implementation-independent requirements. A host QoS manager maps these in turn into implementation-dependent requirements so that necessary resources are allocated to the application by the operation system and network subsystem at the hosts as well as by router QoS managers within the network. QoS managers communicate with their peers or directly with their neighbor managers to coordinate the allocation of resources. Router QoS managers control the allocation of paths within the network by communicating with routing managers, which may use different types of routing protocols to locally or globally build paths that are capable of satisfying the QoS requirements of applications.

208

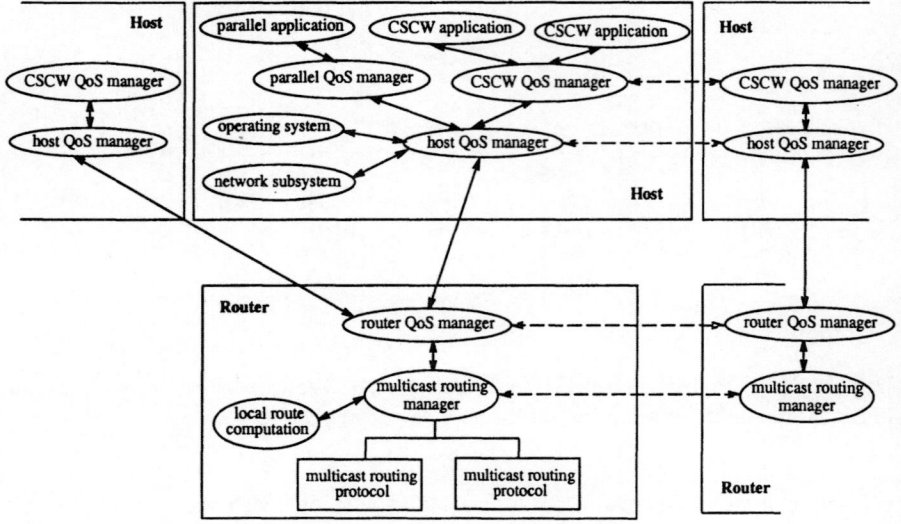

Fig. 11. Integrated QoS architecture.

Such architecture allows for supporting QoS between the endpoints of communication (*i.e.* the applications) in the presence of various dynamics. The QoS is controlled in a manner sensitive to the specifics of the application so that the system can successfully and efficiently deliver targeted QoS. The applications and network are allowed to exchange information so as to adapt mutually to system dynamics. In the following subsections, we elaborate on the delivery of targeted (application-oriented) QoS and the application's capability to be aware of (and adaptive to) system state, and also as an example of network control, we elaborate on building multicast routing trees that are sensitive to QoS.

We finally discuss mechanisms that should be in place in order to scale to large networks and to provide stability and reliability in the presence of changes in the state of the system.

4.1 Application-oriented QoS Mappings

An end-to-end QoS architecture has to deal with mapping the logical view of the applications to a physical allocation of resources. This involves application-specific QoS managers that take as input abstract QoS requests, such as application input graph with various tasks and interdependencies between them. Through a protocol to discover the state and location of physical resources (this knowledge is typically outdated), the QoS manager produces as output the needed physical resources that are likely to satisfy the application's QoS in an efficient manner. See Figure 12. This in turn involves host QoS managers that would communicate with other control entities at hosts and within the network using some signaling protocol to finally allocate the physical resources.

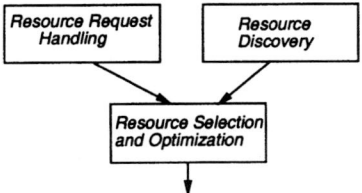

Fig. 12. Application-oriented QoS mapping.

Application-oriented Resource Selection: Given a (typically outdated) view of the system, which physical resources to select to satisfy an application-level QoS request? The answer to this question depends on the nature of the application. For example, consider an application that is real-time in nature and requires the scheduling of real-time tasks on a set of hosts. A real-time task needs to be reserved some amount of CPU cycles to meet a deadline. The host that is selected may end up rejecting the assigned task if the host finds that it does not currently have enough cycles (capacity). For this application, the major objective is to minimize task rejection rate. A traditional load distribution scheme is "load balancing." The goal here is to equalize the load over candidate hosts. However, as shown in Figure 13, this load-balancing strategy may result in capacity fragmentation and so higher rejection. Thus, although load balancing is adequate for providing best-effort QoS (optimizing *average* measures), it is not adequate for providing guaranteed QoS (optimizing *real-time* measures). A more appropriate scheme is *load profiling* [20]. The idea here is to have a more diverse profile of available capacity on the candidate hosts, so we increase the likelihood of finding a feasible host for future requests.

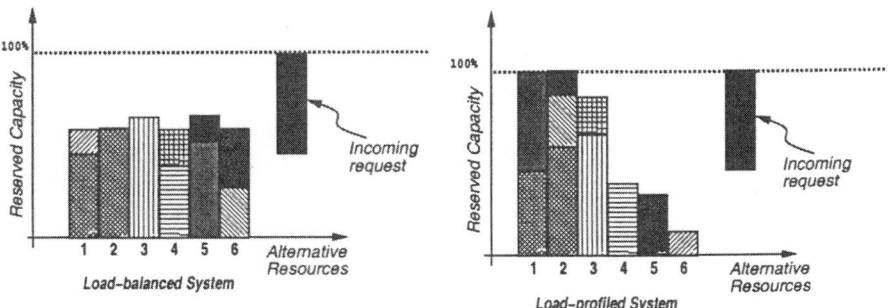

Fig. 13. Load balancing versus load profiling.

Figure 14 shows how after choosing the least-loaded host (with idle capacity of 15) for a class-1 task, we can then only accept 4 consecutive class-2 tasks. On the other hand, if we choose the most-loaded host (with idle capacity of 11), we can then accept 5 consecutive class-2 tasks, as we won't have fragmented capacity in the system. Choosing the most-loaded candidate is a "load packing" strategy, which is only asymptotically optimal for large systems and accurate feedback about system state. In a distributed system with delayed (inaccurate) feedback, a strategy that has the same effect but operates probabilistically is less sensitive to the inaccuracies in feedback information and is more appropriate. We call it "load profiling" strategy.

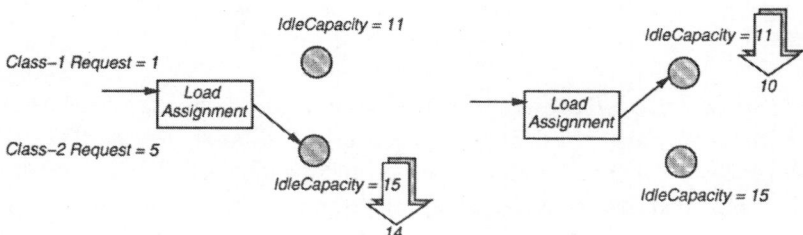

Fig. 14. Example illustrates difference between (a) load balancing and (b) load packing/profiling.

The main idea behind load profiling is illustrated in Figure 15, where the probabilities of selecting each candidate resource are adjusted so we would bring the distribution of QoS requests as close as possible to the distribution of available capacity. This is the well-known supply-demand matching problem.

The gain from load profiling (due to reducing fragmentation) is more significant when we have large requests, which is especially the case when a request represents the aggregate of many micro-requests. This gain over load balancing is also more pronounced as the system becomes more loaded. Extended models that consider the lifetimes of tasks, the costs of migration of tasks, etc. require more careful and more complicated analysis. So in summary, resource selection is an important and difficult problem: how to select resources so as to optimize some application-oriented measure(s) subject to QoS constraints and possibly other constraints on the type of resources, interdependencies between tasks to be assigned, etc. This is a multi-constrained optimization problem that needs fast heuristics that can produce high-quality solutions.

4.2 Network-aware Applications

To make resource selection easier, applications should specify a range of acceptable QoS if possible. This has many other benefits: most importantly is

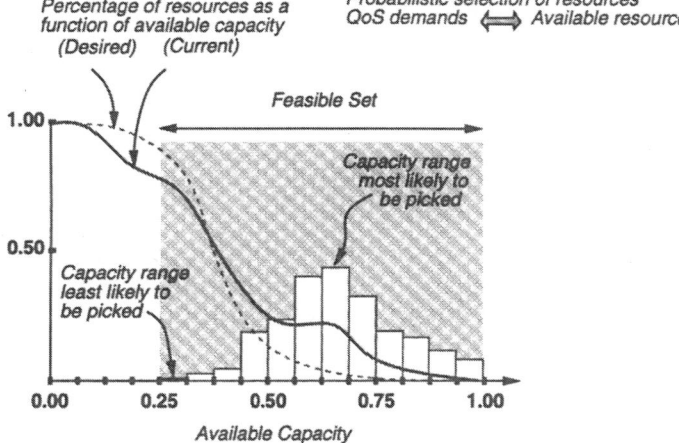

Fig. 15. Maintaining a resource availability profile that matches the characteristics of QoS requests.

that if the requested QoS can not be delivered by the network in a strictly guaranteed manner, then the application can adapt to the currently delivered QoS in a controlled way. For example, the application could send fewer data that have a chance of making it through the network and so the application gets a consistent (although of lesser quality) service. To do this, application-specific QoS managers need to maintain for applications QoS measures of interest to them, and inform applications about the current QoS operating point so that applications adapt accordingly. Applications could also maintain quality by compensating for QoS violations. For example, knowing the loss rate at the receiver, a video source could adjust its FEC (Forward Error Correction) error recovery scheme to compensate for errors. QoS managers could also try to hide QoS violations from applications, for example, by reducing latency through caching or prefetching requested data, by overlapping communication and computation, by migrating processes to where the data resides, etc.

4.3 QoS Multicast Tree Construction

Another important component of an end-to-end QoS architecture is efficient network services that are sensitive to the QoS requested by applications. One important service is multicast routing. A major goal here is to build a delivery multicast tree with minimum cost and which satisfies QoS delivery constraints. Again, this is a multi-constrained optimization problem and we need good and fast heuristics.

An example heuristic is called QDMR (QoS Dependent Multicast Routing) [15]. A nice feature of this heuristic is that it constructs a low-cost tree

212

using a greedy strategy that augments the partially constructed tree with nodes of minimum cost. However, since this can lead to paths that violate QoS delay bounds, the tree construction policy is adapted on the fly to give up some cost savings so as to increase the likelihood of satisfying the QoS delay bound. Figure 16 illustrates the idea.

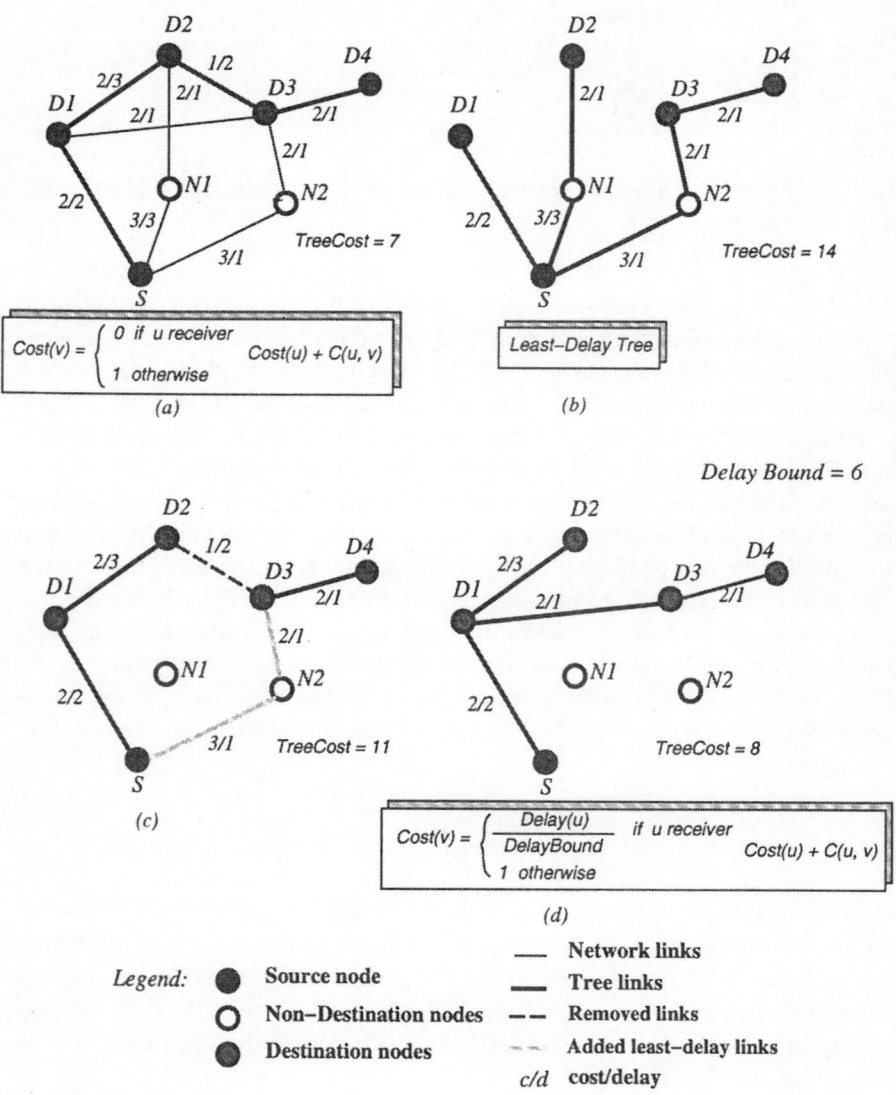

Fig. 16. QoS-aware multicast tree construction.

Figure 16(a) shows a low-cost tree construction policy. The cost of a new node v, denoted by $Cost(v)$, is defined in terms of $Cost(u)$, the cost of node u that is already on the tree, and $C(u,v)$, the cost of the link from u to v. $Cost(u)$ does *not* contribute to $Cost(v)$ if u is a receiver. The idea is to give priority to tree paths going through destination nodes so they are extended to add new nodes (as they would likely have lower cost). By leveraging the cost of reaching a destination to reach other destinations, the total cost of the tree is lowered. This, however, may violate the requested delay bound (which is the case for destination nodes $D3$ and $D4$). In Figure 16(b), the (sender-based) tree of least-delay paths is shown, which would satisfy the delay bound if this were indeed feasible. However, the tree cost is high.

In Figure 16(c), we modify the tree in (a) by replacing the infeasible paths for destinations $D3$ and $D4$ by the corresponding least-delay paths so as to obtain a tree that satisfies the delay bound. However, QDMR can generate a lower cost feasible tree as shown in Figure 16(d). The cost of a new node to be added to the current tree depends on how far we are from violating the delay bound. Paths through destination nodes are no longer given priority as we get closer to violating the delay bound. This makes the tree "bushier" and the delay bound is satisfied at a lower cost.

4.4 Scalability

Scalability is another important aspect of an end-to-end QoS architecture, especially for large wide-area systems. One main goal is to reduce the view a QoS manager has about the state of the system, and based on which it schedules resources (hosts, paths, etc.). This manager could be a sender or a receiver or an agent on behalf of the application, depending on where resource selection/allocation is done. A key to scalability is to separate the way the view is collected from how resources are selected. One way is to have pre-defined classes of applications and collect class statistics to attach to the view, as opposed to statistics about individual applications, for example, the total capacity used by a class rather than the individual capacities used by each application.

Another approach is to have special control entities, called *view-servers* [1], where each view-server maintains only a small view of its surrounding area, as opposed to a full view of the whole system. If a larger area is needed, more than one view-server could be queried and their views merged. Figure 17 illustrates the idea of view-servers.

Another more traditional approach is area-based, where nodes are grouped into level-1 areas, level-1 areas are grouped into level-2 areas, etc. See Figure 18. This is, for example, the scaling approach used in PNNI ATM routing [13]. The idea is that a node has only a detailed view of its own area, and less detailed (summarized or aggregated) views of remote areas, *i.e.* summarized view of level-1 areas in the same level-2 area, of level-2 areas in the same level-3 area, and so on. Different schemes could be used to aggregate an

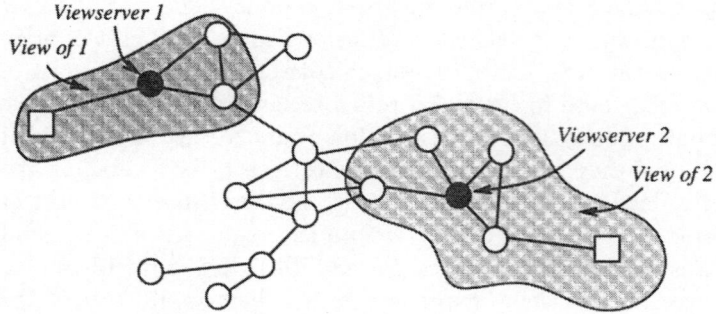

Fig. 17. Viewserver hierarchy.

area. For example, an area may be represented by a fully connected logical graph connecting all B border nodes (those nodes connecting the area to other areas), or by a logical star graph with a virtual node in the center, or by a logical single node. The accuracy of the view, which is presented to the resource selection process, decreases with more aggressive aggregation at the benefit of lesser overhead.

4.5 Robustness

Reliability is another important aspect of an end-to-end QoS architecture. This involves replication of important control entities to survive their failures. It involves avoiding oscillations between alternative configurations that would arise because of the performance interdependencies among the various applications. This could happen if we blindly honor QoS requests, violating existing QoS promises which are then reinstated by again violating other promises. Also, robustness involves reliable switchover to new configurations. For example, switching to a new multicast tree may require keeping transmission over the old tree until the new tree is fully established and the new QoS can be reliably delivered.

5 Conclusion

This article surveys some of the grand challenges in building integrated end-to-end QoS architectures. As we have seen, this involves a plethora of issues in finding fast and good heuristics, defining secure interfaces between different components, investigating the interactions between these components horizontally and vertically, etc. Another important issue is how to develop such a complex software in an easy and reusable way. One recent approach is aspect-oriented programming [18], which differs from the traditional object-oriented in that different aspects of the application or protocol, such as communication, core behavior, structure, etc. are not tangled together, which makes maintenance much easier.

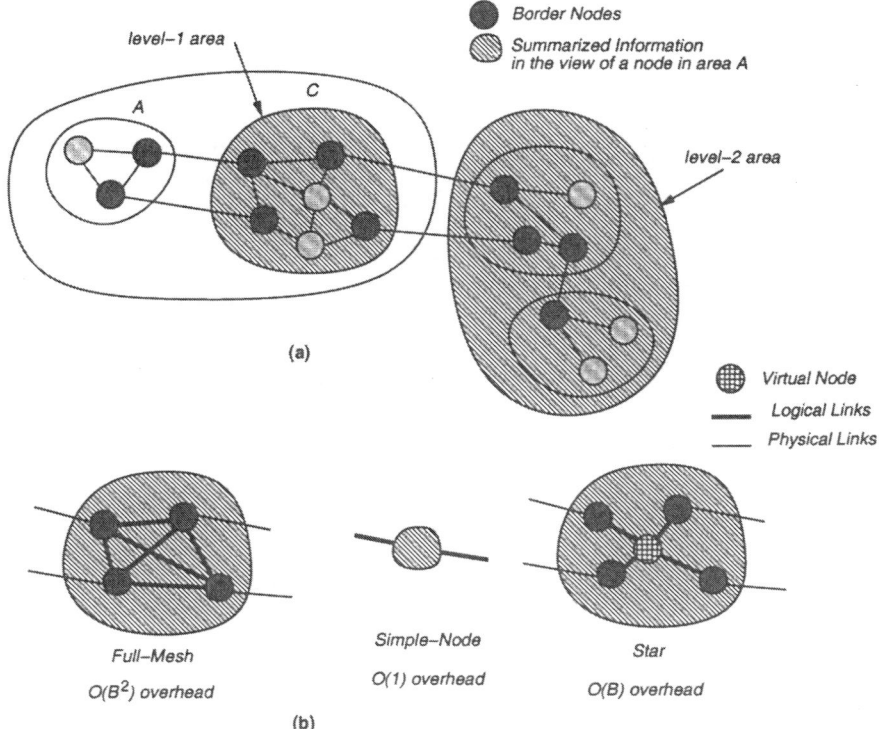

Fig. 18. (a) Area hierarchy, (b) aggregation of area C.

Research and development efforts to build an end-to-end QoS global system are necessarily multi-disciplinary. For such system to become reality, solutions to different problems have to be integrated into a flexible QoS architecture, and the overall performance and cost be evaluated. Initiatives, such as Internet2 [17] and NGI (Next Generation Internet) [16], are providing the infrastructure to deploy and test such advanced architectures.

References

1. C. Alaettinoglu, I. Matta, and A.U. Shankar. A Scalable Virtual Circuit Routing Scheme for ATM Networks. In Proc. *International Conference on Computer Communications and Networks - ICCCN '95*, pages 630–637, Las Vegas, Nevada, September 1995.
2. C. Aurrecoechea, A. Campbell, and L. Hauw. A Survey of QoS Architectures. *ACM/Springer Verlag Multimedia Systems, Special Issue on QoS Architecture*, May 1998.
3. A. Ballardie, P. Francis, and J. Crowcroft. Core Based Trees. In Proc. *SIGCOMM '93*, San Francisco, California, September 1993.
4. S. Blake, D. Black, M. Carlson, E. Davies, Z. Wang, and W. Weiss. An Architecture for Differentiated Services. RFC 2475, December 1998.

5. J-C. Bolot. Adaptive Applications Tutorial. ADAPTS BOF, IETF Meeting, Washington DC, December 1997.

6. B. Braden, D. Clark, and S. Shenker. Integrated Services in the Internet Architecture: An Overview. Internet Draft, October 1993.

7. B. Braden, L. Zhang, S. Berson, S. Herzog, and S. Jamin. Resource ReSerVation Protocol (RSVP) – Version 1 Functional Specification. Internet Draft, March 1996.

8. S. Deering, D. Estrin, D. Farrinacci, V. Jacobson, C. Liu, and L. Wei. Protocol Independent Multicast (PIM): Protocol Specification. Internet Draft, 1995.

9. S. Fischer, A. Hafid, G. Bochmann, and H. de Meer. Cooperative QoS Management for Multimedia Applications. In Proc. *Fourth IEEE International Conference on Multimedia Computing and Systems (ICMCS'97)*, pages 303–310, June 1997.

10. I. Foster and C. Kesselman. Globus: A Metacomputing Infrastructure Toolkit. *Intl. J. Supercomputing Applications*, 11(2):115–128, 1997.

11. M.R. Garey and D. S. Johnson. *Computers and Intractability: A Guide to the Theory of NP-Completeness*. W.H. Freeman and Company, New York, 1979.

12. A. Grimshaw, W. Wulf, and Legion Team. The Legion Vision of a Worldwide Virtual Computer. *Communications of the ACM*, 40(1), January 1997.

13. ATM Forum PNNI Subworking Group. Private Network-Network Specification Interface v1.0 (PNNI 1.0). Technical report, March 1996.

14. ATM Forum Traffic Management Working Group. ATM Forum Traffic Management Specification v4.0. Technical report, 1996.

15. L. Guo and I. Matta. QDMR: An Efficient QoS Dependent Multicast Routing Algorithm. Technical Report NU-CCS-98-05, College of Computer Science, Northeastern University, Boston, MA 02115, August 1998. To appear in *IEEE RTAS '99*.

16. NGI (Next Generation Internet). http://www.ngi.gov.

17. Internet2. http://www.internet2.edu/.

18. G. Kiczales, J. Lamping, A. Mendhekar, C. Maeda, C. Lopes, J-M. Loingtier, and J. Irwin. Aspect-Oriented Programming. In Proc. *European Conference on Object-Oriented Programming*, pages 220–242. Springer Verlag, 1997.

19. J. Kurose. Open Issues and Challenges in Providing Quality of Service Guarantees in High-Speed Networks. *ACM Computer Communication Review*, January 1993.

20. I. Matta and A. Bestavros. A Load Profiling Approach to Routing Guaranteed Bandwidth Flows. In Proc. *IEEE INFOCOM*, March 1998. Extended version in *European Transactions on Telecommunications - Special Issue on Architectures, Protocols and Quality of Service for the Internet of the Future*, February-March 1999.

21. I. Matta and M. Eltoweissy. A Scalable QoS Routing Architecture for Real-Time CSCW Applications. In Proc. *Fourth IEEE Real-Time Technology and Applications Symposium (RTAS'98)*, June 1998.

22. I. Matta, M. Eltoweissy, and K. Lieberherr. From CSCW Applications to Multicast Routing: An Integrated QoS Architecture. In Proc. *IEEE International Conference on Communications (ICC'98)*, June 1998.

23. J. Moy. Multicast Extensions to OSPF. Internet draft, Network Working Group, September 1992.

24. K. Nahrstedt and J. Smith. The QoS Broker. *IEEE Multimedia*, 2(1):53–67, Spring 1995.

25. P. Steenkiste, A. Fisher, and H. Zhang. Darwin: Resource Management in Application-Aware Networks. Technical Report CMU-CS-97-195, Carnegie Mellon University, December 1997.

26. H. Topcuoglu, S. Hariri, D. Kim, Y. Kim, X. Bing, B. Ye, I. Ra, and J. Valente. The Design and Evaluation of a Virtual Distributed Computing Environment. *Cluster Computing*, 1(1):81–93, 1998.

27. D. Waitzman, C. Partridge, and S. Deering. Distance Vector Multicast Routing Protocol. Request for Comments RFC-1175, November 1988.

28. J. Zinky, D. Bakken, and R. Schantz. Architectural Support for Quality of Service for CORBA Objects. *Theory and Practice of Object Systems*, 3(1):55–73, 1997.

A Prototype of a Combined Digital and Retrodigitized Searchable Mathematical Journal

Gerhard O. Michler

Institute for Experimental Mathematics, Essen University, Ellernstr. 29, 45326 Essen, Germany

1 Introduction

Recently many mathematical journals appear in digital and in paper form. The digital version offers many advantages in the future. It is searchable, and in due course it will technically be possible that also a quoted article can be retrieved and partially be shown in a second window on screen if it will be part of a searchable digital library system. Over scientific wide area networks like the German B-Win all digital libraries of the connected universities can be combined to a national distributed research library, and this is about to happen in Germany, and in many other countries. If the legal problems concerning the authentication of the subscribers to a distributed on-line digital library system will be solved, then the authorized members of a German university or research institute will be able to view, read and print the wanted text of digital issues of an available scientific journal at their personal computers. In the future they also want to search in the whole distributed research library, and not only in the recent articles. Therefore the Deutsche Forschungsgemeinschaft (DFG) has provided financial support for the establishment of two Centers for Retrospective Digitization, one at the State and University Library at Göttingen and the other at the Bavarian State Library Munich.

The mathematicians are lucky, because the Göttingen library has been the DFG-Sammelstelle for Mathematics for long, which means that almost all essential mathematical journals and books have been collected there with financial support of the Deutsche Forschungsgemeinschaft. If all the digital versions of the mathematical journals were collected at the digital library in Göttingen, and the document management system AGORA [2] was implemented at the Göttingen Center for Retrospective Digitization, then the authorized German mathematicians could use this searchable digital library from their workstations at their university or research institute.

Of course there is a long way to achieve this ideal situation. Besides financial and difficult legal problems there are also a lot of technical problems which still have to be solved. It is the latter point of view which will be addressed in this article.

2   A prototype mathematical text recognition system

Since 1997 my study group is cooperating with the publisher Birkhäuser (Basel) in order to retrodigitize 6 volumes of the Archiv der Mathematik published by Birkhäuser in the years 1993 till 1996. This is a widely distributed mathematical journal which publishes short articles from all major areas of mathematics. Furthermore, its typesetting is excellent. This offers a chance for the use of optical character recognition (OCR) sytems for the retrospective digitization task.

Many ordinary texts can automatically be retrodigitized by means of commercial OCR systems. In [1] you can read: "With Adobe Acrobat Capture 2.01 software, you can easily turn volumes of paper into searchable Portable Document Format (PDF) libraries. It's perfect for forms, manuals, specifications, books - any important document you need to make accessible on your Web site or intranet." Besides Adobe [1], there are also many other commercial OCR systems, like the one of the World Scientific Publishing Company [4], and FineReader [8].

What is so special with mathematical texts? A convincing answer is given by R. Fateman in his article [5]. There he writes in the introduction: "Conventional OCR programs have low accuracy for mathematics for several reasons. The very sensible heuristics typically used for text recognition include computing the locations of text lines and estimating character sizes using global statistics as well as local processing. These programs may also use language-based statistics (perhaps a spelling dictionary) as tools to improve recognition rates. By contrast, mathematics is not necessarily arranged on lines, its character sizes vary, the letter and symbol frequencies are distinct from normal text, and many other text-oriented heuristics are directly counter-productive. Additionally, even if the mathematics were somehow recognized, conventional OCR programs whose traditional output is (say) ASCII text, need to be substantially augmented with some meta-level language before they can express 'math results' as their output. Although most advanced word-processing programs have some escape mechanism for 'doing' mathematics, there is still no uniform standard for expressing two-dimensional layouts, subscript positioning, variable-sized characters, unusual math operators, etc."

In 1997 the Deutsche Forschungsgemeinschaft has agreed to support the research project "Retrodigitization of the mathematical journal Archiv der Mathematik" for one and then for two more years. Its purpose is to retrodigitize the 6 volumes of this journal which have appeared from 1993 to 1996. Without the legal support of the publisher Birkhäuser this project could not have been started. The outcome will be a prototype of a text recognition system which allows to search in the ordinary text of an originally printed article. Furthermore, it produces a version of the mathematics formulas and symbols that reflects the semantics of the mathematical part of the text such that the retrodigitized formulas are written in tex format. Thus they can be incorporated into new mathematical manuscripts.

Many computer algebra systems like AXIOM, MAPLE, MATHEMAT-
ICA or MAGMA allow symbolic formula manipulation and to display the
calculated formulas as typeset expressions. Unfortunately, at the moment
the computer algebra systems are not able to read in digitized mathematical
formulas. It is hoped that this deficiency will be overcome in the future. Then
the retrospectice digitization text systems will become even more important
for mathematical research.

2.1   Recognition of mathematical expressions and formulas

In [3], [6] and [7] R. Fateman, T. Tokuyasu and coauthors have described
a package of LISP programs for optical mathematical formula recognition
and translation of a scanned mathematical text into digital LISP format. My
former collaborator Dr. J. Rosenboom has trained this special OCR system
so that it became acquainted with the special typesetting of mathematical
formulas used by the printers of the Archiv der Mathematik. Furthermore,
he has extended the recognition algorithms for special mathematical symbols
and the geometry of combined formulas. Inspired by R. Fateman's suggestions
Rosenboom has written a program to parse the digital mathematical LISP
formulas and then to produce an output in tex format. In particular, he
incorporated procedures enabling the LISP program to recognize different
types of fonts like normal, italics, Greek, bold face, etc.

In the LISP code there are several procedures for understanding the layout
of the printed pages of a journal. So it is necessary to have segmentation
procedures for dissecting a page into lines, lines into words, words into letters
or a mathematical formula into mathematical symbols.

Inspired by R. Fateman's article [5] Rosenboom has written a program
which separates ordinary text of a scanned page from areas of the page consisting only of mathematical formulas. Such a separation leads to a substantial
improvement of the retrospective digitization procedure. This was demonstrated by J. Rosenboom in his lecture [14] at the international workshop on
"Retrodigitalization of mathematical journals and automated formula recognition" organized by R. Fateman (Berkeley), E. Mittler (Göttingen) and the
author at the Institute for Experimental Mathematics of Essen University in
December 1997.

The idea of separating the ordinary text of a scanned page from the
remainder lead us to use a commercial OCR system for the recognition of
this part of a scanned page. In our project we use FineReader [8], because it is
very reliable and has a very good application programming interface for C++
and other modern programming languages. This is important, because we do
not have access to the source code of the commercial product FineReader.

The following copy of a Tiff file of a scanned page of an article of the
"Archiv der Mathematik" will be used to explain the different retrodigitizing
procedures.

Arch. Math., Vol. 65, 399–407 (1995)

0003-889X/95/650S-0399 $ 3.30/0
© 1995 Birkhäuser Verlag, Basel

Tameness of biserial algebras

By

WILLIAM CRAWLEY-BOEVEY

In this paper we study finite dimensional associative algebras (with 1) over an algebraically closed field K. An algebra is *biserial* if every indecomposable projective left or right module P contains two uniserial submodules whose sum is the unique maximal submodule of P and whose intersection is either zero or simple. (A module is uniserial if it has a unique composition series.) An algebra A has *tame representation type* if its indecomposable modules lie in one-parameter families, see for example [3]. Here we prove the following result.

Theorem A. *Biserial algebras have tame representation type.*

This appears to have been conjectured for some time; it is already known for so-called special biserial algebras [12] and for many other biserial algebras (see for example [7] and [10, Chapter 2]). The two main ingredients in our proof are Vila-Freyer's structure theorem for basic biserial algebras, which was motivated by this problem, and a modification of Geiß's Theorem that algebras with a tame degeneration are tame, in which we vary the relations instead of the structure coefficients of the algebra. If $\mathrm{Alg}(d)$ is the variety of associative unital algebra structures on an d-dimensional vector space, then Geiß's Theorem [6] states that if the closure of the $\mathrm{GL}(d, K)$-orbit of $A \in \mathrm{Alg}(d)$ contains a tame algebra then A is tame. In our version the algebras may have different dimensions:

Theorem B. *Let A be a finite dimensional algebra, let X be an irreducible variety and let $f_1, \ldots, f_r : X \to A$ be morphisms of varieties (where A has its natural structure as affine space). For $x \in X$ write $A_x = A/\langle f_i(x)\rangle$. Let $x_0, x_1 \in X$. If A_{x_0} is tame and $A_x \cong A_{x_1}$ for general $x \in X$, i.e. for all x in a non-empty open subset of X, then A_{x_1} is tame.*

In practice Geiß's Theorem has usually been used in the form of Theorem B, but in this case one also has to check that all the algebras A_x have the same dimension. Of course Theorem B and Geiß's theorem have a common refinement in which both the algebra A and relations are allowed to vary. The author is supported by the EPSRC of Great Britain.

1. Proof of Theorem B. If an algebraic group G acts on a (not necessarily irreducible) variety Y then the *number of parameters* of G on Y is $\dim_G Y = \max \{\dim Y_{(s)} - s \mid s \geqq 0\}$ where $Y_{(s)}$ is the union of the orbits of dimension s. By a G-stable subset $Z \subseteq Y$ we

2.2 Recognition of ordinary mathematical texts

FineReader can recognize articles written in different languages like English, French, German and Russian. However, it does not like mixtures of languages in a given paper. It achieves recognition by checking its dictionaries of words in a chosen language. For the retrospective digitization of mathematical articles these dictionaries do not suffice, because they do not contain the special mathematical terms, abbreviations, names of authors, quoted scientists or mathematical journals. Therefore additional mathematically oriented dictionaries have been written in C++. They can be read by FineReader over its application interface. Thus the ordinary text of a scanned page of a mathematical article is recognized by FineReader almost perfectly. It can be read on screen and its digital version is written in ASCII. Therefore it is possible to search in this digitized text for words. Those parts of the scanned page which have not been recognized by FineReader like mathematical formulas, geometric pictures or diagrams are defined by our retrospective digitization system to be mathematical formulas. These sections of the scanned page are sent to the LISP program. Its output is an ASCII text in tex format. Also this mathematical contents of the scanned page can be viewed on screen. However, this is done in a different window than the ordinary FineReader text. In section 2.5 it is described how these two parts of the text of a scanned article are linked to each other.

2.3 Getting bibliographic data of an article

In order to recognize the special layout of the first page of a scanned article another special program has been written. It reads any scanned page. By analyzing the first recognized letters of it, it is able to decide whether this page is the first page of the article. If so, then the program recognizes all bibliographic data about this paper: Name of the journal, number of the volume containing the article, first and last page, year of appearance, the international standard serials number (ISSN), owner of copyright (Birkhäuser Verlag), the town of the publisher (Basel).

Moreover, the program recognizes the title, the number of authors and their names together with their first names and initials. From the Tiff file of the example given in section 2.1 it gives the following output.

William Crawley-Boevey, Tameness of biserial algebras, Arch. Math. **65**, 399-407 (1995)

From these data another program produces the following bibliographic data in SGML format:

⟨journal⟩   Arch. Math.   ⟨/journal ⟩
⟨volume⟩   65   ⟨/volume⟩
⟨firstPage⟩   399   ⟨/firstPage⟩
⟨lastPage⟩   407   ⟨/lastPage⟩
⟨year⟩   1995   ⟨/year⟩
⟨authors⟩
⟨author⟩
⟨lastName⟩   Crawley-Boevey   ⟨/lastName⟩
⟨firstName⟩   William   ⟨/firstName⟩
⟨firstInitial⟩   ⟨/firstInitial⟩
⟨secondInitial⟩   ⟨/secondInitial⟩
⟨/author⟩
⟨authors⟩
⟨title⟩   Tameness of biserial algebras ⟨/title⟩

These data written in the standard generalized markup language (SGML) will enable other digital library document management systems like AGORA [2] or MILESS [11] to retrieve these bibliographic records. Such an example is described in [10].

The Archiv der Mathematik always prints the complete addresses of all authors at the end of an article. In recent volumes also their e-mail adresses are given. Both are recognized by our retrodigitizing programs. If necessary these informations can also be produced in SGML format. The end of the last address is also used to mark the end ot the retrodigitized article.

2.4   Recognizing the references of an article

My collaborators Dr. G. Hennecke and Dr. H. Gollan have written another special program for the recognition of the references. It reads any scanned page and decides by itself, whether or not this page contains the beginning or the remaining part of the references. Each reference is digitized in full text, including the abbreviations of the cited journals, volumes, years and page numbers. In the example of section 2.1 the reference [12] mentioned on the first page is recognized as follows.

[12] B. WALD and J. WASCHBÜSCH, Tame biserial algebras, J. Algebra **95**, 480-500 (1985).

From these data the program produces the following SGML file:

⟨referenceNumber⟩	[12]	⟨/referenceNumber⟩
⟨authors⟩		
⟨author⟩		
⟨lastName⟩	Wald	⟨/lastName⟩
⟨firstName⟩		⟨/firstName⟩
⟨firstInitial⟩	B.	⟨/firstInitial⟩
⟨secondInitial⟩		⟨/secondInitial⟩
⟨/author⟩		
⟨author⟩		
⟨lastName⟩	Waschbüsch	⟨/lastName⟩
⟨firstName⟩		⟨/firstName⟩
⟨firstInitial⟩	J.	⟨/firstInitial⟩
⟨secondInitial⟩		⟨/secondInitial⟩
⟨/author⟩		
⟨/authors⟩		
⟨title⟩	Tame biserial algebras	⟨/title⟩
⟨journal⟩	Journal of Algebra	⟨/journal⟩
⟨series⟩		⟨/series⟩
⟨volume⟩	95	⟨/volume⟩
⟨firstPage⟩	480	⟨/firstPage⟩
⟨lastPage⟩	500	⟨/lastPage⟩
⟨year⟩	1985	⟨/year⟩

Using later the functions of a distributed digital library document management system these data allow to search in the digitized volumes of the mathematical journal and retrieve the quoted digital articles. Such an example is described in [10].

2.5 Incorporation of the different digitized texts into a multivalent document system

Many mathematical articles of the Archiv der Mathematik contain pictures, complicated diagrams and tables which cannot be recognized by any OCR system. These parts of a scanned page have to be stored as images. However, they do not contain any information which is necessary for the searchability of a mathematical article.

In order to enable the reader to view the complete content of an article of the retrodigitized mathematical journal on screen each page is scanned with 600 dots per inch, and a TIFF file of the whole page is produced. Applying then the procedures described in sections 2.1, 2.2, 2.3 and 2.4 we so obtain the following separate files:

1) TIFF file of the whole scanned page,
2) ASCII file of its ordinary text,
3) Tex file of its mathematical formulas text,

4) Text and SGML files of the quoted references,

5) SGML file of its bibliographic data.

These different files are only useful for the scientists if they can be linked to each other. This is done by the multivalent document system designed by T.A. Phelps and R. Wilensky [13]. This software system has been produced by T.A. Phelps [12], recently. It is a new general paradigm that regards complex documents as multivalent documents comprising multiple layers of distinct but intimately related content. Phelps and Wilensky write in [13]: "Small, dynamically-loaded program objects, or 'behaviors', activate the content and work in concert with each other and layers of content to support arbitrarily specialized document types. Behaviors bind together the disparate pieces of a multivalent document to present the user with a single unified conceptual document. Examples of the diverse functionality in multivalent documents include: 'OCR select and paste', where the user describes a geometric region on the scanned image of a printed page and the corresponding text characters are copied out."

Therefore this multivalent document system is very useful for our retrospective digitization project. The following picture describes the different layers containing the separate files 1) till 5), and the layer for user annotations.

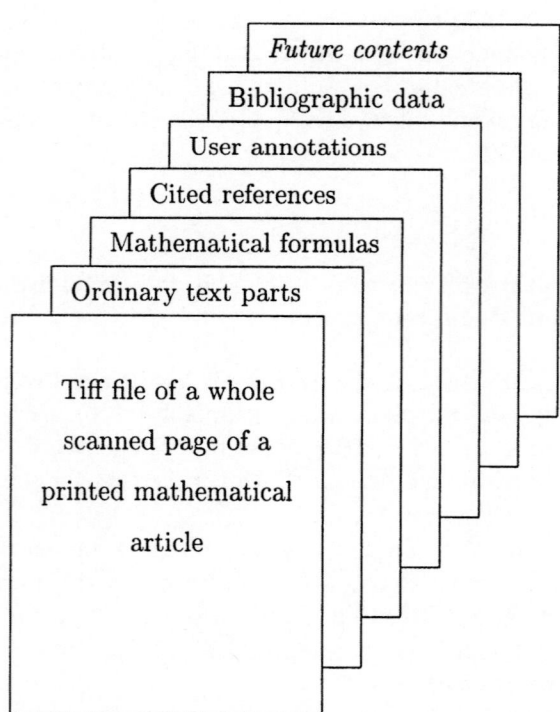

Multiple active semantic layers of the contents of a scanned page

The program of Phelp's multivalent document system (MVD) can automatically read the TIFF files 1) of the scanned pages of an article and show them on screen. Thus the user can have a printout of the whole manuscript. The TIFF file of the mathematical article is contained in layer 1 of its MVD system.

In order to incorporate the ordinary text file 2) into this system another program is written which enables FineReader to produce the digitized ordinary text in Xdoc format. This Xdoc file describes besides the ASCII text also the coordinates of each of its letters. The MVD system now allows to search for a word in the second layer and to show the result on the first layer on screen. Thus the system provides full searchability in the ordinary text of retrodigitized page of a mathematical article. The ordinary text parts of this mathematical article are contained in layer 2 of its MVD system.

As an example we now present the Xdoc file of the ordinary text of W. Crawley-Boevey's article "Tameness of biserial algebras" corresponding to the Tiff file given in section 2.1. It is free of mathematical formulas or expressions.

```
[a;"XDOC.10.0";E;"FrEngine40-CB1"]
[d;"65_5_399.xdc"]
[p;1;P;83;S;0;1666;0;0;3010;4618]
[t;1;1;0;0;A;"";"";"";0;0;0;0;1]
[f;0;"<DEFAULT>";R;q;10;V;0;0;0;10;100]
[f;1;"Courier";R;q;10;V;60;50;10;15;100]
[s;1;88;0;70;p;1]Arch.[h;196;32]Math.,[h;426;28]Vol.[h;569;25]
    65,[h;677;30]399-407[h;965;29](1995)[h;1177;802]0003-889X/95/
    6505-0399[h;2719;28]$[h;2778;23]3.30/0[y;2978;0;70;0;H]
[s;1;1980;0;166;p;1][h;2044;34]1995[h;2209;32]Birkh\"auser
    [h;2565;25]Verlag,[h;2795;32]Basel[y;2978;0;166;1;S]
[s;1;844;0;774;p;1]Tameness[h;1271;41]of[h;1411;29]biserial
    [h;1756;42]algebras[y;2165;0;774;1;S]
[s;1;1461;0;1009;p;1]By[y;1537;0;1009;1;S]
[s;1;1092;0;1148;p;1]WILLIAM[h;1351;28]CRAWLEY-BOEVEY[y;1914;0;
    1148;1;S]
[s;1;108;0;1441;p;1]In[h;176;34]this[h;327;33]paper[h;548;32]
    we[h;667;32]study[h;879;31]finite[h;1076;31]dimensional[h;
    1511;33]associative[h;1902;32]algebras[h;2209;31](with[h;
    2407;37]1)[h;2496;33]over[h;2674;31]an[h;2783;34]alge-
    [y;2973;0;1441;0;H]
[s;1;28;0;1539;p;1]braically[h;315;27]closed[h;545;30]field
    [h;712;36]K.[h;818;31]An[h;944;29]algebra[h;1217;29]is[h;1293;
    30]biserial[h;1558;26]if[h;1632;18]every[h;1826;27]
    indecomposable[h;2393;28]projective[h;2748;27]left[h;2874;30]
    or[y;2975;0;1539;0;H]
[s;1;28;0;1636;p;1]right[h;183;35]module[h;466;39]P[h;547;36]
    contains[h;862;34]two[h;1018;36]uniserial[h;1335;35]submodules
```

[h;1765;36]whose[h;2006;34]sum[h;2174;35]is[h;2257;34]the[h;
2391;35]unique[h;2652;34]maximal[y;2973;0;1636;0;H]

[s;1;27;0;1734;p;1]submodule[h;392;27]of[h;487;24]P[h;553;27]
and[h;702;27]whose[h;934;26]intersection[h;1346;27]is
[h;1421;25]either[h;1635;23]zero[h;1798;28]or[h;1897;25]simple.
[h;2150;27](A[h;2252;26]module[h;2526;26]is[h;2600;27]
uniserial[h;2908;26]if[y;2982;0;1734;0;H]

[s;1;26;0;1832;p;1]it[h;69;41]has[h;218;38]a[h;291;39]unique
[h;556;36]composition[h;1005;40]series.)[h;1262;37]An[h;1395;
40]algebra[h;1679;42]A[h;1772;36]has[h;1917;40]tame[h;2108;37]
representation[h;2612;37]type[h;2783;38]if[h;2869;29]its
[y;2972;0;1832;0;H]

[s;1;26;0;1930;p;1]indecomposable[h;565;38]modules[h;881;39]lie
[h;994;38]in[h;1092;39]one-parameter[h;1621;36]families,[h;
1926;40]see[h;2061;38]for[h;2192;36]example[h;2503;40][[3].[h;
2646;41]Here[h;2847;37]we[y;2972;0;1930;0;H]

[s;1;26;0;2045;p;1]prove[h;213;31]the[h;344;30]following[h;684;
30]result.[y;911;0;2045;1;S]

[s;1;106;0;2224;p;1]Theorem[h;398;24]A.[h;492;36]Biserial[h;772;
30]algebras[h;1073;31]have[h;1245;32]tame[h;1428;31]
representation[h;1927;27]type.[y;2106;0;2224;1;S]

[s;1;106;0;2372;p;1]This[h;250;39]appears[h;545;39]to[h;649;38]
have[h;842;36]been[h;1030;37]conjectured[h;1453;38]for[h;1584;
37]some[h;1791;38]time;[h;1997;36]it[h;2077;38]is[h;2163;37]
already[h;2445;36]known[h;2708;38]for[h;2840;36]so-[y;2972;0;
2372;0;H]

[s;1;27;0;2469;p;1]called[h;217;26]special[h;464;28]biserial
[h;729;26]algebras[h;1030;26][[12][h;1171;28]and[h;1321;25]
for[h;1439;25]many[h;1648;26]other[h;1850;25]biserial[h;2112;
26]algebras[h;2413;24](see[h;2556;25]for[h;2674;24]example
[y;2973;0;2469;0;H]

[s;1;29;0;2581;p;1][[7][h;104;22]and[h;248;22][[10,[h;382;22]
Chapter[h;676;22]2]).[h;804;23]The[h;954;21]two[h;1096;20]
main[h;1281;22]ingredients[h;1668;21]in[h;1750;21]our[h;1886;
18]proof[h;2091;12]are[h;2203;23]Vila-Freyer's[h;2656;19]
structure[y;2972;0;2581;0;H]

[s;1;28;0;2665;p;1]theorem[h;300;26]for[h;419;27]basic[h;611;25]
biserial[h;873;28]algebras,[h;1192;27]which[h;1414;27]was
[h;1563;26]motivated[h;1925;28]by[h;2032;26]this[h;2175;26]
problem,[h;2496;27]and[h;2646;26]a[h;2708;25]modifi[y;2971;0;
2665;0;H]

[s;1;26;0;2763;p;1]cation[h;230;28]of[h;327;17]Gei's[h;552;27]
Theorem[h;878;29]that[h;1037;27]algebras[h;1339;26]with[h;
1509;27]a[h;1572;25]tame[h;1758;27]degeneration[h;2216;27]are

tame, in which we

vary the relations instead of the structure coefficients of the algebra. If is the variety

of associative unital algebra structures on an vector space, then Gei's

Theorem [[6] states that if the closure of the of contains a tame

algebra then A is tame. In our version the algebras may have different dimensions:

Theorem B. Let be a finite dimensional algebra, let be an irreducible variety and

let be morphisms of varieties (where A has its natural structure as

affine space). For write Let If is tame and

for general i.e. for all in a non-empty open subset then is tame.

In practice Gei's Theorem has usually been used in the form of Theorem B, but in

this case one also has to check that

```
[h;1155;20]all[h;1253;22]the[h;1375;18]algebras[h;1669;25]
A^[h;1772;20]have[h;1946;20]the[h;2066;20]same[h;2251;20]
dimension.[h;2630;22]Of[h;2742;11]course[y;2967;0;3884;0;H]
[s;1;20;0;3982;p;1]Theorem[h;319;28]B[h;394;25]and[h;541;26]
Gei's[h;775;26]theorem[h;1073;27]have[h;1253;25]a[h;1314;24]
common[h;1630;26]refinement[h;2008;25]in[h;2093;26]which
[h;2315;25]both[h;2493;26]the[h;2620;24]algebra[h;2888;31]
A[y;2971;0;3982;0;H]
[s;1;20;0;4080;p;1]and[h;141;39]relations[h;467;38]are[h;606;
38]allowed[h;900;38]to[h;1003;38]vary.[h;1201;39]The
[h;1367;38]author[h;1630;36]is[h;1714;38]supported[h;2087;39]
by[h;2205;37]the[h;2342;39]EPSRC[h;2638;39]of[h;2746;29]Great
[y;2966;0;4080;0;H]
[s;1;21;0;4178;p;1]Britain.[y;266;0;4178;1;S]
[s;1;103;0;4375;p;1]1.[h;151;29]Proof[h;372;21]of[h;462;20]
Theorem
[h;775;25]B.[h;865;29]If[h;950;20]an[h;1048;30]algebraic
[h;1379;27]group[h;1604;31]G[h;1685;29]acts[h;1841;29]
on[h;1952;29]a[h;2017;27](not[h;2176;29]necessarily[h;2562;27]
irreducible)[y;2967;0;4375;0;H]
[s;1;20;0;4472;p;1]variety[h;246;31]Y[h;324;24]then[h;491;27]
the[h;618;25]number[h;879;26]of[h;972;15]parameters[h;1346;26]
of[h;1441;20]G[h;1510;27]on[h;1619;31]Y[h;1698;22]is
[h;1769;24]
[s;1;19;0;4570;p;1]where[h;215;43]is[h;436;38]the
[h;574;40]union[h;804;40]of[h;913;30]the[h;1042;41]orbits
[h;1273;40]of[h;1381;29]dimension[h;1754;39]s.[h;1835;42]By
[h;1964;38]a[h;2038;40]G-stable[h;2345;40]subset[h;2590;37]
we[y;2965;0;4570;1;S]
[g;1666;0;0;3010;4618]
```

In order to incorporate the mathematical formula latex file 3) into the MVD system the software of the MVD has been extended such that it can read in file 3) as its third layer. To each mathematical formula or expression its coordinates are attached. This is necessary in order to keep the connection between the ordinary text of the second layer and the mathematical text of the third layer. Here the reader can mark a relevant mathematical formula and show it on screen. Another program will be written enabling the viewer to transfer the obtained mathematical formula into a new mathematical latex manuscript.

As an example we now printout the mathematical formula text of Crawley-Boeyey's article "Tameness of biserial algebras".

Alg(d) [h; 2537; 19]

d-dimensional [h; 2080; 35]

$GL(d, K)$-orbit [h; 1967;24] of $A \in [h; 2188; 23]$Alg(d) [h; 2426; 20]

Theorem [h; 394; 25] **B.** [h; 484; 37] Let [h; 627; 38] A [h; 715; 31]

X [h; 1980; 38]

$f_1, \ldots, f_r : [h; 434; 31]X[h; 515; 30] \to [h; 607, 33]A$ [h; 691; 36]

$x \in X$ [h; 881; 58]

$A_x = [h; 1320; 32]A/(f_i(x)).$ [h; 1654; 59]

$x_0, x_1 \in X$ [h; 2222; 59]

A_{x_0} [h; 2494; 52]

$A_x[h; 107; 29] \cong [h; 187; 39]A_{x_1}$ [h; 325; 33]

$x[h; 774; 22] \in [h; 833; 28]X$ [h; 934; 26]

x [h; 1324; 26]

X [h; 2406; 26]

A_{x_1} [h; 2696; 29]

$\dim_G[h; 1961; 21]Y[h; 2030; 21] = [h; 2105; 27]\max[h; 2272; 23]$
$\{\dim[h; 2450; 22]Y_{(s)}[h; 2564; 27] - [h; 2645; 28]s[h; 2700; 16] \mid [h; 2743; 18]$
$s[h; 2788; 27] \geq [h; 2867; 28]0\}[y; 2962; 0; 4472; 0; H][s; 1; 19; 0, 4570, p; 1]$

$Y_{(s)}$ [h; 349; 39]

$Z[h; 2682; 27] \subseteq [h; 2762; 36]Y$ [h; 2848; 29]

We remark that **Theorem B** has only be put into this printout of the mathematical formula text in order to help the reader finding the corresponding text in the Xdoc file of the ordinary text and in the Tiff file.

The fourth layer contains file 4) with the bibliographic data of the quoted articles and their SGML formats.

The layer with the user annotations allows the communication between the author or library and the possible readers of a retrodigitized article. So misprints, comments and suggestions for further improvement of the article and its presentation in the digital library can be mentioned here.

The sixth layer is used for the preparation of the searchability between the different articles of the retrodigitized volumes of the mathematical journal.

Therefore it contains all the bibliographic data about the given article in SGML format.

Since our retrospective digitization programs can recognize the special layout of the first page and of the references, the MVD system of the prototype allows to mark and call a reference of a retrodigitized article contained in the digital library within the MVD format.

3 Combining retrodigitized and recent digital volumes

All the programs described in the second section will be applied to retrodigitalize the 6 volumes of the Archiv der Mathematik published in the years 1993 till 1996. The result can be considered to be a prototype for retrospective digitization of general mathematical journals. Of course for each other periodical the training systems of the OCR programs have to be modified. However, the developed and applied technologies can be adjusted.

It is therefore even more important to combine the retrodigitized volumes of the Archiv der Mathematik with the recent digital issues of this periodical. Since 1997 this journal is published in paper and in digital form. The digital articles can be received from the Springer Link in Heidelberg over the internet by the authorized members of the universities and research institutes, because the Springer data base contains them in portable document format (PDF). This format allows to view each page of the received article on screen. Furthermore, the reader can produce a printout of it. Also the full text can be searched for words, but not for mathematical formulas or expressions. At the Springer Link the bibliographic records and abstracts of the articles are stored in SGML format.

The publisher Birkhäuser has agreed to extend our present collaboration in order to produce a software allowing to connect the retrodigitized and the recent digitized volumes of the Archiv der Mathematik in one experimental digital library system. For that he will provide the digital texts of the recent 6 volumes in PDF format together with bibliographic records about their articles in SGML format.

Since we then will have the bibliographic data of the retrodigitized and of the digital volumes in SGML format, we can use the MILESS system [11] to produce a common platform for the both parts of the digital articles of the Archiv der Mathematik. MILESS is a library server, developed at Essen University in a joint research project of the Computer Center and the Central Library. It uses the IBM DB2 Digital Library product [9] to provide access to digital and multimedia documents in a reliable and systematic way. As described in [10] MILESS allows the storage of such material in any format such as audio, video, HTML, XML, PDF. Since both the retrodigitized and the new digital articles of the Archiv der Mathematik have an SGML-description of their bibliographic data, we can use these to provide MILESS with the necessary information about any given article. For the old retrodigitized articles we will use the produced MVD system described in the second section as the storage format. The new articles will be stored as PDF-files, as it is done in the Springer LINK. So both parts of the digital journal can be linked within the MILESS library system. Furthermore, the search functions of MILESS allow now to search in the bibliographic data and also for words in ASCII-versions of the articles of both parts of the Archiv der Mathematik.

This part of the project will be pursued in due course. The director of the Essen Computer Center has agreed to provide access to the MILESS digital library program. We do not need any access to the source code of the IBM DB2 Digital Library product which is called by the useful public domain programs of MILESS.

There is no doubt that this experiment will be successful. It is the last corner stone for building a prototype of a combined digital and retrodigitized searchable mathematical journal. Instead of MILESS the publisher could also use the digital library data base of the Springer LINK. Since the Center for Retrospective Digitization of the Göttingen State and University Library is about to introduce AGORA, a new document management system [2], it is very likely that also by means of that library system retrodigitized and recent digital volumes of a mathematical journal can be linked. The director of the Göttingen State and University Library has agreed to support such experiments with our prototype in another joint project.

4 Future improvements

As the reader will have observed so far the multivalent document system (MVD) does not produce the ordinary mathematical text parts of scanned mathematical article in PDF format. On the other side the programs used at the Springer LINK do not provide any access to the mathematical formulas. They are also not able to call the text of a quoted article in the references which is available in the publishers digital library data base.

In order to show that referencing within the digital articles of the Archiv der Mathematik is possible one could also retrodigitize the recent digital texts and put the obtained ordinary text part of a page into the multivalent document system (MVD) in Xdoc format. Such an experiment is planned, and will have the support of the publisher Birkhäuser. However, it will only prove that it is very promising to start a new joint project, in which the software of the present prototype is extended in such a way that the retrodigitized and the recent digital volumes will be stored in a common PDF or postscript format. Furthermore, the bibliographic and referencing records have to be written in a common SGML or XML format. Such a future project requires joint efforts of the software developers for the publisher, for the MILESS project and for the Essen retrodigitization project.

If successful, then the resulting new prototype of one combined retrodigitized and digital mathematical journal will show that such a software system can also be used for retrospective digitization and combining its results with the digital volumes of other mathematical journals contained in a distributed digital research library. In particular, it will then be possible to retrieve and read quoted articles of different journals as long as they are part of a digital library system. Furthermore, such a software will allow to mark a retrodigitized or recent digital mathematical formula on screen and have it transferred into a new mathematical manuscript.

Acknowledgements

The author kindly acknowledges financial support by DFG grant III N 2 - 542 81(1) Essen BIB45 ENug 01-02.

He is very grateful to the Birkhäuser Verlag for its generous technical and legal support. The author also thanks Professor R. Fateman, T.A. Phelps Ph.D and Professor D. Wilensky of the University of California (Berkeley) for their advice and their programs. Finally, he owes his thanks to his former collaborators Dr. G. Hennecke, Dr. J. Rosenboom, and present collaborators C. Begall, Dr. H. Gollan and Dr. R. Staszewski who have done or do all the programming and hard work.

References

1. Adobe Acrobat Capture 2.01 for Windows 95 and Windows NT(R), http://www.adobe.com/prodindex/acrobat/capture.htm
2. Agora - Digitales Dokumentenmanagementsystem für die Inhalte Ihrer Bibliothek, http://www.agora.de
3. Benjamin P. Berman, Richard J. Fateman, Nicholas Mitchell and Taku Tokuyasu, "Optical character recognition and parsing of typeset mathematics", *Journal of Visual Communication and Image Representation*, vol. 7 (1996), 2-15.
4. D. Blostein and A. Grbavec, "Recognition of Mathematical Notation", Chapter 22 in P.S.P. Wang and H. Bunke (eds.), *Handbook on Optical Character Recognition and Document Analysis*, World Scientific Publishing Company, 1996.
5. R. J. Fateman, "How to find mathematics on a scanned page", Preprint 1997, Univ. Calif. Berkeley.
6. R. Fateman and T. Tokuyasu, "A suite of programs for document structuring and image analysis using Lisp", UC Berkeley, technical report, 1996.
7. R. Fateman and T. Tokuyasu, "Progress in recognizing typeset mathematics", 1997, Univ. California Berkeley.
8. FineReader OCR Engine, Developer's Guide, ABBYY Software House (BIT Software), Moscow, 1993-1997.
9. IBM DB2 Digital Library, http://www.software.ibm.com./is/dig-lib/
10. H. Gollan, F. Lützenkirchen, D. Nastoll, "MILESS - a learning and teaching server for multi-media documents", Preprint.
11. MILESS - Multimedialer Lehr- und Lernserver Essen, http://miless.uni-essen.de

12. T. A. Phelps, "Multivalent Documents: Anytime, Anywhere, Any Type, Every Way User-Improvable Digital Documents and Systems", dissertation, UC Berkeley, 1998.
13. T. A. Phelps, R. Wilensky, "Multivalent Documents: Inducing Structure and Behaviors in Online Digital Documents", in *Proceedings of the 29th Hawaii International Conference on System Sciences* Maui, Hawaii, January 3-6, 1996.
14. J. Rosenboom "A prototype mathematical text recognition system", Lecture at the international workshop on "Retrodigitalization of mathematical journals and automated formula recognition", Institute for Experimental Mathematics, Essen University, 10 - 12 December 1997.

Gigabit Networking in Norway
Infrastructure, Applications and Projects

Thomas Plagemann

UniK - Center for Technology at Kjeller
University of Oslo
http://www.unik.no/~plageman

Abstract. Norway is a country with large geographical dimensions and a very low number of inhabitants. This combination makes advanced telecommunication services and distributed multimedia applications, like telemedicine and distance education, very important for the Norwegian society. Obviously, an appropriate networking infrastructure is necessary to enable such services and applications. In order to cover all important locations in Norway, this network represents a very large Wide-Area Network (WAN) within a single nation. This paper describes the Norwegian academic networking infrastructure, and gives an overview of Norwegian research institutions, programs, and projects. Furthermore, we describe in two case studies one examplary multimedia application and one ongoing research project in the area of gigabit networking and multimedia middleware.

1   Introduction

Traditionally, Norway is a very advanced country in the area of networking and data communications. For example, in 1975 Kjeller (which is situated 30 kilometers north-east of Oslo) and London were the only european nodes in the former internet, called ARPANET. Today, Norway is one of the leading countries in the world with respect to access and usage of the internet. According to the Norwegian Gallup institute that is specialized in interview based market analysis, 46% of all Norwegians have access to the internet, 33% are regularly using the internet, and 24% of all Norwegian households are connected to the internet [8]. There are two further facts about Norway that make a study of advanced networking in Norway quite interesting:

- Norway has approximately 4.3 million inhabitants. Consequently, there are only a few universities and research institutions.
- Norway stretches over 2000 kilometers from south to north. This geographical dimension combined with the low number of inhabitants makes advanced telecommunication services and distributed multimedia application, like telemedicine and distance education, very important for the Norwegian society. Obviously, an appropriate networking infrastructure is necessary to enable such services and applications. In order to cover all important locations in Norway, this network represents a very large Wide-Area Network (WAN) within a single nation.

This paper has two main goals: (1) to give a general overview of Norwegian research activities in the area of gigabit networking and to provide appropriate references; and (2) to give a more detailed description of two typical examples for research projects and the usage of the Norwegian networking infrastructure. In this survey, we consider also distributed multimedia systems and applications. These systems typically operate only with several Mbit/s bandwidth per user, but the potentially large number of concurrent users imposes considerable requirements onto gigabit networks.

The first part of this paper describes the Norwegian academic research network infrastructure, the connected research institutions, research programs, and relevant projects. In the second part, we present two case studies. The first case study describes the electronic classroom system that is used for teaching regular university courses. The second case study presents the ongoing MULTE (Multimedia Middleware for Low-Latency High-Throughput Environments) project.

2 National Networking Infrastructure

Since 1987, the organization UNINETT has the responsibility for the academic networking infrastructure in Norway. This includes [17]:

- to develop and maintain the national data network for research and higher education,
- to propagate the usage of open standards, and
- to stimulate research and development that is important in the context of UNINETT's activities.

It is a strategic goal for Norway, to keep up with research and development on new network services, like *Internet 2* and *Next Generation Internet*, in the USA; and to actively participate in the 5th Framework of the European Union. In this context, multimedia and real-time services, like IP telephony, digital libraries, distance education, and virtual reality, play an important role and require a considerable amount of bandwidth. Table 1 summarizes the bandwidth requirements UNINETT estimates for the periode 1998 - 2003 for the Norwegian backbone network, regional networks, access networks, and internal educational networks [18].

In order to meet the current and future requirements, UNINETT, the Norwegian Research Council, and Telenor officially opened in September 1998 the *National Research Network*. This network comprises two parts: the *research network* and the *test network*. The research network is a stable network to be used for productive services and for new (multimedia) applications. All Norwegian Universities, four Engineering Schools (Mo i Rana, Stavanger, Grimstad, and Halden) and research institutions at Kjeller are connected by the research network. Additionally, Lillehammer will be connected during 1999.

Table 1. Estimated bandwidth requirements in Mbit/s [18]

	1998	1999	2000	2003
Backbone network	40-150	100-300	300-600	2000-4000
Regional networks	0,25-30	10-60	20-150	150-600
Access networks	0,1-10	0,5-20	2-150	150-300
Internal educational networks	2-10	10-40	10-100	80-300

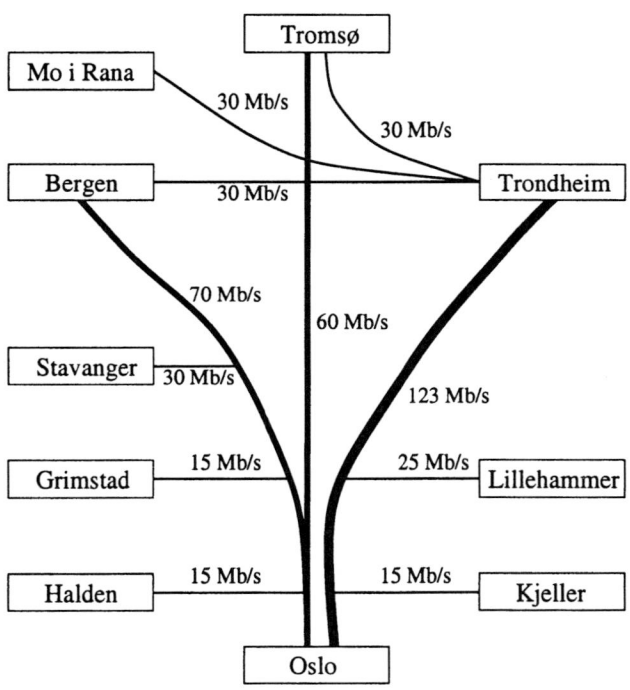

Fig. 1. Topology of the Norwegian test network (according to [19])

Figure 1 specifies the links and the available amount of bandwidth between these institutions.

In contrast, the test network enables academic research institutions to experiment with new network protocols, e.g., IPv6 and RSVP, and applications in a Wide-Area Network (WAN) without interfering with the productive services in the research network. Only the leading (academic) research institutions in the area of networking and distributed systems, i.e., the four Universities, UniK - Center for Technology at Kjeller and Telenor Research and Development have access to the test network.[1] In particular, the test network offers an infrastructure to [18]:

- realize the national IPv6 infrastructure,
- experiment with protocol mechanisms to support QoS on top of ATM, IPv6, or other internet based services,
- introduce reliable multicast, and
- perform experimental research with new protocols and services.

Both networks - the research network and the test network - are based on the commercial 155Mbit/s ATM/SDH WAN from Telenor, called *Nordicom*. In order to manage and control the access to the research network, UNINETT connects each node[2] in the National Research Network with an ATM switch (Cisco LightStream A1010) to Nordicom. Based on Virtual Paths (VPs) in Nordicom, these ATM switches establish Virtual Circuits (VCs), by using the Private Network-to-Network Interface (PNNI) signalling protocol. Thus, the Cisco switch supports the User-Network Interface (UNI) signalling protocol. Additionally, the ATM switches from UNINETT are connected to IP routers that route IPv4 packets in the research network (and IPv6 packets in the test network) over the VCs towards their destination. Research institutions with local ATM networks can choose whether they want to use IP services or directly ATM services. Figure 2 illustrates the basic architecture at a node.

Fig. 2. Node architecture

3 Overview of Research Activities

3.1 Research Institutions

The main academic research institutions in the area of gigabit networking and related areas are:

---

[1] Commercial institutions can apply for access to the test network on a per-project basis.

[2] A node corresponds to Gigabit Point of Presence (GigaPOP) in the Internet2 terminology.

- At the University of Bergen, the Department of Information Science does research in the area of information systems and the Department of Informatics in the areas of algorithms, bioinformatics, code theory, numerical analysis, program development, and optimization.

- The Department of Informatics at the University of Oslo is actively working in the areas: computer science, microelectronics, mathematical modeling, systems development, and image processing. Relevant activities include: Swipp (Switched Interconnection of Parallel Processors), SCI (Scalable Coherent Interface), Multimedia Communication Laboratory (MMCL), and ENNCE (Enhanced Next Generation Networked Computing Environment).

- At the Norwegian University of Science and Technology in Trondheim (NTNU), the Department of Telematics is working in the areas of distributed systems, traffic analysis, and reliability. The Department of Computer Science and Information Science is doing research in the areas of artificial intelligence, image processing, human-computer interfaces and systems development, information management, algorithms, and database systems. Important projects include PaP (plug-and-play) project and WIRAC (Wideband Radio Access).

- UniK is a foundation at which faculty members are either affiliated with the University of Oslo or the NTNU. Areas of research interest at UniK are: distributed multimedia systems, telecommunications, opto-electronics, and mathematical modeling. Relevant activities include: OMODIS (Object-Oriented Modeling and Database Support in Distributed Systems), INSTANCE (Intermediate Storage Node Concept), ENNCE/MULTE (Multimedia Middleware for Low-Latency High-Throughput Environments).

- The Department of Computer Science at the University of Tromsø is focussing its activities on distributed operating systems and open distributed systems. Relevant activities include: TACOMA, MacroScope, Vortex; ENNCE/MULTE.

- The Norwegian Computing Center performs applied research in the fields of information technology. Selected activities include: LAVA (Delivery of video over ATM), IMiS (Infrastructure for Multimedia Services in Seamless Networks), and ENNCE.

- SINTEF Telecom and Informatics performs research in the areas of computer science, telecommunications, electronics, and accoustics. Relevant activities include IMiS and OMODIS.

- NORUT IT is working in the areas of earth observation, information and communication technology. Selected activities include NorTelemed (telemedicine applications) and LAVA.

- The Norwegian Defence Research Establishment (FFI) is a state operated, civilian research establishment reporting directly to the Norwegian Ministry of Defence. FFI is an interdisciplinary establishment representing most of the engineering fields, as well as biology, medicine, political science, and economics. Relevant and not classified research activities include ENNCE/MULTE.

The leading commercial research intitution is Telenor Research and Development (R&D), a part of the former national PTT. Main areas of interest include service development and network solutions. Interesting projects include: Project I (next generation IP protocols and applications) and DOVRE (Distributed Object-oriented Virtual Reality Environment). Further companies that perform research in the scope of this paper are Ericsson, Alcatel, and Thomson-CSF Norcom.

3.2 Research Programs and Projects

National research projects are mainly founded by the Norwegian Research Council (NFR). NFR is organized in areas, like the area of Natural Sciences and Technology (NT) which in turn are organized in research programs and activities. Ongoing research programs of interest in NT include:

- The Distributed IT Systems (DITS) lasts from 1996 to 2000. This program has a budget of 70 Million Norwegian Kronws (MNOK). The Program supports basic research within the following three main areas: (1) construction and usage of distributed IT systems, (2) methods for construction and maintenance of systems and applications for distributed information handling, and (3) basic software and hardware technology for distributed IT systems.
- Main goal of the Basic Telecommunication Research Program (GT) is to support strategic and basic telecom research at universities and research institutes in the following four areas: (1) mobile systems, (2) broadband systems, (3) transport networks and end-systems, and (4) telecommunication systems for people with special needs. The program has a budget of 78 MNOK for the period 1997 - 2001.
- Main goal of the Super Computing program is to provide access to national super computer resources to scientific research projects. These resources include: one Cray J90 and one Cray T3E at NTNU, Silicon Graphics Origin2000 at the University of Bergen, and an IBM SP2 at the University of Oslo. NFR contributes 110 MNOK to the budget in the program periode 1999 - 2003.

Furthermore, universities and industry are financially supporting research projects. Following, we briefly describe some research projects in three prominent areas: cluster technology, middleware, and multimedia applications. A list of these and further research projects, and references to online documentation can be found in the Appendix.

Cluster technology: The SCI (Scalable Coherent Interface) research group at the University of Oslo studies how cluster software and hardware can be created, analyzed, efficiently utilized, and maintained. Especially, I/O and network access within SCI and ATM, and performance studies of SCI are of interest in this context [13]. At the University of Tromsø, two projects are concerned with multicomputer systems, clusters, and distributed operating systems. The primary goal of the MacroScope project is to design and build a multicomputer via a distributed operating system based on distributed shared memory. The experimental hardware development platform consists of eight Hewlett-Packard NetServers, each equipped with four Intel Pentium Pro processors running at 166 MHz. Each NetServer has 128 MB memory and two peer PCI buses. The NetServers are connected via a Myrinet interconnect from Myricom. The Myrinet has a peek bandwidth of 2x1.28 Gb/s, and hosts one megabyte of SRAM on the network interface. The Vortex operating system is currently running on uniprocessors, 2, 4 and 8-way Pentium II/Pentium Pro based multiprocessors. The current implementation includes support for multithreaded processes, virtual memory, network communication over UDP/IP (100 Mbit ethernet), gigabit network communication (Myrinet), a RAM based file system, basic synchronization (mutexes, semaphores, etc.), APIC symmetric I/O mode, a Plan9 like namespace for resources, and other features.

Middleware: The TACOMA project (Tromsø And COrnell Moving Agents) focuses on operating system support for agents and how agents can be used to solve problems traditionally addressed by other distributed computing paradigms, e.g., the client/server model [10]. The plug-and-play project is financed by GT and will specify and explore aspects of a self-configuring and self-adapting architecture with plug-and-play functionality for transport and teleservice components. The goal is to develop the technology and to demonstrate the ideas central to this plug-and-play concept. The objective of OMODIS is to create basic research results within the domain of modeling for distributed multimedia systems with emphasis on object-oriented modeling and Quality of Service (QoS) modeling, based on a distributed persistent object architecture [9]. OMODIS is financed by DITS in the periode 1996 - 2001. The ENNCE/MULTE project is described in more detail in Section 5.

Applications: The NorTelemed project finishes in 1999. New services for telemedicine, like remote diagnostics and remote consultation, are developed, tested, and evaluated in this project. The focus of the main LAVA project (1995 - 1996) was the delivery of video and audio over ATM, including video compression technology, transport protocols for ATM, and multimedia databases. Results of LAVA include a MPEG System Stream player application and a server for delivery of streams [5]. In 1998, a LAVA extension was started, called LAVA Education, which focuses on the use of interactive multimedia systems for educational purposes. DOVRE is a project at Telenor R&D. DOVRE is a software platform for developing networked real-time 3D applications. The primary goal of DOVRE is to provide a platform for work, education, entertainment and co-operation in distributed digital 3D worlds [3]. The Electronic Classroom is described as case study in the following Section.

4 Case Study 1: The Electronic Classroom

The first case study we present in this paper discusses an example of an advanced application that is used on top of the research network to provide a reliable service.

In the MUNIN project [4] and the MultiTeam project [1], the Center for Information Technology Services at the University of Oslo (USIT), Telenor R&D, and the Center for Technology at Kjeller (UniK) have developed the so-called electronic classroom for distance education. Since 1993, the two electronic classrooms at the University of Oslo are used for teaching regular courses to overcome separation in space by exchanging digital audio, video, and whiteboard information. Currently, four electronic classrooms are established in Norway: two at the University of Oslo, one at the University of Bergen, and one at the Engineering School of Hedmark. Since 1997, the electronic classroom system is commercially available from New Learning AS.

The following sections give an overview of the application and describe the system architecture. A detailed analysis of QoS aspects in this system can be found in [15].

4.1 Application

The main goal of the distributed electronic classroom is to make the teaching situation in a distributed classroom as similar as possible to an ordinary classroom. Thus, the number of seats for students is limited to maximal 20 in each classroom. During a lecture, at least two electronic classrooms are connected. Teacher and students can freely interact with each other, this is not dependent on whether they are in the same or in different classrooms. This interactivity is achieved through the three main parts of each electronic classroom: electronic whiteboard, audio system, and video system. All participants can see each other, can talk to each other, and may use the shared whiteboard to write, draw, and present prepared material from each site. The electronic whiteboard, audio, and video system in turn consist of several components. Figure 3 shows the students in the classroom with the teacher and Figure 4 shows the two whiteboards (one with the picture of the teacher) in the remote classroom.

In addition to the ordinary classroom structure that is visible on these pictures, i.e., student and teacher area, a technical back room is located behind the classroom. Figure 5 illustrates the basic layout of an electronic classroom.

Fig. 3. Electronic classroom with teacher

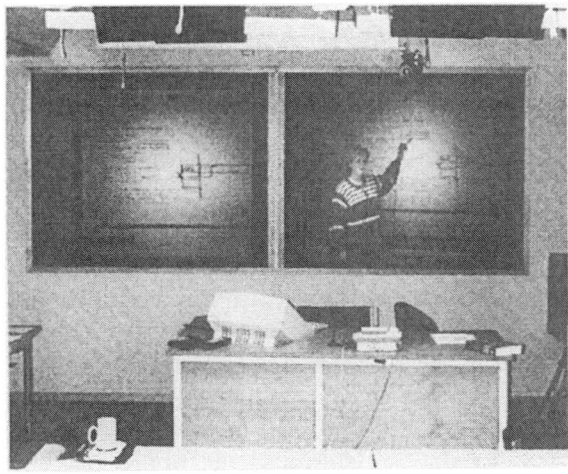

Fig. 4. Remote electronic classroom with students only

The electronic whiteboard is a synonym for a collection of software and hardware elements to display and edit lecture notes and transparencies that are written in Hypertext Markup Language (HTML). The whiteboard itself is a 100" semi-transparent shield that is used together with a video canon and a mirror as a second monitor of an HP 725/50 workstation in the back room. A light-pen is the input device for the whiteboard. A distributed application has been developed that can be characterized as World-Wide Web (WWW) browser with editing and scanning features. When a WWW page is displayed, lecturer and students in all connected classrooms can concurrently write, draw, and erase comments on it by using the light-pen. Thus, floor control

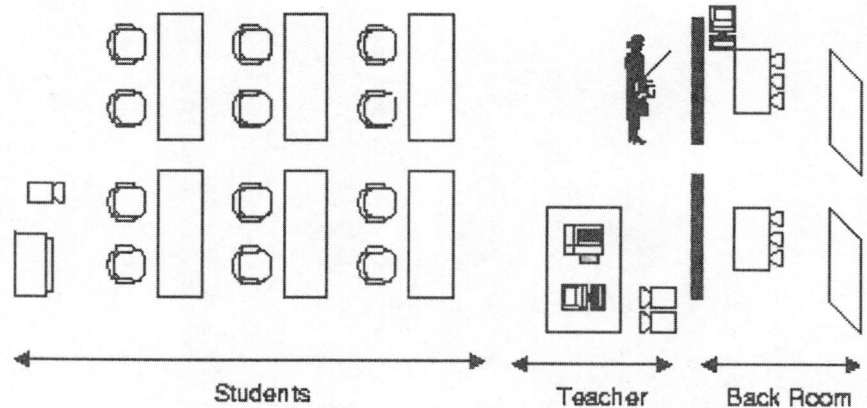

Fig. 5. Basic layout of an electronic classroom

is achieved through the social protocol - as in an ordinary classroom - and is not enforced by the system. Furthermore, a scanner can be used to scan and display on the fly new material, like a page from a book, on the shared whiteboard. The entire application can be managed from a workstation in the classroom.

The video system comprises three cameras, a video switch, a set of monitors, and a H.261 coding/decoding device (codec) to generate a compressed digital video stream. One camera is automatically following and focusing on the lecturer. The other two cameras capture all events happening in the two slightly overlapping parts of the student area in the classroom. The audio system detects the location in the classroom with the loudest audio source, i.e., a student or the teacher that is talking. A video switch selects one of the three cameras that captures this location and person in order to produce the outgoing video signal. Two control monitors are placed in the back of each classroom. The upper monitor displays the incoming video stream, i.e., pictures from the remote classroom, and the lower monitor displays the outgoing video stream, i.e., video information from the local classroom. Thus, the teacher can see the students in the remote classroom and can control the outgoing video information while facing the local students. The students in turn can see the remote classroom on a second large screen which is also assembled out of a whiteboard, a video canon that is connected to the output of the H.261 codec, and a mirror in the back room.

The audio system includes a set of microphones that are mounted on the ceiling. The microphones are evenly distributed in order to capture the voice of all the participants and to identify the location of the loudest audio signal in the classroom. Furthermore, the teacher is equipped with a wireless microphone. To generate a digital audio stream, two codecs are available: the audio codec from the workstation and the audio codec in the H.261 codec. Thus, one of the three coding schemes can be selected: 8 bit 8 Khz PCM coding (64 Kbit/s), 8 bit 16 Khz PCM coding (128 Kbit/s), and 16 bit 16 Khz linear coding (256 Kbit/s). Speakers are mounted at the ceiling to reproduce the audio stream from the remote site.

4.2 Platform

The aim of the electronic classroom system is to be an open system. Therefore, standardized internet protocols have been used as far as possible (see Figure 6). There are four streams, which are using IPv4 as network protocol: management, audio, video, and whiteboard stream. The management part of the classroom, e.g., setting up a session, is performed in a point-to-point manner and utilizes the reliable TCP protocol. The data exchange (audio, video, and whiteboard stream) during a lecture requires multicast capable protocols, because more than two classrooms can be interconnected. Therefore, UDP is used on top of IP multicast. The audio and video streams have stringent timing requirements, and audio and video packets are time-stamped with the RTP protocol. For both streams, software modules are used in the application to adapt the streams from the codecs to the RTP protocol, i.e., fill a certain number of samples or parts of video frames into a protocol data unit (PDU). In contrast to audio and video, which tolerate errors, the whiteboard application cannot tolerate errors and is therefore placed on top of a proprietary multicast error control protocol (based on retransmissions) on top of UDP.

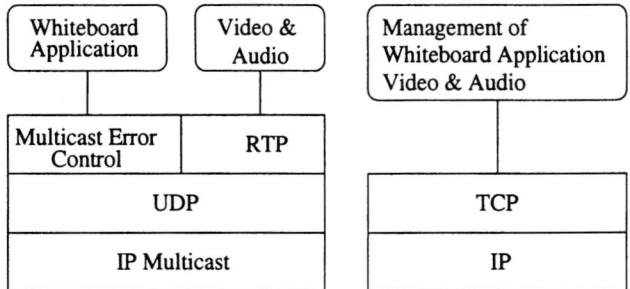

Fig. 6. Protocol stacks used in the electronic classroom

The network topology between the electronic classrooms is basically defined by the research network. The classrooms are connected either via a local ATM switch (ForeRunner 200) or via dedicated ethernet, routers and a FDDI ring to the research network. Addressing and routing of traffic on the

IP layer is mainly performed from the workstations in the back rooms and the routers that are directly attached to the research network. As a backup solution, six ISDN lines can be used to interconnect two classrooms.

5 Case Study 2: The ENNCE/MULTE Project

In this section, we discuss the ENNCE/MULTE project, because it establishes a Metropolitan-Area Network (MAN) with Gigabit/s capacity and utilizes the particular features of the test network.

5.1 Overview

The ENNCE/MULTE project is a collaboration project between the University of Tromsø, University of Oslo, FFI, Thomson-CSF Norcom, and UniK. The project is funded by the Norwegian Research Council under the Basic Telecommunication Research Program in the periode 1997 - 2001.

The need for QoS, real-time behavior, and high performance in distributed multimedia applications like the electronic classroom, or command-and-control systems is the starting point for the ENNCE/MULTE project. On a first glance, it seems that the necessary technology to build an appropriate system platform for such applications is already commercially available: ATM networks that offer high bandwidth and (guaranteed) QoS to higher layer protocols, and implementations of the distributed object computing middleware standard Common Object Request Broker Architecture (CORBA) from the Object Management Group (OMG). Object Request Brokers (ORBs) represent the heart of CORBA and enable the invocation of methods of remote objects, despite their location, underlying network and transport protocols, and end-system heterogeneity [20]. However, nearly all CORBA implementations (e.g., IONA's Orbix and Visigenic's VisiBroker) are based on the communication protocols TCP/IP. It is well known that TCP/IP is not able to support the wide range of multimedia requirements, even if it runs over high-speed networks like ATM. Furthermore, CORBA itself is not well-suited for performance sensitive multimedia and real-time applications, because it lacks streams support, standard QoS policies and mechanisms, real-time features, and performance optimizations [16].

The main hypothesis of the ENNCE/MULTE project is that satisfying the broad range of requirements of current and future distributed multimedia applications requires flexible and adaptable middleware that can be dynamically tailored to specific application needs. A further hypothesis is that a flexible protocol system is an appropriate basis for the construction of flexible and adaptable middleware. Based on these hypotheses, the MULTE project breaks down into the following areas of concern:

- Analysis of application requirements, based on multimedia applications that are developed from the project partners: video journaling at the

University of Tromsø, command-and-control systems on naval vehicles at FFI, and the electronic classroom at UNIK [15].

- Low latency high throughput transmission is based on the Gigabit service offered from the Gigabit ATM Network Kit from the Washington University in St. Louis, an SCI network, and Gigabit Ethernet.
- Stream binding and enhanced interoperable multicast for heterogeneous environments requires appropriate abstractions at the upper API and configuration and management of filters at intermediate and end-systems [7].
- Flexible connection management that comprises mechanisms to adapt connection set-up and release mechanisms to QoS requirements and to make them independent of the particular protocol functionality.
- Flexible protocol systems that perform the communication tasks of the ORB-core.

In the following subsections, we describe the architecture of the first prototype of a flexible multimedia ORB and the MULTE Gigabit network over which the ORB will be used.

5.2 Flexible Multimedia ORB

A flexible protocol system allows dynamic selection, configuration and reconfiguration of protocol modules to dynamically shape the functionality of a protocol to satisfy specific application requirements and/or adapt to changing service properties of the underlying network. The basic idea of flexible end-to-end protocols is that they are configured to include only the necessary functionality required to satsify the application for the particular connection. This might even include filter modules to resolve incompatibilities among stream flow endpoints and/or to scale stream flows due to different network technologies in intermediate networks. The goal of a particular configuration of protocol modules is to support the required QoS for requested connections. This will include point-to-point, point-to-multipoint, and multipoint-to-multipoint connections. As a starting point, we use the Da CaPo (Dynamic Configuration of Protocols) system [14] to build a flexible multimedia ORB.

Overview of Da CaPo: Da CaPo splits communication systems into three layers denoted A, C, and T. End-systems communicate via the transport infrastructure (layer T), representing the available communication infrastructure with end-to-end connectivity (i.e., T services are generic). In layer C the end-to-end communication support adds functionality to T services such that at the AC-interface, services are provided to run distributed applications (layer A). Layer C is decomposed into protocol functions instead of sublayers. Each protocol function encapsulates a typical protocol task like error detection, acknowledgment, flow control, encryption/decryption, etc. Data dependencies between protocol functions are specified in a protocol graph. T layer modules and A layer modules terminate the module graph of a module configuration. T modules realize access points to T services and A modules realize access points to layer C services. Both module types "consume" or "produce" packets. For example, in a distributed video application a frame grabber and compression board produces video data. Applications specify their requirements within a service request and Da CaPo configures in real-time layer C protocols that are optimally adapted to application requirements, network services, and available resources. This includes determining appropriate protocol configurations and QoS at runtime, ensuring through peer negotiations that communicating peers use the same protocol for a layer C connection, initiates connection establishment and release, and handles errors which cannot be treated inside single modules. Furthermore, Da CaPo coordinates the reconfiguration of a protocol if the application requirements are no longer fulfilled. The main focus of the Da CaPo prototype is on the relationship of functionality and QoS of end-to-end protocols as well as the corresponding resource utilization. Applications specify in a service request their QoS requirements in form of an objective function. On the basis of this specification, the most appropriate modules from a functional and resource utilization point of view are selected. Furthermore, it is ensured that sufficient resources are available to support the requested QoS without decreasing the QoS of already established connections (i.e., admission control within layer C).

Integration of Da CaPo in COOL: At UniK, we develop a new multi-threaded version of Da CaPo on top of the real-time micro-kernel operating system Chorus, that takes full advantage of the real-time support of Chorus [11]. Furthermore, we integrate Da CaPo into the CORBA implementation COOL such that the COOL-ORB is able to negotiate QoS and utilizes optimized protocol configurations instead of TCP/IP [2]. Figure 7 illustrates the architecture of the extended COOL-ORB on top of Chorus.

The COOL communication subsystem is split in two parts to separate the message protocol, i.e., the Inter-ORB Protocol (IIOP) and the proprietary COOL Protocol, from the underlying transport protocols, i.e., TCP/IP and Chorus Inter-Process Communication (IPC). A generic message protocol provides a common interface upwards, thus generated IDL stubs and skeletons are protocol independent. A generic transport protocol provides a common interface for the different transport implementations.

There are to alternatives to integrate Da CaPo in this architecture: (1) Da CaPo represents simply another transport protocol. This alternative is our first prototype implementation for Da CaPo in COOL, accompanied with an extended version of IIOP called QoS-IIOP, or QIOP. QIOP encapsulates QoS information from application level IDL interfaces and conveys this information down to the transport layer and performs at the peer system the reverse operation. Da CaPo uses this information for configuration of protocols.

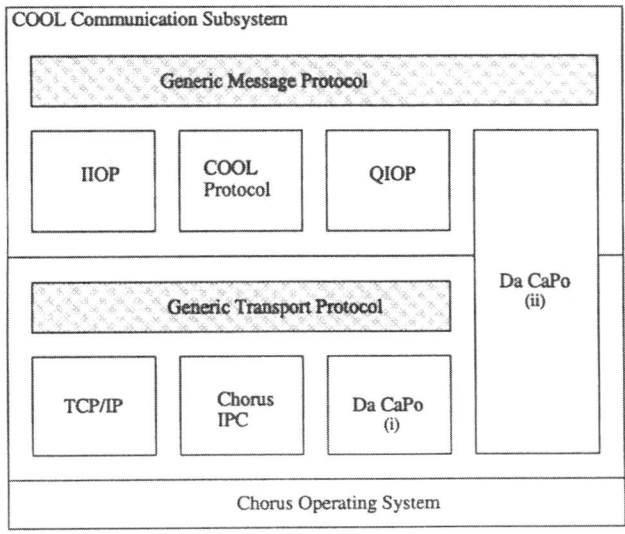

Fig. 7. Integration of Da CaPo into COOL and Chorus

The next step is to implement the second alternative, where Da CaPo additionally configures a message protocol. The message protocols are then Da CaPo modules formatting requests for marshaling and demarshaling in stubs and skeletons.

5.3 MULTE Gigabit Network

At FFI and UNIK, we currently establish a metropolitan area network that combines traditional network technologies, like 100 Mbit/s Ethernet and 155 Mbit/s ATM, with the following Gigabit network technologies [12]:

- The Gigabit ATM Network Kits are based on technology that has been developed at the Washington University in St. Louis [6]. The ATM switches support several different link speeds up to 2.4 Gb/s. The ATM Network Interface Cards (NICs) operate at up to 1.2 Gb/s.
- SCI is standardized by ANSI-IEEE (Std 1596-1992). SCI provides distributed shared memory to a cluster of nodes, e.g., workstations, memory, disks, high speed network interfaces etc. Hardware-supported shared memory can be used in various applications, ranging from closely synchronized parallel programming to LAN support. The aggregated bandwidth of the SCI ring used at FFI is 1.2 Gbit/s.
- Gigabit Ethernet supports data transfer rates of 1 Gbit/s and is standardized in the IEEE 802.3z standard.

Figure 8 illustrates the network topology and infrastructure. Two Gigabit ATM switches, one at FFI and one at UNIK, that are connected with a 1.2

Gigabit/s link build the core of the network. At FFI, five PCs are connected with PCI cards to this switch and additionally to an SCI ring. At UNIK, we connect SunUltra workstations and PCs to the Gigabit switch, to Gigabit Ethernet, and to the local 155 Mbit/s ATM network from ForeSystems. The available access to different types of gigabit networks in one end-system enables us to directly compare these technologies. Especially, we will experimentally evaluate with the flexible multimedia ORB the possibility to select an appropriate network service on the fly. The gigabit network is directly connected to the test network via the Fore ATM switch. This enables the University of Tromsø to access the gigabit network in Kjeller, and we are currently using the flexibility of Da CaPo and the possibilities of the test network to study the influence of various protocol configurations combined with different network reservations onto the QoS of streamed video transfer between Kjeller and Tromsø (distance of approx. 1500 km).

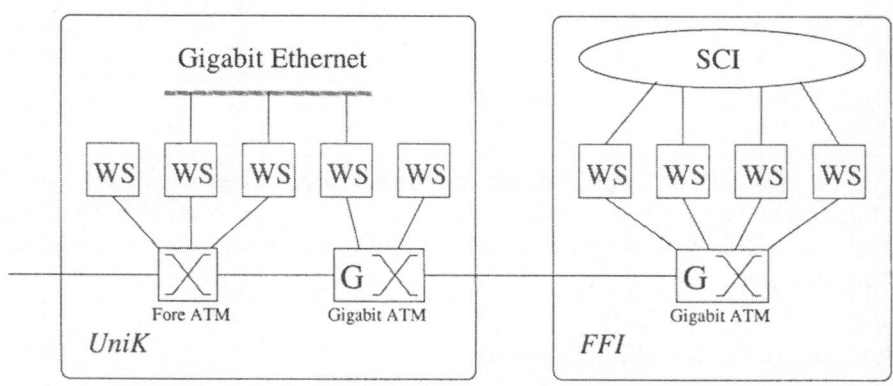

Fig. 8. Gigabit network at Kjeller

6 Concluding Remarks

The aim of this paper is twofold, on the one hand we intend to give an overview of gigabit networking and related areas in Norway. Thus, the first part desribes the Norwegian academic networking infrastructure, research institutions, programs, and projects. On the other hand, we want to provide a more detailed description of two exemplary activities, the electronic classroom and the ENNCE/MULTE project. In this context, the definition of *relevant*, *interesting*, and *important* activities is always based to a certain amount on subjective measures, even if it is intended to present an objective selection of activities and projects. Therefore, we apologize if we have missed important activities.

Acknowledgements: I wish to thank Tom Kristensen for a lot of help in preparing the paper. Furthermore, I would like to acknowledge Petter Kongshaug and Olav Kvittem from UNINETT for providing details about the research and test network.

References

1. Bakke, J. W., Hestnes, B., Martinsen, H. (1994) Distance Education in the Electronic Classroom. Technical Report Telenor Research and Development, TF R 20/94, Kjeller (Norway)
2. Blindheim, R. (1999) Extending the Object Request Broker COOL with flexible Quality of Service Support", Master Thesis at the University of Oslo, Department of Informatics, February 1999
3. Bottar, E. (1997) Telepresence through Distributed Augmented Reality. Scientific Report R&D 44/97 Telenor
4. Bringsrud, K., and Pedersen, G. (1993) Distributed Electronic Classrooms with Large Electronic White Boards. Proceedings of 4th Joint European Networking Conference (JENC 4), Trondheim (Norway), May 1993, 132–144
5. Bryhni, H., Lovett, H., Maartmann-Moe, E., Solvoll, T., Sørensen, T. (1996) On-demand regional television over the Internet. Proceedings of the ACM Multimedia 96 Conference, Boston, November 1996, 99–108
6. Chaney, T., Fingerhut, A., Flucke, M., Turner, J. (1997) Design of a Gigabit ATM Switch. Proceedings of IEEE Infocom, April 1997
7. Eliassen, F., Mehus, S. (1998) Type Checking Stream Flow Endpoints. Proceedings of Middleware'98, Chapman & Hall, Sept. 1998
8. Gallup (1998) Intertrack December 1998 (in Norwegian). available at: http://www.gallup.no/menu/internett/default.htm
9. Goebel, V., Plagemann, T., Berre, A.-J., Nygård, M. (1996) OMODIS - Object-Oriented Modeling and Database Support for Distributed Systems. Proceedings of Norsk Informatikk Konferanse (NIK'96), Alta (Norway), November 1996, 7–18
10. Johansen, D., Schneider, F. B., van Renesse, R. (1998) What TACOMA Taught Us. To appear, Mobility, Mobile Agents and Process Migration - An edited Collection, Milojicic D., Douglis, F., Wheeler, R. (Eds.), Addison Wesley Publishing Company
11. Kristensen, T. (1999) Extending the Object Request Broker COOL with flexible Quality of Service Support (in Norwegian), Master Thesis at the University of Oslo, Department of Informatics, in progress
12. Macdonald, R. (1998) End-to-end Quality of Service Architecture for the TDF. ENNCE/WP2 Technical report TR02/98F
13. Omang, K. (1998) Performance of a Cluster of PCI Based UltraSparc Workstations Interconnected with SCI. Proceedings of Network-Based Parallel Computing, Communication, Architecture, and Applications, CANPC'98, Las Vegas, Nevada, Jan/Feb 1998, Lecture Notes in Computer Science, No.1362, 232–246.
14. Plagemann, T. (1994) A Framework for Dynamic Protocol Configuration", PhD Thesis, Swiss Federal Institute of Technology Zurich (Diss. ETH No. 10830), Zurich, Switzerland, September 1994
15. Plagemann, T., Goebel V. (1999) Analysis of Quality-of-Service in a Wide-Area Interactive Distance Learning System. To appear in Telecommunication Systems Journal, Balzer Science Publishers

16. Schmidt, D. C., Gokhale, A. S., Harrison, T. H., Parulkar, G. (1997) A High-Performance End System Architecture for Real-Time CORBA, IEEE Communications Magazine, Vol. 35, No. 2, February 1997, 72–77
17. UNINETT (1993) Research Network and Internet2 (in Norwegian), UNINyTT nr. 1 1993, electronically available at: http://www.uninett.no/UNINyTT-1-93.html
18. UNINETT (1998) Research Network and Internet2 (in Norwegian), UNINyTT nr. 1/2 1998, electronically available at: http://www.uninett.no/UNINyTT/1-98
19. UNINETT (1998) Research Network Estabished! (in Norwegian), UNINyTT nr. 3 1998, electronically available at: http://www.uninett.no/UNINyTT/3-98
20. Vinkoski, S. (1997) CORBA: Integrating Divers Applications Within Distributed Heterogeneous Environments, IEEE Communications Magazine, Vol. 35, No. 2, February 1997, 46–55

A Appendix

Table 2. Institutions

University of Bergen, Department of Information Science	http://www.ifi.uib.no/index.html
University of Bergen, Department of Informatics	http://www.ii.uib.no/index_e.shtml
University of Oslo, Department of Informatics	http://www.ifi.uio.no/
UniK - Center for Technology at Kjeller	http://www.unik.no
University of Tromsø, Department of Computer Science	http://www.cs.uit.no/EN/
NTNU, Department of Telematics	http://www.item.ntnu.no/index-e.html
NTNU, Department of Computer Science and Information Science	http://www.idi.ntnu.no/
Norwegian Defence Research Establishment	http:www.ffi.no
Norwegian Computing Center	http://www.nr.no/ekstern/engelsk/www.generelt.html
SINTEF Telecom and Informatics	http://www.informatics.sintef.no/
NORUT IT	http://www.norut.no/itek/
Telenor Research and Development	http://www.fou.telenor.no/english/
Ericsson	http://www.ericsson.no
Alcatel	http://www.alcatel.no/telecom/
Thomson CSF Norcom	http://www.thomson-csf.no/
UNINETT	http://www.uninett.no/index.en.html

Table 3. Programs, projects, and activities

DITS	http://www.ifi.uio.no/ dits/translate/ index.html
GT	http://www.sol.no/forskningsradet/program/profil/gtele/
Supercomputing program	http://www.sol.no/forskningsradet/program/tungregn/index.htm
Telecom 2005 - Mobile communication	http://www.item.ntnu.no/~tc2005/
ADAPT-FT	http://www.ifi.uio.no/~adapt/
CAGIS	http://www.idi.ntnu.no/~cagis/
ClustRa	http://www.clustra.com/
DOVRE	http://televr.fou.telenor.no/html/dovre.html
ENNCE	http://www.unik.no/~paal/ennce.html
ENNCE/MULTE	http://www.unik.no/~plageman/multe.html
GDD	http://www.gdd.cs.uit.no/gdd/
GOODS	http://www.ifi.uio.no/~goods/
INSTANCE	http://www.unik.no/~plageman/instance.html
IMiS	http://www.informatics.sintef.no/www/prosj/imis/imis.htm
LAVA	http://www.nr.no/lava/
MacroScope	http://www.cs.uit.no/forskning/DOS/MacroScope
Mice-nsc	http://sauce.uio.no/mice-nsc/
MMCL	http://www.ifi.uio.no/~mmcl/
Multimedia Databases	http://www.idi.ntnu.no/grupper/DB-grp/projects/multimedia.html
Network Management	http://www.item.ntnu.no/~netman/
Plug and Play (PaP)	http://www.item.ntnu.no/~plugandplay/
project I	http://pi.nta.no/indexe.html
SCI activities at the University of Oslo	http://www.ifi.uio.no/~sci
Swipp (Switched Interconnection of Parallel Processors)	http://www.ifi.uio.no/~swipp/
TACOMA	http://www.tacoma.cs.uit.no/
Vortex	http://www.vortex.cs.uit.no/vortex.html
Wirac	http://www.tele.ntnu.no/wirac/

Low speed ATM over ADSL and the Need for High Speed Networks
A case study in Göttingen

Gerhard J. A. Schneider

Gesellschaft für wissenschaftliche Datenverarbeitung Göttingen, Am Fassberg, D-37077 Göttingen, Germany

Abstract. The use of modern technology from non-standard sources allows IT centres to find temporary solutions for operational needs, for example to provide access quickly to a networking infrastructure for the local scientific community. This paper describes the experiences of GWDG with ADSL equipment. Although originally a consumer technology it can also be used in a scientific environment. Despite of having some clear limitations, ADSL technology can be used quite well if other means are not readily available.

In addition various networking issues that arise in a scientific environment are discussed, using the situation in Göttingen as an example.

1 Introduction

GWDG, the *Gesellschaft für wissenschaftliche Datenverarbeitung Göttingen*, is the joint IT centre of the University of Göttingen and the Max Planck Society. Five major research institutes of the Society are situated in the Göttingen area. While four of them are within the city boundaries, the fifth is some 30 kms away. In order to provide adequate access to the network infrastructure also for this institute, a dark fibre link has been installed in cooperation with the local water supply company. It turned out that this solution was cheaper than a 3 year lease of a high speed link from any of the telecommunication carriers.

Apart from doing research in applied computer science, the major tasks of GWDG are to provide strategic services for its customers, as well as the operation of local midrange parallel computers and of the high speed data backbone GöNET in the Göttingen area.

2 The WAN infrastructure

The Internet connectivity for the German science community is provided by the *Deutsches Forschungsnetz* DFN ([1]). It operates B-WiN, a nation-wide ATM backbone (which is physically part of the network of *Deutsche Telekom*) with access points of 34 Mbit/s and 155 Mbits/s. This network will migrate to Gigabit speed in the year 2000 and it will then offer access speeds of up

258

to 622 Mbit/s and later even more. The main B-WiN nodes are currently housed on the premises of the *Deutsche Telekom*. Customers have either own access points (pricing is independent of the location and varies from EUR 400000 p.a. for 34 Mbit/s to EUR 600000 p.a. for 155 Mbit/s) or connect to the nearest access point via leased or private lines. While the second option allows the sharing of an access point, it may add the overhead of the cost for leased lines. Although prices for leased lines have started to fall since the liberalisation of the telecommunication market in early 1998, they still provide a problem for remote sites.

The ATM backbone is primarily used to transport IP traffic between member sites. Thus PVCs exist between the various routers in the main nodes. For detailed information about the current infrastructure of the network, including an up-to-date map, see [2]. The map in Fig. 1 reflects the situation in April 1999.

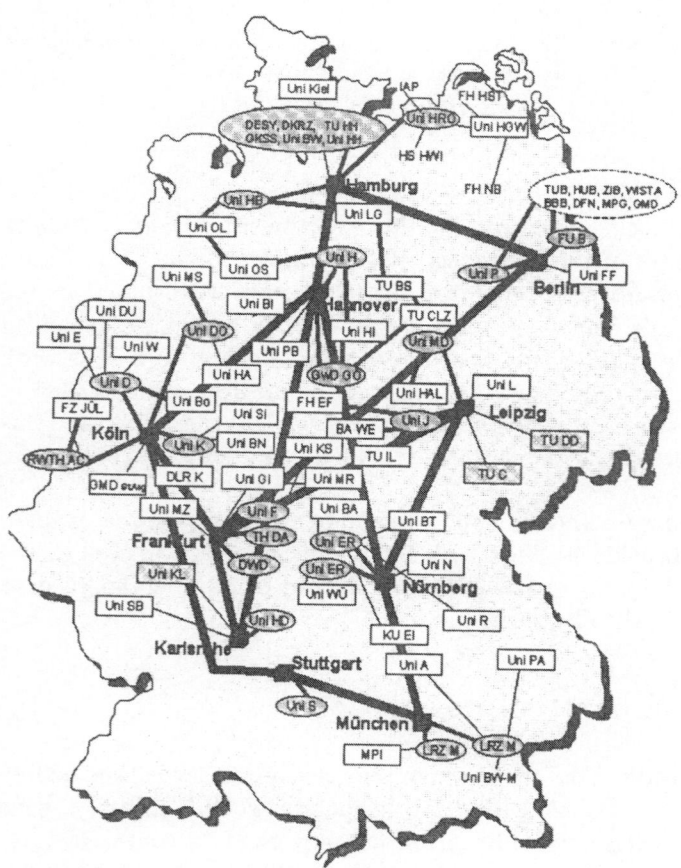

Fig. 1. High-speed network B-WiN of the *Deutsches Forschungsnetz*

In addition to plain IP traffic, it is also possible to order PVCs and quite recently SVCs between individual sites. Prices for such connections are very moderate and therefore are not prohibitive. The reasons for these quality of service connections can be special demands from research groups – like priority access to supercomputers – or video conferences. It should be added that the majority of such individual PVCs also carry only IP traffic, but in a guaranteed environment. Thus many aspects of the current discussion on quality of service connections over the internet have successfully been solved within the German science network by using appropriate transport technologies. While *ATM to the desktop* may still be discussed and perhaps never arrive, ATM is certainly well suited as a backbone technology, especially with respect to quality of service.

The B-WiN also offers connectivity to the US networks, currently at 155 Mbit/s, with another upgrade to 310 Mbit/s due in July 1999, via its Hannover node.

The US connectivity highlights various political problems many European network providers to the scientific communities are facing: although there is a significant flow of data from Europe to the US, commercial US providers reject the idea of cofunding transatlantic lines. In addition the diverse provider infrastructure in the United States basically forces Europeans to build their own distribution network infrastructure in the US to allow for adequate connectivity with different leading IP subnetworks, in order to achieve decent throughput rates to US universities and other research partners. So in essence, while US sites are benefitting from European data sources, the European networks have to pay for this.

DFN's B-WiN is also part of the European ATM network TEN-155 which provides interconnectivity between the different European science networks. Rather than relying on an obscure peering via CIXes, the ATM network allows for bilateral agreements between the various institutions. It seems that this model is superior to the US model with its shortcomings described above, at least from the point of view of international access.

Exchange with the commercial Internet in Germany is ensured via a 34 Mbit/s link [1] to the DE-CIX in Frankfurt. Although load on this link is heavy, many commercial providers currently seem unwilling or unable to allow an increase of the link speed, as their networks may be unable to cope with the flow of data requested by their users from the servers in the science community. As a result, some commercial Internet providers ordered direct links to the German Science Network DFN.

3   The Landeswissenschaftsnetz Nord

In the past centuries it was customary to found universities off the main political centres, to keep the influence of riotous students away from politics.

---

[1] this link will be upgraded to 68 Mbit/s in May 1999

In fact, Göttingen is a perfect example, since it was located at the far southern end of the Kingdom of Hannover in 1737. Similarly, in the mid seventies in this century many newly founded universities and polytechnics were placed in remote areas, mainly to provide some local infrastructure in otherwise poor areas.

Smaller sites which do not have the bandwidth requirements or the financial power to justify a B-WiN node now face the additional cost for leased lines to access the nearest B-WiN site and the internet. Thus the two German states of Lower Saxony (Niedersachsen) and Bremen joined forces to improve the situation for such remote sites by providing a statewide infrastructure for teleteaching and video conferences as well as telemedicine. It still seems unclear whether teleseminars will be the choice of the future for university education. In any case it is necessary to train students to use these tools, which no doubt will become part of their working life later on. In addition the export of knowledge from universities to companies may become an additional challenge. Teleteaching methods could be used to enable direct transfer of ideas and modern developments to industry in order to provide a competitive advantage. Similar arguments hold for the medical sector. The dense population in Germany may not require the needs for teleoperations, but developing and providing the necessary tools and methods may eventually be important for the export oriented medical industry.

The new state network LWN (*LandesWissenschaftsnetz Nord*) became operational in March 1999. It consists of a 155 Mbit/s ATM ring (see fig. 2) connecting the major institutions as well as access lines for smaller remote sites operating with at least 2 Mbit/s. Thus access to Internet technology at rates required by modern developments is now guaranteed for all state institutions in higher education. Since more and more local schools connect to the nearest university or polytechnic (at their own cost) the availability of appropriate connectivity also has a positive effect on secondary education.

The LWN is fully compatible with DFN's Science Network B-WiN and interconnects via the three sites in Bremen, Hannover and Göttingen. Each interconnect operates at 155 Mbit/s. The price structure of the DFN resulted in part of the funding of the State network coming from merging various 34 Mbit/s access points, by benefitting from the economy of scale. Thanks to the compatibility, PVCs via LWN and B-WiN do not present any problems. In particular this ensures that participants of the LWN are not cut off from forthcoming developments but on the contrary can participate much better than before.

The ring infrastructure allows better use of the available resources as there are always two paths between any two institutions. In addition it provides an obvious fault tolerance in accessing the three interconnected sites. In particular, in Göttingen the lines for LWN and B-WiN physically arrive at different locations on campus.

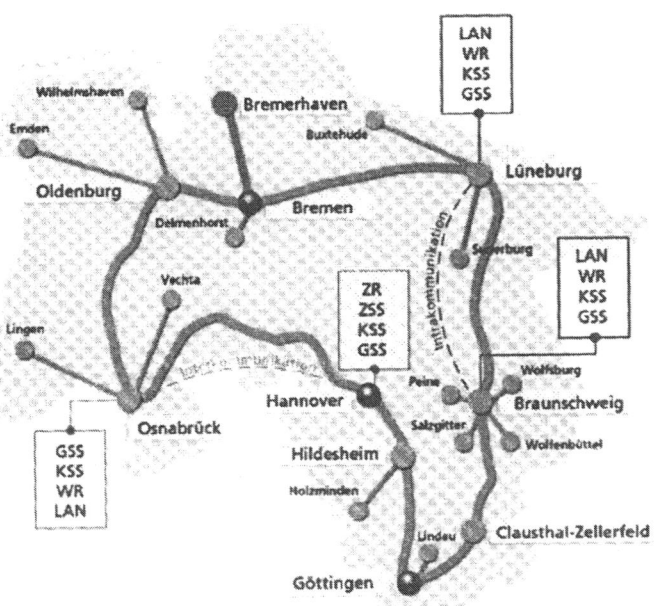

Fig. 2. Landeswissenschaftsnetz Nord

4 Dial-in

The classical situation with respect to providing dial-in support for users required universities to purchase appropriate equipment and to lease access lines from *Deutsche Telekom*. The ongoing liberalisation forces carriers to generate traffic to compensate for declining revenue. As a result *Deutsche Telekom* is now placing routers on university premises and is providing the necessary lines at no cost for the scientific institutions. Thus the dial-in capacity has been boosted significantly in the past months. As the Telekom infrastructure is very modern this means that rather than large modem banks, a few S_{2m} ISDN trunk lines provide the required capacity both for ISDN as well as for analogue connections, including V.90. Since connection charges are identical for ISDN and analog users, more and more users switch to ISDN because of its faster and more reliable performance. Connecting to the university for one hour at 64 kbit/s currently costs less than EUR 1 during off-peak. Since the basic ISDN S_0 subscriber access offers two (virtual) 64 kbit/s lines, connections at 128 kbit/s are also possible, but cost twice as much. Since connections are charged based on time intervals (which may be up to 4 minutes long), demand-driven automatic dialing of the second connection of the S_0 trunk may not be wise in all cases.

Although 64 kbit/s don't seem to be much in these days of high speed networking, especially with respect to the 155 Mbit/s WAN technology, the number of connections may present a challenge. GWDG is currently operating some 10 S_{2m} dial-in trunks which may result in a demand of up to 20 Mbit/s on WAN performance, in addition to the LAN traffic. Fortunately dial-in demand is typically at its peak in the evening and night hours and compensates nicely daytime LAN demand. The dial-in characteristics are best shown by the following diagram – see fig. 3 – which seems typical for a German science institution, but also reflects the pricing structure of the carrier (currently rates for local calls are cheaper after 18:00 and cheapest after 21:00). It also shows that scientists and students tend to work late.

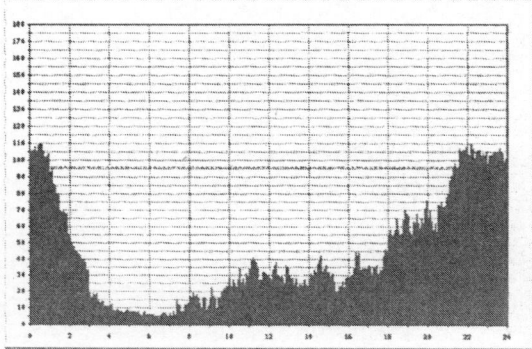

Fig. 3. usage of dial-in lines over 24 hours

5 LAN

While access to WAN technologies can now be aquired at relatively short notice due to the abundance of fibre optic cables with the long distance carriers, bringing the local LAN onto modern technologies is a time and money consuming exercise. Shortage of funds in the public sector mean that many construction plans have to be postponed.

Göttingen is a nice in-town University, with many departments housed in old and picturesque buildings, including C. F. Gauss' original observatory. Although this does boost the academic atmosphere and makes the University very attractive, it turns into a nightmare for networkers. Connecting a building to the backbone not only means installing the cabling locally while conforming to the requirements of conservation laws but also digging across roads and public premises. Only recently has the legal framework been liberalised in this respect.

As a result the backbone infrastructure in Göttingen is up-to-date with a 622 Mbit/s ATM backbone as well as an FDDI ring connecting the various central points of the university. Most science faculties now have either 100 Mbit/s or 10 Mbit/s access to the backbone.

Arts and Social Sciences are typically not placed high on the list of priorities since the need for networking was not obvious for these disciplines when priorities were set 10 years ago. Thus GWDG is now faced with rising demands from a new group of users but with little or no extra funds to meet this specific demand.

However the University owns a large and extensive copper network which was installed together with the PABX in the late 1960s. Basically each office is on this telephone network and there are plenty of spare wires into each building. While the PABX itself is now more of historic interest and up for replacement, the copper network still seems to be in excellent condition. Although classical modem connections provide a first way to access the backbone over these wires, speed is not adequate even for modest requirements from science.

6 ADSL

In early 1998, GWDG teamed up with *Ericsson* to investigate the possibility of providing higher speeds over this copper network. The then newly released *Ericsson* ANxDSL equipment was to be in the centre of the investigation.

Analysis of the equipment showed at a very early stage that it offered some interesting advantages over other solutions in the ADSL sector which made it particularly interesting for deployment in a LAN environment. The most appealing feature is that *Ericsson*'s ANxDSL is delivering native ATM to the customer premises. The network terminating equipment offers two native ATM plugs (ATM 25 Mbit/s as a matter of fact) as well as an Ethernet port to carry IP LAN traffic over ATM. This port is bridge-tunneled according to RFC 1483. Therefore it is very easy to transport at least two different LANs, e.g. a VLAN for administrative purposes as well as the standard LAN infrastructure for science. Thus the typical paranoia of the administrative sector with respect to IP traffic can be overcome at no extra cost. Other institutions in the State were forced to install a separate administrative LAN, and since funds are available only once this meant that essentially scientific needs had to be sacrificed to accommodate administrative demands.

The main ATM hub for the ADSL equipment was placed next to a main ATM switch on the campus network. Thus a seamless integration became possible.

Since copper lines are readily available it became possible to deliver a connection to the GöNET to many sites almost immediately. The time for waiting for a LAN connection was reduced from several years to several days.

Although ADSL is primarily a consumer technology, offering high bandwidth to the customer site and comparatively little bandwidth in the opposite direction, it turns out that it offers a fully functional solution also in an high performance LAN environment. It enables users to experience the possibilities of high speed internet so that skills will have been developed when the proper connection will be installed in the future.

The actual ADSL system consists of two parts:

• At the customer site a network terminating device ANxDSL-NT (fig. 4)is installed. This device has 3 ports: two ports offering native 25.6 Mbit/s ATM access and a twisted pair port for direct Ethernet access.

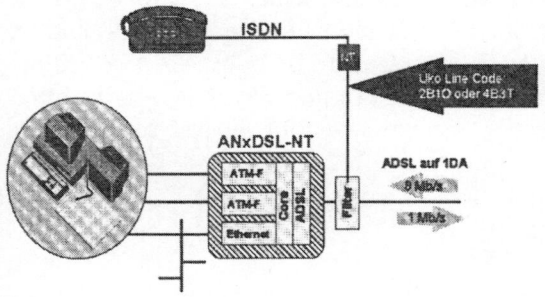

Fig. 4. network termination at customer site

A filter in the NT allows the splitting of plain old telephony service (POTS) and ADSL on the customer premises. A similar splitter for the European ISDN infrastructure will soon be available. Thus POTS and ISDN are not affected by a power failure, while the NT requires electrical power to operate.

• At the central site a line terminating device ANxDSL-LT (fig. 5) is installed. This consists of a shelf holding up to 15 cards (two ports each, at the time of writing 4-port-cards were about to be released), connected via the backplane to a 155 Mbit/s STM-1 interface.

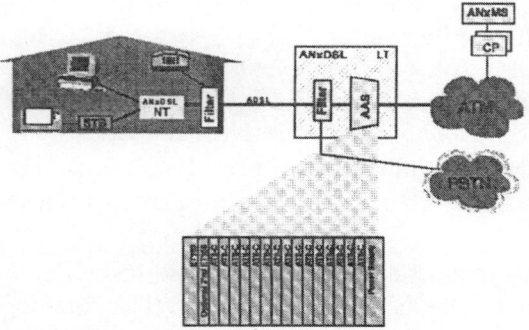

Fig. 5. AnXDSL equipment overview

Ericsson also provides a concentrator allowing the connection of 16 such shelves onto **one** 155 Mbit/s STM-1 interface. In principle, up to 480 ADSL lines can thus be connected to the WAN, depending on traffic and performance requirements.

The original setup at GWDG consisted of one ANxDSL-LT located close to the central PBAX of the university and 30 ANxDSL-NT devices. Now almost 50 lines are in operation. The telephone functionality was not tested.

It is clear, however, that current ADSL technology cannot be seen as a replacement for traditional LAN technology, especially with respect to multimedia applications and high end central services, like a centralized backup for large data.

In particular the core system is not designed to handle massive LAN traffic but rather to support the occasional dial-in from many users at different times with a resulting moderate demand on networking. The coupling of LANs via ADSL however tends to create a continuous demand for bandwidth, especially when student dorms are on the net. In our trial, the 30 lines generated a theoretical bandwidth of 240 Mbit/s and a practical peak demand of almost 100 Mbit/s.

6.1 Experiences

Provided the available copper wire is of reasonable quality, the ADSL technology works amazingly well and is very easy to set up. Under reasonably good conditions, the bandwidth achieved with this technology is indeed up to the promises made in the manuals. Yet ADSL puts a serious demand on a network and it seems that carriers who are thinking of offering the technology to consumers may underestimate the necessary upgrades in the backbone.

Most of the performance problems observed during the trials originated from previously unnoticed defects in the copper wires.

Apart from private copper lines, GWDG also experimented with a 64 kbit/s leased line from *Deutsche Telekom* (in fact a cheap copper wire) to run ADSL, which at the beginning was not (officially) known to *Deutsche Telekom*. The results were just as promising. As a consequence *Deutsche Telekom* is now monitoring the progress in Göttingen and is about to sign a contract with GWDG. This contract will enable GWDG to rent additional copper lines at a very moderate cost (well below the traditional cost of leased lines) to connect remote sites as well as the homes of some University staff members to GöNET. Although *Deutsche Telekom* is about to launch their own ADSL pilots in various German cities, this contract will allow for a special infrastructure in Göttingen, offering more functionality and options because of the scientific interest behind the setup.

To highlight the experiences the following table 1 gives an overview of some of the speeds obtained over the network. Length of cable is certainly one limiting factor, but there are obviously others which we could not discover due to the lack of measuring equipment.

Table 1. Line lengths and link speeds of ADSL:

Location	line length (meter)	downlink speed (kbit/s)	uplink speed (kbit/s)
Kolosseum	1700	7968	640
Studentendorf	1900	7968	640
Neuer botanischer Garten	2750	7712	832
Gerichtsmedizin	5400	2304	640
Studienzentrum Subtropen	2000	7968	800
Studienzentrum *U. of California*	3600	4320	608
Medizinische Physik	2000	7456	640
Botanischer Garten	2900	6688	608
ZENS	2000	7264	736
Volkskunde	3500	6592	544
Völkerkundemuseum	3200	6560	224
Sprachlehrzentrum	2900	5824	704
Anthropologie	5100	3872	640
Ibero-Amerika-Institut	2600	6720	672
Umweltgeschichte	2700	7968	160
Heizkraftwerk	800	4352	448
Akademie der Wissenschaften	3650	5536	544
Restaurierungsstelle	3500	4736	832

7   City of Göttingen interconnected

In another attempt to speed up the building of the GöNET, talks started with the City of Göttingen with the aim of jointly using the available infrastructure. Both ATM and secure end-to-end encryption provide the technology to run LANs with contradicting security requirements over the same cable infrastructure, thus providing the potential for the cost-effective sharing of resources.

The City of Göttingen also owns a fibre optic network (used to control traffic lights and to connect sites like public swimming pools to the central administration in the City Hall) as well as a copper network. The copper network is of interest to offer permanent internet connections to primary and secondary schools, currently at modem speed.

It turned out that at some specific location GöNET and the City network are just 30m apart. As a result, the University, the City and GWDG signed a contract in late 1998 and decided to join forces.

After the networks became physically connected in March 1999, various university buildings can now easily be reached via existing fibres of city network as some of them are close to public sites or traffic lights. The positive experiences at GWDG with ADSL equipment has triggered the decision to connect all local schools in Göttingen via ADSL to a central site in the City Hall and from there via a 2 Mbit/s PVC over ATM directly to the German Science Network. Access to GöNET will be at a higher speed, so that local schools may also gain insight into the paradigm changes caused by high speed networking.

8 Summary

Modern telecommunication systems allow the rapid deployment of currently adequate bandwidth to a large number of sites. Protocols like ATM as well as encryption permit the operation of different LANs over the same infrastructure. In addition, issues concerning quality of services like guaranteed or restricted bandwidth can be solved easily with ATM, both locally and on a nationwide basis.

The sudden decline in prices for WAN connection leads to an inverse networking pyramid: While backbone and WAN are capable of delivering the bandwidth required by modern communication, the local infrastructure both to and in the buildings does not keep up with this development, due to funding issues. ADSL provides a way to quickly connect sites at reasonable speed and to bridge the time gap until fibre is installed.

In fact the Deutsche Forschungsgemeinschaft (DFG) has acknowledged this inverted networking phenomenon and is working on a memorandum to highlight the need for additional resources for local networks.

References

1. Verein zur Förderung eines deutschen Forschungsnetzes DFN, Berlin, http://www.dfn.de
2. B-WiN-Karte, http://www.dfn.de/b-winkarte.html
3. ADSL-Projekt der GWDG, http://www.gwdg.de/adsl

The NRW Metacomputing Initiative *

Uwe Schwiegelshohn[1] and Ramin Yahyapour[1]

Computer Engineering Institute, University Dortmund, 44221 Dortmund, Germany

Abstract. In this paper the Northrhine-Westphalian metacomputing initiative is described. We start by discussing various general aspects of metacomputing and explain the reasons for founding the initiative with the goal to build a metacomputer pilot. The initiative consists of several subprojects that address metacomputing applications and the generation of a suitable infrastructure. The latter includes the components user interface, security, distributed file-systems and the management of a metacomputer. At last, we specifically discuss the aspect of job scheduling in a metacomputer and present an approach that is based on a brokerage and trading concept.

1   The Need for High Performance Computing

High Performance Computing (*HPC*) has become an important tool for research and development in many different areas [2,14]. Originally, supercomputers were mainly used to address problems in physics. Then, the term *Grand Challenges* has been introduced in the eighties to describe a variety of technical problems which require the availability of significant computing power. Today the number of fields, which are not in need of computers, are rapidly decreasing. In addition to the core areas of physics and engineering, high performance computer equipment is essential for e.g. the design of new drugs, an accurate weather forecast [13] or the creation of new movies. Other new applications especially in the field of education are currently under development.

But HPC is not just a necessity for a few companies or institutions. For instance, many companies in various areas of engineering are faced today with the task to constantly design new complex systems and bring them into the production line as soon as possible. *Time to market* has become an important parameter for high tech companies trying to grow in the global market. Therefore, many of those companies use system simulation as a key element for rapid prototyping to reduce development cycles. On the other hand the complexity of many products in the fields of telecommunication and information technology makes the availability of large computing resources indispensable. Hence, the access to high performance computing for a broad range of users may be a key factor to further stimulate innovation [10].

---

* Supported by the NRW Metacomputing grant

In recent years the computer industry constantly increased the computing power of their products. While a top-of-the-line PC was equipped with a 33MHz processor and 4MB DRAM in 1991 [7] a similar computer in 1998 included a 450MHz processor and 128MB memory [8]. Note that this does not consider the additional technical advances in the architecture of the processor. But the demand is growing at an even faster pace. For instance, the number of computers in an average company has increased significantly from 1991 to 1998. The same development can also be observed for HPC. We claim that no matter how much technology advances there will always be a non negligible number of applications which require the fastest computing equipment available.

Unfortunately, the mentioned short development cycles in information technology equipment result in severe drawbacks for a HPC provider. By definition, HPC uses technology at the leading edge. Therefore, HPC components are expensive. In addition, todays HPC equipment will certainly not fall into this category five years from now unless it is frequently upgraded during this time span [19]. This results in high costs and a significant maintenance effort. In order to balance those costs a high degree of utilization is a must for HPC resources. For instance, few commercial owners of supercomputers can afford to see their machines idle during the night or the weekend as it is common place with most PCs.

Small companies may therefore face a dilemma. While the access to HPC resources is necessary for the development of new products they do not have enough applications to efficiently use such equipment. It certainly does not pay off to run secretary programs like word processing on a supercomputer. Those users would need a company or an institution where they can easily get access to HPC resources when needed. In the academic environment the computer centers assume such a role for the various research labs within a university. As it cannot be expected that all potential users are living in close proximity from each other and from HPC resources, powerful networks are an essential component of a suitable HPC infrastructure.

Further, there are significant architectural differences between todays high performance computers [12], like parallel vector computers (e.g. Fujitsu VPP700 [20]), large symmetric multiprocessors (e.g. SGI PowerChallenge [16]), tightly coupled parallel computers with distributed memory (e.g. IBM RS/6000 SP [3]) and large clusters of workstations (e.g. Beowulf, [4,17]). Also, machines from different manufacturers typically require or support different software. On the other hand, some HPC applications need a specific architecture as

- they are not portable for historic reasons,
- they are optimized for this machine or
- they can make best use of the available architectural properties.

It is therefore unlikely that a single supercomputer will be sufficient for all potential users in a region. But if no other HPC resources are locally available some users face the choice to run their application on the local equipment or to ask for an account at another location. The first approach results in a decreased efficiency while the second approach is typically quite cumbersome.

2 Metacomputing Infrastructure

Such a HPC infrastructure may be based upon a single HPC center which provides all required resources, that is all HPC equipment is concentrated and maintained at one location. Access to resources is rented to users for their applications. Therefore, HPC users only pay for their actual usage while they are not forced to care for support and maintenance of the system. Unfortunately, this approach also has a few drawbacks:

- All HPC use needs network bandwidth. Therefore, large investments in a dedicated network structure are necessary.
- The center may either be a potential single point of failure or special care must be taken to prevent situations like the disruption of the whole infrastructure by a single power failure.
- The center is completely decoupled from the applications. This may be a disadvantage for some users like e.g. those designing new applications.

In addition, a single HPC center requires central planning and may show little flexibility.

Alternatively, the concept of a distributed heterogeneous supercomputer can be used. Such an infrastructure is also called a metacomputer. It consists of geographically distributed HPC installations which are linked by an efficient network. The location of a HPC component will depend of the demand of local users. A suitable distribution of HPC resources allows a significant reduction of the network load in comparison to the central approach. Further, HPC resources from different providers may be included into the infrastructure and can compete for customers. This absence of a single institution controlling all HPC resources may be a significant advantage especially for commercial users. In addition, the failure of any single component will not lead to a breakdown of the whole metacomputer.

While metacomputing offers a variety of promising prospects it is not clear whether this concept is actually feasible. To this end several questions must be addressed:

- What are the technological requirements for metacomputing?
- Will this concept find acceptance in the user community including potentially new users from industry?
- Which problems will arise in the management of a metacomputer?
- What will be the performance of a metacomputer in comparison to a large installation of a supercomputer?
- What are the costs for building and maintaining a metacomputer?

2.1 Metacomputing Scenarios

Before finding a method to answer those questions it is necessary to precisely define the use of a metacomputer. In general there are three scenarios with different degrees of user involvement and with different system requirements.

Single Site Application In this scenario each job is executed on a single HPC component in the metacomputer. If any component has not enough resources, like e.g. processors, to execute a job completely, it will also not run parts of that job. Of course, a job may be assigned in parallel to several HPC components for reasons of performance, that is to increase the probability that the job will be completed at a given deadline. But in this case all copies of the job are independent from each other. For single site applications the maximum job size for the metacomputer is determined by the size of the largest component.

The user need not modify any of his applications. It is only necessary to specify the execution requirements of his job like e.g. the amount of memory or the minimal number of processors or the necessary software. Taking the requirements and possibly additional restrictions into account the metacomputer picks its best suited component for the execution of the job (*location transparency*). Even if all HPC resources in the metacomputer are working at full load, the metacomputer can increase overall efficiency by running jobs on the HPC component best suited for them.

Homogeneous Multi Site Applications In addition to single site applications, some jobs may also be executed in parallel on different HPC components of the same type, e.g. several IBM RS/6000 computers are combined to jointly run a large job. In a large metacomputer this scenario significantly expands the number of HPC resources which are potentially available to a single job by a forming a *virtual supercomputer*. As the cost for most types of supercomputers grows superlinearly with the size, this approach may be an interesting option for all cases where such big jobs must only be executed once in a while.

However, multi site applications require the concurrent availability of several HPC components. This includes the network that links the compute components. Therefore, management of such a system becomes more difficult. In addition, the user will not receive the same communication performance as on a single large supercomputer. Hence, she must design her applications accordingly. Further, some problems with large random communication patters may not run as a multi site application or take a huge performance hit. Nevertheless, there are numerous applications that require limit communication overhead and can therefore benefit from a multi site execution. This is especially true when applications are developed with a metacomputing system in mind.

Heterogeneous Multi Site Applications This scenario further expands the homogeneous multi site concept by allowing that potentially all HPC resources of a metacomputer are used for the execution of a single job. However, it is not necessary that all those components are actually running the same executable. It is also possible that a job is automatically piped from one set of HPC components to the next. Nevertheless, this will result in a substantial coordination effort. Further, the *workflow* of the job must be carefully planned taking into consideration various resource constraints like network bandwidth or the size of different machines in the metacomputer. This requires a new programming paradigm and significant additional user effort. On the other hand, the prospect of a gigantic virtual supercomputer may be well worth the work.

2.2 Requirements for a Metacomputing Pilot Project

The best approach to answer the previously posed questions is the establishment of a pilot project including some applications. This will help to determine the actual technical problems and suitable solutions for them. Early user participation will provide a helpful feedback for the system designers. Such a close cooperation between developers and users of a metacomputer is an essential element of a pilot project. Unfortunately, any new HPC installation leads to high initial costs. It is therefore highly advisable to select locations for this pilot study where most if not all of the required HPC equipment is already in place.

Taking all these requirements into consideration the German state of Northrhine Westphalia (NRW) offers an excellent basis for the realization of the project. It hosts the largest concentration of universities and other research institutions in Germany. Most of these institutions already own HPC equipment in their computing centers which operate independently. This equipment includes almost all common HPC platforms like e.g. Sun Enterprise 10000, IBM RS/6000 SP, Cray T3E, SGI Origin and others. The inclusions of many different platforms guarantees the desired degree of flexibility. It also makes the new metacomputer attractive for a wide range of users as almost everyone can find her favorite HPC hardware in the system. In addition, the state has a powerful network infrastructure which links these institutions. It is presently based on an ATM backbone which allows Quality-of-Service features and virtual channels (PVC/SVC). This all together constitutes a suitable system infrastructure for metacomputing.

Note further that the large number of research institutions from many areas also guarantees a diversity of research projects requiring HPC. Finally, several technology centers with small high-tech companies are also located in NRW resulting in a large pool of potential users. Therefore, all requirements for a metacomputing pilot project are met in Northrhine Westphalia. On the other hand, the setup of a metacomputer may provide significant benefits for the economy and research projects in Northrhine Westphalia.

3 The NRW Metacomputing Initiative

Based on these thoughts, the NRW Metacomputing Initiative was proposed by A. Bachem, B. Monien and F. Ramme in 1996 ([1]). It started in July 1996 and is planned to conclude in June 1999. The project is coordinated by B. Monien of University Paderborn. It is jointly funded by the state of Northrhine-Westphalia and the participating research institutions which are named below:

- Paderborn Center for Parallel Computing (PC^2, University of Paderborn)
- University of Cologne
- University of Dortmund
- Technical University (RWTH) Aachen
- Central Institute for Applied Mathematics (ZAM), Forschungszentrum Jülich
- GMD National Research Center for Information Technology, Bonn

Besides generating a working metacomputer the initiative has the goal to find answers to the following specific questions:

- What are the system requirements for HPC components in a metacomputer?
- Does the metacomputer generate a need for a new type of HPC component or for significant modifications of the existing ones?
- Which applications can benefit most from a metacomputer?

The initiative consists of several system and application projects that work on different aspects of metacomputing, see Fig. 1.

3.1 Application Projects

The inclusion of application projects from the beginning had the goal of supporting a constant communication process between users and system designers. These applications can further be used for test and evaluation of the metacomputer pilot. This includes functionality and performance aspects. The first user projects may also give a indication about the characteristic properties of future multi site application regarding

- communication patterns,
- network requirements, and
- software adaptations.

The subjects of the application projects are Molecular Dynamic Simulation, Traffic Simulation, and Weather Forecast. However, in this description we will primarily focus on the system design of the metacomputer and therefore not go into the details of those projects.

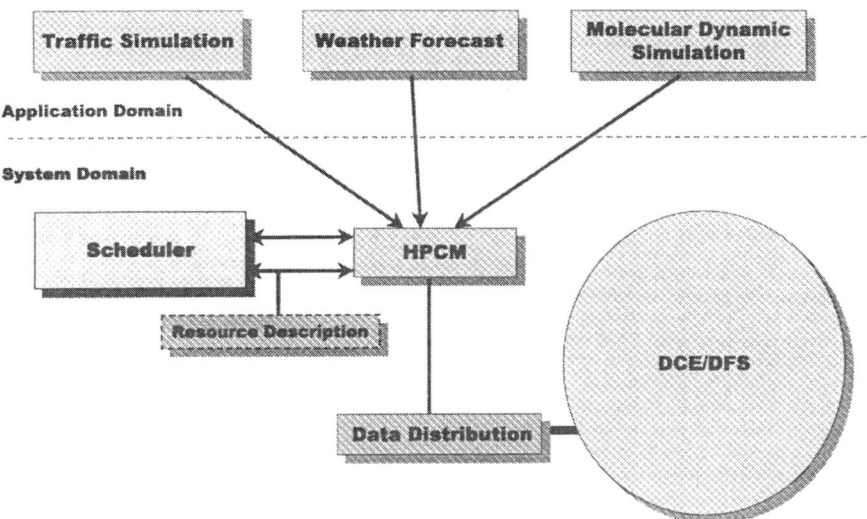

Fig. 1. Projects of the Initiative

3.2 System Projects

As already mentioned the metacomputer uses existing HPC installations. This includes both hardware and system software (operating systems, local management software). In order to combine those resources into a working metacomputer the following problems must be addressed:

- Coordinated management
- Interfaces
- Security

These problems became the subject of several projects in the initiative. In the next sections all those system projects are briefly described while the project *Schedule* is discussed in more detail.

4 Data Distribution and Authentication with DCE/DFS

As metacomputing in this initiative is done over the public Internet, insecure channels are used for communication. Also, computers of different political administration domains are part of the metacomputer. This requires authentication of remote users. Finally, hardware and software of the HPC components must be protected from unauthorized access. Hence, there is need for secure authentication and secure communication. On the other hand, it is important to limit the resulting overhead for users and administrators to achieve a high degree of acceptance and participation.

In the initiative, it was decided to use the standardized Distributed Common Environment (DCE) as an existing and proven software solution. DCE allows secure authentication and communication as well as cross authentication between cells, that is separate administrative domains. Therefore, user login or job startup is possible without the need to supply a password for every machine. Furthermore, the Distributed Filesystem System (DFS) is used to generate a shared file system that provides a dedicated home or project directory on every platform. As DFS uses DCE features for encryption and authentication, system and user files are secured. DCE/DFS has further the advantage of being available for most common platforms.

This system project has the goal to set up DCE/DFS cells for various NRW institutions and to provide cross authentication between them for metacomputing users. Further, mechanisms are developed to enter the authentication cells from outside the DCE/DFS framework. This allows job submission from machines that are not using DCE. The project further includes performance measurements for DFS and the available network infrastructure. The results show a significant speedup in comparison to NFS. Nevertheless, data prefetching is still beneficial for data intensive applications.

5 Metacomputing User Interface

This project deals with the development of a user interface to the metacomputer [21]. To achieve transparency and a high degree of usability the interface should be unique and available for all platforms. Thus, the interface is written in Java and is able to run over the net on all common Java Virtual Machines in e.g. web browsers. Therefore, new versions of the interface are instantly available to all users who download it from the web on startup as a Java applet.

The interface allows the setting of mandatory and voluntary parameters for a job. It provides status information about jobs and available machines. To maintain security for passwords and jobs, the communication is encrypted via third-party software (Cryptix). Signed applets ensure that only the authorized applet from the original site is used.

The Java User interface connects to the HPCM management of the NRW metacomputer. It transmits job requests and authentication information. If the user is not working from a DFS enabled hosts, the applet can upload application data into the DFS cell.

6 Management Architecture HPCM

The HPCM project provides the infrastructure for the metacomputing management. It consists of a management daemon and several coupling modules which communicate platform specific information to the HPCM layer. The management daemon executes on the HPCM server machine and receives

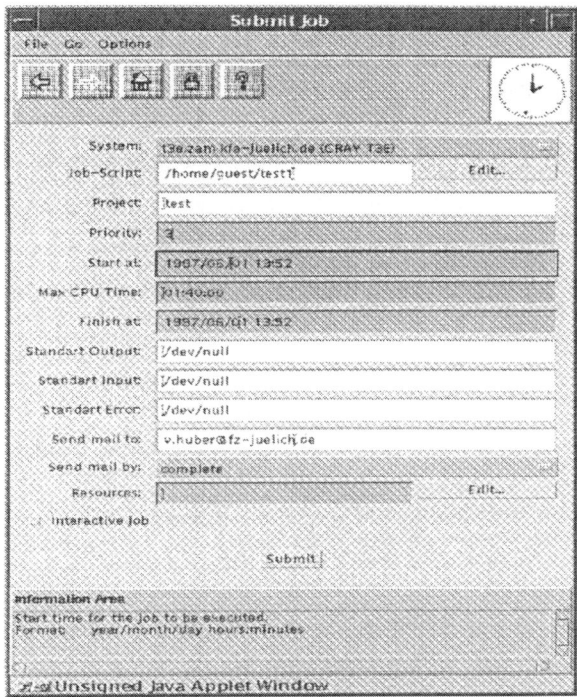

Fig. 2. Screenshot of the Java User Interface for HPCM

requests from the Java user interface. It is the administrative instance that generates the global view of the metacomputer with its components and the participating users. Note that similar multi-tier architectures can be found in other metacomputing projects, as e.g. in Globus [6].

The coupling module is an interface to various computing platforms. Besides abstracting the available information and access methods from the management it interacts with the available local management of the HPC component. The current implementations of coupling modules in the initiative range from NQE [5], CCS [9] to LoadLeveler. Additional modules as e.g. for LSF [22] can easily be derived from the existing implementations.

7 Metacomputing Scheduling

Typically, owners of HPC installations are only willing to include their resources into a metacomputer, if the performance of their components will not degrade in the new environment. This is especially true for commercial owners. Similarly, users expect a better performance for their jobs. Note that the expression *performance* has not been defined as different people may attach a different meaning to it. In most cases however, an owner wants a high system load for her machine while a user is interested in a short response time for his job or at least a *fair* resource allocation.

Therefore, job scheduling and resource allocation are one of the core problems in the metacomputing architecture. As existing system software is used, the metacomputer scheduler must interact with the local schedulers on all HPC components. To avoid the bottleneck of a centralized scheduler and to increase flexibility a distributed approach is employed. Therefore, the paradigms for job schedulers of parallel computers and for metacomputer schedulers differ significantly, see Table 1.

Table 1. Different Scheduling Paradigms

Parallel Computer Scheduling	Metacomputer Scheduling
The network is ignored.	The network is a resource.
Load information is instantly available.	Load information must be collected.
Homogeneous system environment	Heterogeneous system environment
Mostly first-come-first-serve scheduling	Resource reservation for future time frames
Central scheduler	Distributed scheduler

To implement a distributed metacomputing scheduler we use an architecture which is based upon so called *MetaDomains*. All MetaDomains of a metacomputer form a redundant network. Typically, a MetaDomain is associated with local HPC resources. That is all HPC resources at one site are connected to a single MetaDomain. For local HPC access any user can choose to either submit her job directly to the local component or to use the local MetaDomain.

Therefore, the metacomputer does not require exclusive access to a HPC resource. The logical structure of such a scheduler is described in Fig. 3. This network can be dynamically extended or altered. Such a property is advantageous as, for instance, individual HPC components may be temporarily unavailable due to maintenance or new HPC resources may be introduced into the metacomputer. The presented architecture guarantees a high degree of flexibility.

MetaDomains communicate among one another by transmitting or requesting information about resources and jobs. To this end a MetaDomain inquires local schedulers about system load and job status. A MetaDomain can also allocate local HPC resources to requests. The distributed scheduling itself is based upon a brokerage and trading concept which is executed between the MetaDomains. In detail, a MetaDomain tries to

- satisfy local demand if possible,

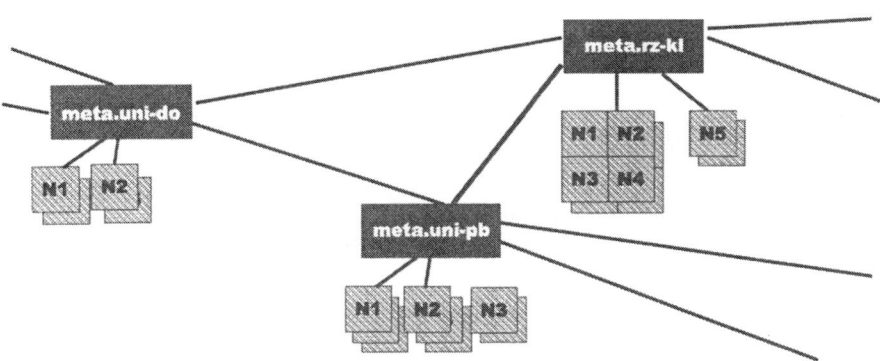

Fig. 3. Logical Infrastructure

- ask other MetaDomains for resources, if the local demand cannot be satisfied,
- offer local HPC resources to other MetaDomains for suitable remote jobs, and
- act as an intermediary for remote requests.

Note that we did not address the actual job submission. This process is not necessarily a task of the scheduler. Once a suitable allocation of HPC resources (including network resources) to a job has been found the actual submission is independent of the scheduler. Also, the scheduling objectives are not specified. As already mentioned there may not be a single scheduling objective in a metacomputer. Each HPC component can define its specific objectives. Similarly, each user may associate specific constraints with his job like a deadline or a cost limit. It is the task of the trading system to find matches between requests and offers. This way not all users and all components are forced to fit into a single framework as it is usually done in conventional scheduling. Now, it is their responsibility to define their own objectives. The implementation of the metacomputing scheduler must only provide the framework for such a definition and it must be able to compare any request with any offer to find a match.

In our metacomputer scheduling concept only the local HPC scheduler is responsible for the load distribution on the corresponding HPC resource. Therefore, it can also accept jobs from sources other than the metacomputer. The metacomputer scheduler only addresses the load imbalance between different HPC resources. To execute multi site applications however, the concurrent availability of different HPC resources and sufficient network bandwidth between them becomes necessary as already described in Sec. 2.1. For reasons of efficiency this requires resource reservation for future time frames and the concept of guaranteed availability. Although most HPC schedulers do not presently support such an approach it can be implemented by using preemption (a *checkpoint-restart* facility) while still maintaining a high system load.

In the project *SCHEDULE* [18] of the initiative a metacomputer scheduler was designed using *CORBA* [15] to allow transparent and language independent access to distributed management instances. For the evaluation of different scheduling methods a simulation framework has further been implemented. It is used to compare different scheduling algorithms regarding their applicability for a metacomputing network. The benefit of possible technology enhancements, like for example preemption, to the quality of the schedule is also determined with the help of the simulator. As already mentioned communication between resources during a multi site job execution must be taken into account as well. To this end the available network must be considered as a limited resource that is managed by the schedulers in the MetaDomains. The inclusion of this objective into the scheduler is part of the future work.

8 Status

The NRW-Metacomputing initiative has developed a functioning management system that has been deployed in a pilot installation to connect the parallel computers of the participating institutions, that are namely a Cray T3E in Jülich, IBM RS6000/SP in Dortmund, Sun Enterprise 10000 in Cologne and a Parsytec CC in Paderborn.

In this test-phase the application projects of the initiative has been used to show the benefits of such a metacomputer. This projects represents typical examples of problems that are well suited for metacomputing. They can easily be ported to different architectures and provide a small network communication footprint. The developed HPCM management software is going into production use in 1999, providing public access to all users.

The Scheduling project provides a working interface to the mentioned system types. Currently only single site applications are supported as the present implementation does not include reservation. However, simulations have been executed to evaluate whether the backfilling strategy [?] can be used to consider reservations. These simulations have yielded promising results. As many commercial local schedulers are already using backfilling, only small changes to these schedulers are required.

9 Conclusion

It is the primary goal of metacomputing to provide users with easy access to more HPC resources. This includes a platform independent simple user interface. The user herself has the ability to exploit the flexibility of the system to her advantage by clearly specifying the resource requirements of her job. In order to benefit from multi site computing she may need to apply new programming paradigms.

The owner of HPC components in a metacomputer must only focus on the maintenance of a single platform. He must not strive to satisfy all local users with a limited budget as some specific demands can be forward to other HPC installations within the network. Also, there is no need to maintain a separate user interface. With an independent user interface the integration of new resources will become easier. On the other hand the owner may face some pressure to increase standardization. Although the approach guarantees a high degree of flexibility he may also lose some control over the allocation of local resources.

The manufacturers of HPC resources may see a decrease in sales of really big machines, while it will become more common to buy midsize systems and integrate them into an infrastructure of existing resources. The overall user community will increase as more users will gain access to these resources. In todays systems manageability and open interfaces to other management systems are not a strong selling argument. This may change if more large systems become part of a heterogeneous metacomputing environment.

References

1. A. Bachem, B. Monien, F. Ramme. Der Forschungsverbund NRW-Metacomputing – verteiltes Höchstleistungsrechnen (1996), http://www.uni-paderborn.de/pc2/nrw-mc/html_rep/html_rep.html
2. High Performance Computing and Communication (1997). NSTC. http://www.hpcc.gov/pubs/blue97
3. IBM RS/6000 SP Product Line. http://www.rs6000.ibm.com/hardware/largescale/
4. D. Becker, T. Sterling, D. Savarese, J. Dorband, U. Ranawak, C. Packer. Beowulf: A parallel workstation for scientific computation (1995), Proceedings, International Conference on Parallel Processing
5. Introducing NQE. (1998), Cray Research Publication, Silion Graphics, Inc.
6. I. Foster, C. Kesselman. Globus: A metacomputing infrastructure toolkit (1997), The International Journal of Supercomputer Applications and High Performance Computing, 11(2), pp 115–128
7. Intel Microprocessors, Volume II (1991) Handbook. Intel Corporation
8. Pentium II Xeon[tm] Processor Technology Brief (1998), Intel Corporation
9. A. Keller, A. Reinefeld. CCS Resource Management in Networked HPC Systems (1998), In Proceedings Heterogenous Computing Workshop (HCW) at IPPS/SPDP'98
10. S. Karin, S. Graham. The High Performance Continuum (Nov 1998), Communications of the ACM, pp. 32 - 35
11. D.A. Lifka. The ANL/IBM SP Scheduling System, Springer LNCS 949, Proceedings of the Job Scheduling Strategies for Parallel Processing Wokshop, IPPS'95, pp. 295 - 303
12. P. Messina, D. Culler, W. Pfeiffer, W. Martin, J. Oden, G. Smith. Architecture (Nov 1998), Communications of the ACM, pp. 36 - 44
13. Robert C. Malone, Richard D. Smith, and John K. Dukowicz. Climate, the Ocean, and Parallel Computing (1993), Los Alamos Science, No.21

14. Grand Challenges, National Challenges, and Multidisciplinary Challenges (1998), NSF Report
    http://www.cise.nsf.gov/general/workshops/nsf_gc.html
15. Object Management Group Document. The Common Object Request Broker: Architecture and Specification (1998), Revision 2.2
16. SGI PowerChallenge XL Product Line
    http://www.sgi.com/remanufactured/challenge, SGI
17. T. Sterling. Applications and Scaling of Beowulf-class Clusters (1998), Workshop on Personal Computers based Networks Of Workstations, IPPS'98
18. U. Schwiegelshohn, R.Yahyapour. Resource Allocation and Scheduling in Metasystems, Springer LNCS 1593, Proceedings of the Distributed Computing and Metacomputing Workshop, HPCN'99, Amsterdam, pp. 851 - 860
19. J. Dongarra, H. Meurer, E. Strohmaier. (Nov. 1998), TOP500 Supercomputing Sites, http://www.top500.org
20. Fujitsu VPP700E,
    http://www.fujitsu.co.jp/index-e.html
21. V. Sander, D. Erwin, V. Huber. High-Performance Computer Management Based on Java (1998), Proceedings of High Performance Computing and Networking Europe (HPCN), Amsterdam, pp. 526 - 534
22. S. Zhou. LSF: load sharing in large-scale heterogeneous distributed systems (1992), In Proceedings Workshop on Cluster Computing

Design and Evaluation of ParaStation2

Thomas M. Warschko and Joachim M. Blum and Walter F. Tichy

Institut für Programmstrukturen und Datenorganisation, Fakultät für Informatik, Am Fasanengarten 5, Universität Karlsruhe, D-76128 Karlsruhe, Germany

Summary. ParaStation is a communications fabric to connect off-the-shelf workstations into a supercomputer. This paper presents ParaStation2, an adaption of the ParaStation system (which was build on top of our own hardware) to the Myrinet hardware. The main focus lies on the design and implementation of ParaStation2's flow control protocol to ensure reliable data transmission at network interface level, which is different to most other projects using Myrinet.

One-way latency is $14.5\mu s$ to $18\mu s$ (depending on the hardware platform) and throughput is 50 MByte/s to 65 MByte/s, which compares well to other approaches. At application level, we were able to achieve a performance of 5.3 GFLOP running a matrix multiplication on 8 DEC Alpha machines (21164A, 500 MHz).

1. Introduction

ParaStation2 is a communication subsystem on top of Myricom's Myrinet hardware [BCF⁺95] to connect off-the-shelf workstations and PCs into a parallel supercomputer. The approach is to combine the benefits of a high-speed MPP network with the excellent price/performance ratio and the standardized programming interfaces (e.g. Unix sockets, PVM, MPI) of conventional workstations. Well-known programming interfaces ensure portability over a wide range of different systems. The integration of a high-speed MPP network opens up the opportunity to eliminate most of the communication overhead.

ParaStation was originally developed for the ParaStation hardware, a self-routing network with autonomous distributed switching, hardware flow-control at link-level combined with a back-pressure mechanism, and a reliable and deadlock-free transmission of variable sized packets (up to 512 byte). This base system is now being adopted to the Myrinet hardware, which has a fully programmable network interface and a much better base performance than the classic ParaStation hardware (see section 2.). The major difference is the absence of reliable data transmission, which has to be implemented at network interface level on the Myrinet hardware (see sections 3. and 4.).

ParaStation offers as programming interfaces well-known and standardized protocols (e.g. TCP and UDP Unix sockets) and programming environments (e.g. MPI and PVM) at a reasonable and acceptable performance level (see section 5.) rather than squeezing the most out of the hardware using a communication layer with nonstandard semantics.

2. ParaStation vs. Myrinet

Table 2.1 presents a brief comparison between the ParaStation [WBT96] and the Myrinet [BCF+95] hardware.

Table 2.1. Comparison between ParaStation and Myrinet

	ParaStation	Myrinet
Technology	PCI-Bus adapter	PCI-Bus adapter & Switches
Topology	2D-Torus	hierarchical crossbar
Bandwidth	128 Mbit/s	1.28 Gbit/s
Flow control	Link level	Link level
Flow control policy	back-blocking	back-blocking & discard
Error detection	Parity	CRC
Error management	Fatal	implementation dependent
Interface	FIFO	SRAM
Processor	none (FPGA)	32bit RISC (LanAI)

ParaStation with it's two incoming and two outgoing links naturally uses a 2D torus as its network topology. The necessary switching elements (between the X and Y dimension) are located on each ParaStation adapter and therefore no central switch is needed. Myrinet instead uses cascadable switching elements (8 or 16 way crossbars) and therefore has no limitations on the network topology built. In terms of transmission speed there is also a clear advantage for Myrinet. Both systems implement flow control at link level but with different policies. Whereas ParaStation implements a strict back-blocking mechanism between all nodes, Myrinet only blocks for a while and then starts discarding packets. This behaviour helps to keep the network alive even in case of faulty components, but also forces the implementation of a higher-level flow control protocol to guarantee reliable transmission. ParaStation simply blocks if there is not enough buffer space in the next node on the way to the final destination and waits until the receiver starts accepting messages. Reference [War98] proves, that this behaviour is deadlock free and reliable, as long as the receiver keeps consuming packets. As a consequence ParaStation does not need any higher-level flow control mechanism.

Another major difference between ParaStation and Myrinet is the programming interface. ParaStation provides a simple FIFO interface to send and receive messages along with some flags describing the status of the incoming and outgoing FIFO. Prior to a send operation the sender checks the flags to ensure that the sender FIFO can accept a complete packet[1]. If there is enough space, it writes the complete packet into the FIFO and ParaStation's flow control mechanism ensures that the packet will eventually make its way to the receiver. On the receiving side the status flags indicate whether a complete packet has arrived in the receiver FIFO. Thus, the receiver is

---

[1] A packet is up to 512 byte long.

able to receive the whole packet at once rather than polling for individual flits. Writing to and reading from the transmission FIFO is done by the CPU (PIO[2]) rather than using the DMA[3] engines. The Myrinet board uses a 32bit RISC CPU called *LanAI*, fast SRAM memory (up to 1 MByte) and three programmable DMA engines – two on the network side to send and receive packets and one as interface to the host. The LanAI is fully programmable (in C / C++) and the necessary development kit (especially a modified gcc compiler) is available from Myricom. The kit opens up the opportunity to implement and test a much broader design space for high speed transmission protocols than with the ParaStation system. In fact this capability in addition to the high performance of the Myrinet hardware was the main criteria to choose Myrinet as the hardware platform for ParaStation2.

3. Design considerations for ParaStation2

The major questions to answer is how to interface the Myrinet hardware to the rest of the ParaStation software, especially the upper layers with their variety of implemented protocols (Ports, Sockets, Active Messages, MPI, PVM, FastRPC, Java Sockets and RMI). There are three different approaches:

1. Emulating ParaStation on the Myrinet adapter: Simulating ParaStation's transmission FIFO with a small LanAI program running on the Myrinet adapter would not be a problem. But as ParaStation is using programmed I/O to receive incoming packets this approach would lead to unacceptable performance (see [BRB98b]).
2. Emulating ParaStation at software level: As the ParaStation system already has a small hardware dependent software layer called HAL (*hardware abstraction layer*), this approach allows the use of all Myrinet specific communication features as well as a simple interface to the upper protocol layers of the ParaStation system.
3. Designing a new system: This approach would lead to an ideal system and probably the best performance, but we would have to rewrite or redesign most parts of the the ParaStation system.

Because of its simplicity, we choose the second strategy to interface the existing ParaStation software to the Myrinet hardware. The second question to answer is how to guarantee reliable transmission of packets with the Myrinet hardware. As said before, the original ParaStation hardware offers reliable and deadlock free packet transmission as long as the receiver keeps accepting packets. Myrinet instead discards packets (after blocking a certain amount of time) which may happen when the receiver is running out of resources or is unable to receive packets fast enough. Additionally the Myrinet hardware

---

[2] Programmed I/O
[3] Direct Memory Access

seems to lose packets under certain circumstances, e.g. in heavy bidirectional traffic with small packets. The upper layers of the ParaStation system rely on a reliable data transmission, so a low level flow control mechanism – either within the Myrinet control program running on the LanAI processor or as part of the HAL interface – is required.

4. Implementation of ParaStation2

The goal of this section is to explain the basics of the ParaStation2 protocol. Most parts of the protocol are implemented in a Myrinet control program (MCP) running on the Myrinet adapter. The protocol guarantees a reliable data transmission so that only minor changes to the HAL have to be made and it is possible to use all upper layers of the ParaStation system without any changes.

4.1 Basic operation

Figure 4.1 shows the basic operation during message transmission of the ParaStation2 protocol. The basic protocol has four independent parts: (a) the interaction between the sending application and the sender network interface (NI), (b) the interaction between the sending and the receiving NI, (c) the interaction between the receiving NI and the receiving host, and (d) the interaction between the receiving application and the host.

First, the sender checks if there is a free send buffer (step 1). This is accomplished by a simple table lookup in the host memory, which reflects the status of the buffers of the send ring located in the fast SRAM of the network interface (Myrinet adapter). If there is buffer space available, the sender copies (step 2) the data to a free slot of the circular send buffer located in the network interface (NI) using programmed I/O. Afterwards the NI is notified (a descriptor is written) that the used slot in the send ring is ready for transmission and the buffer in host memory is marked as in transit. A detailed description of the buffer handling is given in section 4.2. In step (3), the NI sends the data to the network using its DMA engines.

When the NI receives a packet (step 4) it stores the packet in a free slot of the receive ring using its receive DMA engine. The flow control protocol ensures that there is at least one free slot in the receive ring to store the incoming packet. Once the packet is received completely and if there's another free slot in the receive ring, the flow control protocol acknowledges the received packet (step 5). The flow control mechanism is discussed in section 4.3. As soon as the sender receives the ACK (step 6), it releases the slot in the send ring and the host is notified (step 7) to update the status of the send ring.

In the receiving NI the process of reading data from the network is completely decoupled from the transmission of data to the host memory. When

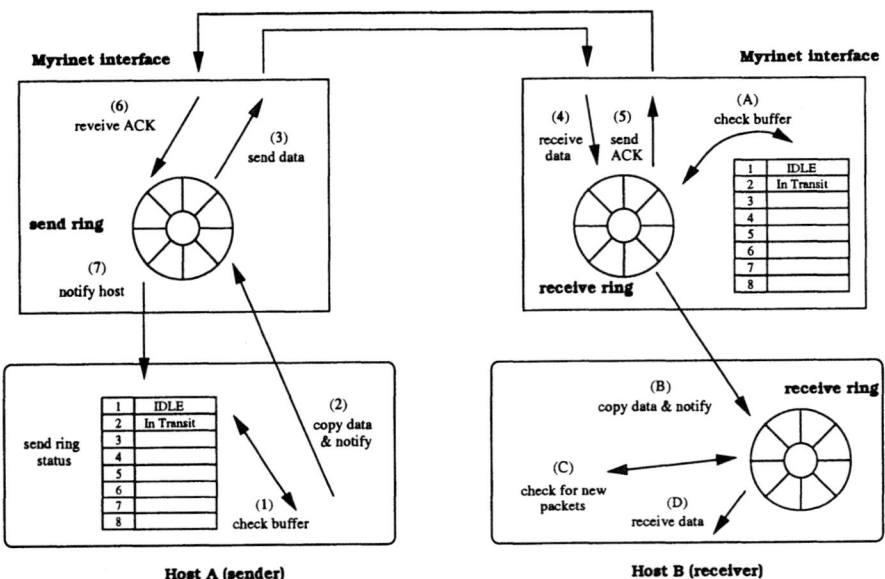

Fig. 4.1. Data transmission in ParaStation2

a complete packet has been received from the network, the NI checks for a free receive buffer in the host memory (step A). If there is no buffer space available, the packet will stay in the NI until a host buffer becomes available. Otherwise the NI copies the data into host memory using DMA and notifies the host about the reception of a new packet by writing a packet descriptor (step B). Concurrently, the application software checks (polls) for new packets (step C) and eventually, after a packet descriptor has been written in step (B), the data is copied to application memory (step D).

Obviously, the data transmission phases in the basic protocol (step 2, 3, 4, and B) can be pipelined between consecutive packets. The ring buffers in the NI (sender and receiver) are used to decouple the NI from the host processor. At the sender, the host is able to copy packets to NI as long as there is buffer space available although the NI itself might be waiting for acknowledgements. The NI uses a transmission window to allow a certain amount of outstanding acknowledgements which must not necessarily equal the size of the send ring. At the receiver the NI receive ring is used to temporarily store packets if the host is not able to process the incoming packets fast enough.

4.2 Buffer handling

Each buffer or slot in one of the send or receive rings can be in one of the following states:

IDLE: The buffer is empty and can be used to store a packet (up to 4096 byte).

INTRANSIT: This buffer is currently involved in a send or receive operation, which has been started but which is still active.

READY: This buffer is ready for further operation either a send to the receiver NI (if it's a send buffer) or a transfer to host memory (if it's a receive buffer).

RETRANSMIT: This buffer is marked for retransmission, because of a negative acknowledgement or a timeout (send buffer only).

Figure 4.2 shows the state transition diagrams for both send and receive buffers in the network interface.

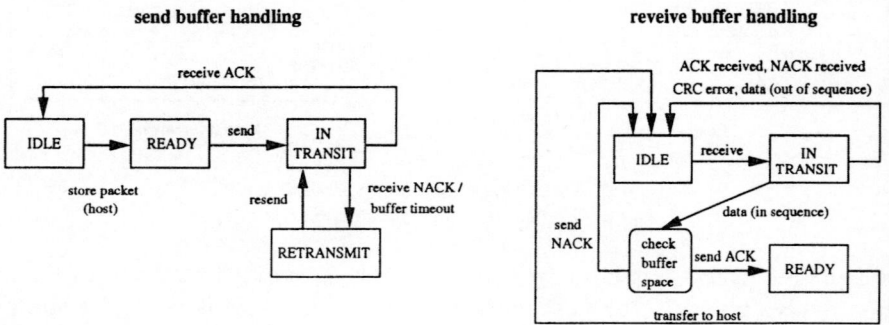

Fig. 4.2. Buffer handling in sender and receiver

At the sender the NI waits until a send buffer becomes **READY**, which is accomplished by the host after it has copied the data and the packet descriptor to the NI (step 2 in figure 4.1). After the buffer becomes **READY** the NI starts a send operation (network DMA) and marks the buffer **INTRANSIT**. When an acknowledgement (ACK) for this buffer arrives (step 6 in figure 4.1), the buffer is released (step 7) and marked **IDLE**. If a negative acknowledgement (NACK) arrives or the ACK does not arrive in time (or gets lost) the buffer is marked for retransmission (**RETRANSMIT**). The next time the NI tries to send a packet it sees the **RETRANSMIT** buffer and resends this buffer, changing the state to **INTRANSIT** again. This **RETRANSMIT** – **INTRANSIT** cycle may happen several times until an ACK arrives and the buffer is marked **IDLE**.

At the receiver the buffer handling is quite similar (see figure 4.2). As soon as the NI sees an incoming packet it starts a receive DMA operation and the state of the associated buffer changes from **IDLE** to **INTRANSIT** (see step 4 in figure 4.1). Assuming that the received packet contains user data, is not corrupted, and has a valid sequence number[4] the NI checks for another

---

[4] For a discussion of the ACK/NACK protocol see section 4.3.

free buffer in the receive ring. If there is another free buffer it sends an ACK back to the sender and the buffer is marked **READY**. Otherwise a NACK is sent, the packet discarded and the buffer released immediately (marked **IDLE**). The check for a second free buffer in the receive ring ensures that there is at least one free buffer to receive incoming packets anytime, because any packet eating up the last buffer will be discarded. When the received packet contains protocol data (ACK or NACK), the NI processes the packet and releases the buffer. In case of an error (CRC) the buffer is marked **IDLE** immediately without further processing. If the received data packet does not have a valid sequence number, the packed is discarded and the sender is notified by sending a NACK back. Thus, the receiver refuses to accept data out of sequence and waits until the sender will resend the missing packet.

4.3 Flow control protocol

ParaStation2 uses a flow control protocol with a fixed sized transmission window and 8 bit sequence numbers (related to individual sender/receiver pairs), where each packet has to be acknowledged (either with a positive or a negative acknowledgement) in combination with a timeout and retransmission mechanism in case that an acknowledgement gets lost or does not arrive within a certain amount of time. The protocols assumes the hardware to be **unreliable** and is able to deal with any number of corrupted or lost packets (containing either user data or protocol information). Table 4.1 gives an overview of possible cases within the protocol, an explanation of each case as well as the resulting action initiated.

Table 4.1. Packet processing within receiver

packet type	sequence check	explanation	resulting action
DATA	<	lost ACK	resend ACK
	=	ok	check buffer space (see fig 4.2)
	>	lost data	ignore & send NACK
ACK	<	duplicate ACK	ignore packet
	=	ok	release buffer
	>	previous ACK lost	ignore packet
NACK	none		mark buffer for retransmission
CRC	none	error detected	ignore packet

When a data packet is received, the NI compares the sequence number of the packet with the assumed sequence number for the sending node. If the numbers are equal, the received packet is the one that is expected and the NI continues with its regular operation. A received sequence number smaller than expected indicates a duplicated data packet caused by a lost or late ACK. Thus the correct action to take is to resend the ACK, because the sender expects one. Is the received sequence number larger than expected, the packet with the correct sequence number has been corrupted (CRC) or lost. As the protocol does not have a selective retransmission mechanism the packet is simply discarded and the sender is notified with a negative acknowledgement (NACK). Thus, this packet will be retransmitted later, either because the sender got the NACK, or because of a timeout. As the missing packet also causes a timeout at the sending side, the packets will eventually arrive in the correct order.

On the reception of an ACK packet, the NI also checks the sequence number and if it is ok it continues processing and releases the acknowledged buffer. If the received sequence number is smaller than assumed, we've received a duplicated ACK because the sender ran into a transmission timeout before the correct ACK was received and the receiver has resent an ACK upon the arrival of an already acknowledged data packet[5]. The response in this case is simply to ignore the ACK. A received sequence number larger than what is expected indicates that the correct ACK has been corrupted or lost. Thus the action taken is to ignore the ACK, but the associated buffer is marked for retransmission to force the receiver to resend the ACK. The buffer associated with the assumed (and missing) ACK will timeout and be resent which also forces the receiver to resend the ACK.

A received NACK packet does not need sequence checking; the associated buffer is marked for retransmission as long as it is in the INTRANSIT state. Otherwise the NACK is ignored (the buffer is in RETANSMIT state anyway). In case of a CRC error the packet is dropped immediately and no further action is initiated, because the protocol is unable to detect errors in the protocol header.

The resulting protocol is able to handle any number of corrupted or lost packets containing either user data or protocol information, as long as the NI and the connection between the incorporated nodes is working. The protocol was developed to ensure reliability of data transmission at NI level, not to handle hardware failures in terms of fault tolerance. The protocol itself can be optimized in some cases (e.g. better handling of ACK's with a larger sequence number), but this is left to future implementations. In comparison to existing protocols, this protocol can roughly be classified as a variation of the TCP protocol.

5. Basic performance of the protocol hierarchy

In table 5.1, performance figures of all software layers in the ParaStation2 system are presented[6]. The various levels presented are the hardware abstraction

---

[5] This case may sound strange, but we've seen this behaviour several times.

[6] For a detailed discussion of the ParaStation2 protocol hierarchy, have a look at our paper called *ParaStation User Level Communication* in this proceedings.

layer (HAL), which is the lowest layer of the hierarchy, the so called *ports* and
TCP layers, which are build on top of the HAL, and standardized communication libraries such as MPI and PVM, which are optimized for ParaStation2
and build on top of the ports layer. Latency is calculated as round-trip/2 for
a 4 byte ping-pong and throughput is measured using a pairwise exchange
for large messages (up to 32K). N/2 denotes the packet size in bytes when
half of the maximum throughput is reached. The performance data is given
for three different host systems, namely a 350MHz Pentium II running Linux
(2.0.35), a 500MHz and a 600MHz Alpha 21164 system running Digital Unix
(4.0D).

Table 5.1. Basic performance parameters of ParaStation2

System	Measurement		HAL	Ports	TCP	MPI	PVM
				Programming interface			
Pentium II	Latency	[μs]	14.5	18.7	20.2	25	
350 MHz	Throughput [MByte/s]		56	48	51	43	
	N/2	[Byte]	256	500	500	400	
Alpha 21164	Latency	[μs]	17.5	24	24	30	29
500 MHz	Throughput [MByte/s]		65	55	57	50	49
	N/2	[Byte]	512	500	500	500	1000
Alpha 21164	Latency	[μs]	18.0	24	25	25	28
600 MHz	Throughput [MByte/s]		64	56	59	51	48
	N/2	[Byte]	350	700	700	500	700

The latency at HAL level of 14.5μs to 18μs is somewhat higher than for
comparable systems such as LFC (11.9μs) or FM (13.2μs) [BRB98a]. This
is because neither LFC nor FM copies the data it receives to the application
and second, both LFC and FM incorrectly assume Myrinet to be reliable.
The maximum throughput with 56 MByte/s to 65 MByte/s of ParaSta-
tion2 is between the throughput of FM (40.5 MByte/s) and LFC (up to
70 MByte/s). If LFC or FM starts copying the received data to the application (as Para/-Station does) the throughput decreases for large messages to
30 – 35 MByte/s [BRB98a] whereas ParaStation2's throughput keeps quite
stable close to maximum level ($\sim$ 50 MByte/s).

Switching from a single-programming environment (HAL) to multi-programming environments (upper layers) results in a slight performance degradation regarding latency as well as throughput. The reason for increasing
latencies is due to locking overhead to ensure correct interaction between
competitive applications. The decreased throughput is caused by additional
buffering and a complex buffer management.

6. Performance at application level

Focusing only on latency and throughput is too narrow for a complete evaluation. It is necessary to show that a low-latency, high-throughput communication subsystem also achieves a reasonable application efficiency. For this reason we installed the widely used and publicly available ScaLAPACK[7] library [CDD+95], which uses both BLACS[8] [DW95] and MPI as communication subsystem on ParaStation2. The benchmark we use is the parallel matrix multiplication for general dense matrixes from the PBLAS library, which is part of ScaLAPACK. Table 6.1 shows the performance in MFLOP's running on our 8 processor DEC-Alpha cluster (500 MHz, 21164A).

Table 6.1. Parallel matrix multiplication on ParaStation2

Problem size (n)	Uniprocessor MFlop (Eff.)	1 Node	2 Nodes	4 Nodes	6 Nodes	8 Nodes
		Performance in MFlop (Efficiency)				
1000	782	731	1276	2304	3243	3871
	(100%)	(93.5%)	(81.6%)	(73.6%)	(69.1%)	(61.9%)
2000	785	743	1359	2546	3582	4683
	(100%)	(94.6%)	(86.6%)	(81.1%)	(76.1%)	(74.6%)
3000	790	755	1396	2700	3908	4887
	(100%)	(95.6%)	(88.4%)	(85.4%)	(82.4%)	(77.3%)
4000	772		1398	2694	4044	**5337**
	(100%)		(90.5%)	(87.2%)	(87.3%)	(86.4%)

First, we measured the uniprocessor performance of a highly optimized matrix multiplication (cache aware assembler code), which acts as a reference to calculate the efficiency of the parallel versions. A uniprocessor performance of 772 to 790 MFLOP on a 500 MHz processor proves that the program is highly optimized (IPC of more than 1.5). Obviously the parallel version executed on an uniprocessor has to be somewhat slower, but the measured efficiency of 93.5% to 95.6% is very high. Using more nodes, the absolute performance in MFLOP increases steadily while the efficiency decreases smoothly. The maximum performance achieved was 5.3 GFLOP using 8 nodes which is quite good compared to the 10.1 GFLOP of the 100 nodes Berkeley NOW cluster[9].

7. Related work

There are several approaches which use Myrinet as a hardware interconnect to build parallel systems: Active Messages and the Berkeley NOW cluster, especially Active Messages-II [CMC97], Illinois Fast Messages (FM) [PLC95], the

---

[7] Scalable Linear Algebra Package.
[8] Basic Linear Algebra Communication Subroutines
[9] see http://now.cs.berkeley.edu

basic interface for parallelism from the University of Lyon (BIP) [PT97], the link-level flow control protocol (LFC) [BRB98a] from the distributed ASCI supercomputer, PM [TOHI98] from the Real World Computing Partnership in Japan, the virtual memory mapped communication VMMC and VMMC-II [DBLP97] from Princeton University, Hamlyn [BJM+96], the user-level network interface U-Net [vEB+95], and Trapeze [YCGL97].

The major difference between these projects and ParaStation2 is twofold. First, ParaStation2 focuses on a variety of standardized programming interfaces, such as UNIX sockets (TCP and UDP), MPI, PVM, and Java sockets and RMI with a reasonable performance at each level rather than a single purpose, nonstandard, proprietary interface which squeezes the most out of the hardware for a specific application.

The second difference due to reliability assumptions of the Myrinet hardware (see figure 7.1).

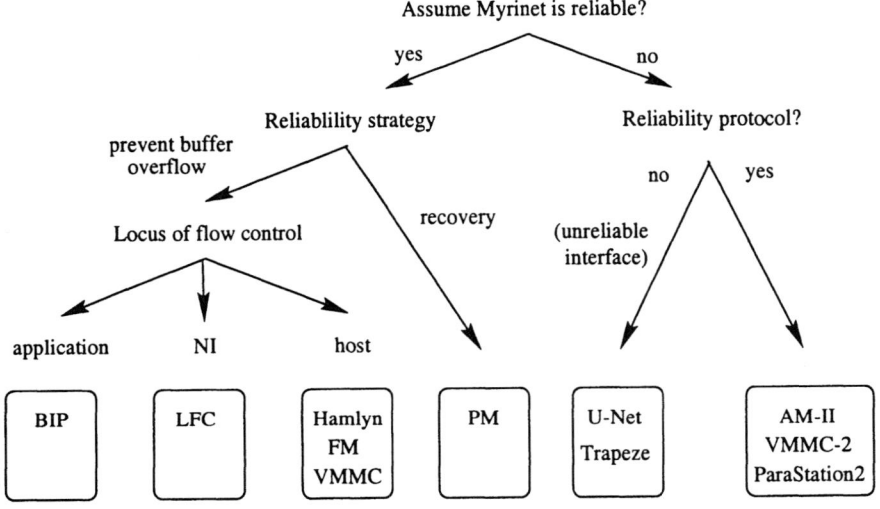

Fig. 7.1. Myrinet and Reliability (from [BRB98b])

Most approaches assume Myrinet to be reliable or pass the unreliability on to the application layer. Only AM-II, VMMC-2 and ParaStation2 accept the unreliability of Myrinet and provide mechanisms to ensure reliable data transmission. The reason why most projects assume Myrinet to be reliable is mainly due to the rather low error rate at hardware level. We have observed that the link-level flow control mechanism seems to fail by overwriting or dropping complete packets under certain circumstances. The only way to detect this behaviour is to count packets or to use sequence numbers within packets, because the hardware neither blocks the transmission nor signals any error. Furthermore, the hardware does not distinguish between data and control packets while dropping one of these. Thus, a simple flow control protocol to prevent buffer overflow assuming that control packets will be delivered reliably is not sufficient to ensure reliable transmission. Although AM-II [CMC97] and VMMC2 [DBLP97] do not explicitly state problems with the Myrinet hardware, they introduced a protocol to ensure reliable communication as they switched from AM to AM-II or VMMC to VMMC2 respectively. The same holds for ParaStation2 which also started using strict back-blocking until serious problems arose.

8. Conclusion and further work

In this paper we've presented the design of ParaStation2, especially the ACK/NACK retransmission protocol to ensure reliable data transmission at network interface level. The advantage of this approach was that we could reuse the ParaStation code with minor changes and getting the complete functionality of the ParaStation system (especially the variety of standardizes and well-known interfaces) for free.

The evaluation of ParaStation2 shows that ParaStation2 compares well with other approaches in the cluster community using Myrinet. ParaStation2 is not the fastest systems in terms of pure latency and throughput, but in contrast to most other approaches it offers the reliable interface which is – in our experience – more important to the user than an ultra high-speed, but unreliable interface. At the level of application performance the 5.3 GFLOPs has not been achieved before with this small number of nodes.

The future plans for ParaStation2 are to optimise the interface between the software and the Myrinet hardware to get even more performance out of the system. Second, ports to other platforms such as Sparc/Solaris and Alpha/Linux are on the way.

References

[BCF+95] Nanette J. Boden, Danny Cohen, Robert E. Felderman, Alan E. Kulawik, Charles L. Seitz, Jarov N. Seizovic, and Wen-King Su. Myrinet: A Gigabit-per-Second Local Area Network. *IEEE Micro*, 15(1):29–36, February 1995.

[BJM+96] G. Buzzard, D. Jacobson, M. MacKey, S. Marovich, and J. Wilkes. An Implementation of the Hamlyn Sender-Managed Interface Architecture. In *The 2nd USENIX Symp. on Operating Systems Design and Implementation*, pages 245–259, Seattle, WA, October 1996.

[BRB98a] Raoul A. F. Bhoedjang, Tim Rühl and Henri E. Bal. *LFC: A Communication Substrate for Myrinet*. Fourth Annual Conference of the Advanced School for Computing and Imaging, June 1998, Lommel, Belgium.

[BRB98b] Raoul A. F. Bhoedjang, Tim Rühl and Henri E. Bal. *User-Level Network Interface Protocols*. IEEE Computer, 31(11), pp. 52 – 60, November 1998.

[CDD+95] J. Choi, J. Demmel, I. Dhillon, J. Dongarra, S. Ostrouchov, A. Petitet, K. Stanley, D. Walker, and R. C. Whaley. ScaLAPACK: A portable linear algebra library for distributed memory computers – design issues and performance. Technical Report UT CS-95-283, LAPACK Working Note #95, University of Tennessee, 1995.

[CMC97] B. Chung, A. Mainwaring, and D. Culler. Virtual Network Transport Protocols for Myrinet. In *Hot Interconnects'97*, Stanford, CA, April 97.

[DBLP97] C. Dubnicki, A. Bilas, K. Li, and J. Philbin. Design and Implementation of Virtual Memory-Mapped Communication on Myrinet. In *11th Int. Parallel Processing Symposium*, pages 388–396, Geneva, Switzerland, April 1997.

[DW95] J. Dongarra and R. C. Whaley. A user's guide to the blacs v1.0. Technical Report UT CS-95-281, LAPACK Working Note #94, University of Tennessee, 1995.

[PLC95] S. Pakin, M. Lauria, and A. Chien. High Performance Messages on Workstations: Illinois Fast Messages (FM) for Myrinet. In *Supercomputing '95*, San Diego, CA, December 1995.

[PT97] L. Prylli and B. Tourancheau. Protocol Design for High Performance Networking: A Myrinet Experience. Technical Report 97-22, LIP-ENS Lyon, July 1997.

[TOHI98] H. Tezuka, F. O'Carrol, A. Hori, and Y. Ishikawa. Pin-down Cache: A Virtual Memory Management Technique for Zero-copy Communication. In *12th Int. Parallel Processing Symposium*, pages 308–314, Orlando, FL March 1998.

[YCGL97] K. Yocum, J. Chase, A. Gallatin, and A. Lebeck. Cut-Through Delivery in Trapeze: An Exercize in Low-Latency Messaging. In *The 6th Int. Symp. on High Performance Distributed Computing*, Portland, OR, August 1997.

[vEB+95] T. von Eicken, A. Basu, V. Buch, and W. Vogel. U-Net: A User-Level Network Interface for Parallel and Distributed Computing. In *Proc. of the 15th Symp. on Operating System Principles*, pages 303–316, Copper Mountain, CO, December 1995.

[WBT96] Thomas M. Warschko, Joachim M. Blum, and Walter F. Tichy. The ParaStation Project: Using Workstations as Building Blocks for Parallel Computing. In *Proceedings of the International Conference on Parallel and Distributed Processing, Techniques and Applications (PDPTA'96)*, pages 375–386, Sunnyvale, CA, August 9–11, 1996.

[War98] Thomas M. Warschko. *Effiziente Kommunikation in Parallelrechnerarchitekturen*. Dissertation, Universität Karlsruhe, Fakultät für Informatik. *Published as:* VDI Fortschrittberichte, Reihe 10: Informatik / Kommunikationstechnik Nr. 525. ISBN: 3-18-352510-0. März 1998.

Broadcast Communication in ATM Computer Networks and Mathematical Algorithm Development

Michael Weller

Institute for Experimental Mathematics, Ellernstr. 29, 45326 Essen, Germany

Abstract. This article emphasizes the importance of collective communication and especially broadcasts in mathematical algorithms. It points out that algorithms in discrete mathematics usually transmit higher amounts of data faster than typical floating point algorithms. It describes the o.tel.o ATM Testbed of the GMD National Research Center for Information Technology, St. Augustin, and the Institute of Experimental Mathematics, Essen, and the experiences with distributed computing in this network. It turns out that the current implementations of IP over ATM and libraries for distributed computing are not yet suited for high performance computing. However, ATM itself is able to perform fast broad- or multicasts. Hence it might be worthwhile to design a message passing library based on native ATM.

1 Broadcasts in mathematical algorithms

Distributed and parallel computing plays an important role in chemistry, physics, engineering and thus numerical mathematics. However, there are parts of pure and discrete mathematics like cryptography, computational group theory and representation theory which can also benefit from using a computer. In general, the algorithms transform the original problem to a huge problem in linear algebra over a finite field [2].

Dense equation systems in 300,000 or more variables [3,4] or the computation and enumeration of hundreds of millions of vectors or even subspaces [5,6] show up easily. However, an entry of a matrix or vector can often be realized by a small number of bits. Operations on the machine words representing small vectors of such entries are either done by integer additions, logical bit operations, or (in the more complex cases) by table lookups.

For the hardware involved in solving such problems this means that there are many trivial (hence fast) arithmetical operations one has to perform. In return, the problems are big themselves and parallelization only makes sense if each computing node receives a substantial amount of data to deal with. For the communication in a distributed application this means that it usually has to transfer a huge amount of data fast. Therefore the computing nodes must have large amounts of memory with a high memory bandwidth to satisfy the speed requirements of the CPU handling many trivial integer operations. If table lookups are involved in the program, a cache is often unable to hide the lack of actual memory bandwidth from the application program.

Numerical applications typically involve slower operations on floating point numbers and no table lookups and reduce the unbalance between memory, communication and CPU speed. However, there are no problems with numerical stability in discrete mathematics.

There are many interesting computational problems in discrete mathematics requiring access to parallel supercomputers. Since they are rare, it seems interesting to couple smaller parallel machines and workstation clusters of different institutions over a WAN to perform these computational tasks.

Many of the linear algebra algorithms use broadcasts and other group communications like the algorithm sketched in Fig. 1.

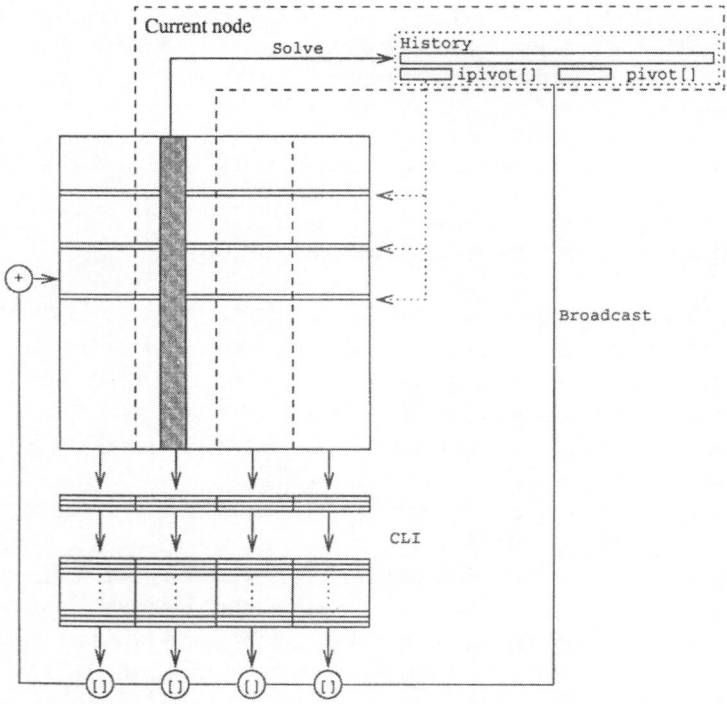

Fig. 1. Data flow of parallel gaussian elimination over a finite field: One of the processes solves a few columns of the equation system residing in its memory. It broadcasts instructions to the other nodes which perform the same operations as the sender on the remaining columns. Then control passes to the next node. The processes perform the pre-solving and broadcasting in a round-robin fashion to achieve a better load distribution. The algorithm is described in [4] in more detail and was also used in a world record computation achieved in [3]

Special networks designed for parallel computers like the CM-5 by Thinking Machines are able to perform broadcasts and reduction operations in hardware. Currently it seems not possible to perform reductions using standard network hardware. Also, broadcasts appear more often in our problems and are even used during the initialization phase of algorithms which do not use them afterwards.

2 Implementation of Broadcasts

Ideally, the application programmer should not care about the implementation of broadcasts. This ought to be done by a communication library. Typical implementations are shown in Fig. 2. Often a tree like implementation is used. Each node which already received a copy of the data helps distributing it. While this reduces the broadcast to $\log_2(n)$ steps of point to point communications there is more than one node sending at the same time. This causes collisions in any non switched network connecting the nodes and leads to network congestion, if, for example, several hosts in the same cluster send data to hosts in a cluster at another location over a WAN connection.

Tree structured broadcast: Simple (PVM) broadcast: Cyclic broadcast:

Non-switched Network / Ethernet:

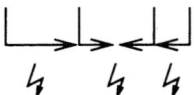

Fig. 2. Different implementations of broadcasts. From left to right: Elaborate implementation using a tree like distribution of data in $\log_2(n)$ steps but causing collisions in non switched networks (some MPI, POE on old IBM SP), simple implementation for non switched networks (PVM), cyclic communication done by the application

Just sending the data in a loop to all $n - 1$ recipients avoids such congestion, but is much slower. It is worth noting that this way is still faster than using the tree broadcast on a shared media Ethernet. Not only must the tree broadcast degenerate to a sequential scheme here, but the collisions reduce the efficiency even more. We tried several types of broadcast in our experiment shown in Fig. 1. Using sequential broadcast from the beginning on shared media 10Mbps Ethernet required 8h 43m where the attempt to use a tree broadcast required 11h 44m.

In that experiment we obtained good results using a cyclic pattern for the sequential broadcast. The master only sends the data to a successor, then resumes his computational work. This has benefits as the master already had to do some precomputation to find the data to be broadcasted. That is, in a sense he is already behind the other nodes and must not be delayed any further. Its successor then sends the data to the next node and so on, until all nodes received the data. As we are using a round robin method to move the master node in this algorithm anyway this interacts nicely with this algorithm. This observation was also made with numerical solvers of floating point equation systems.

3   ATM and distributed computing

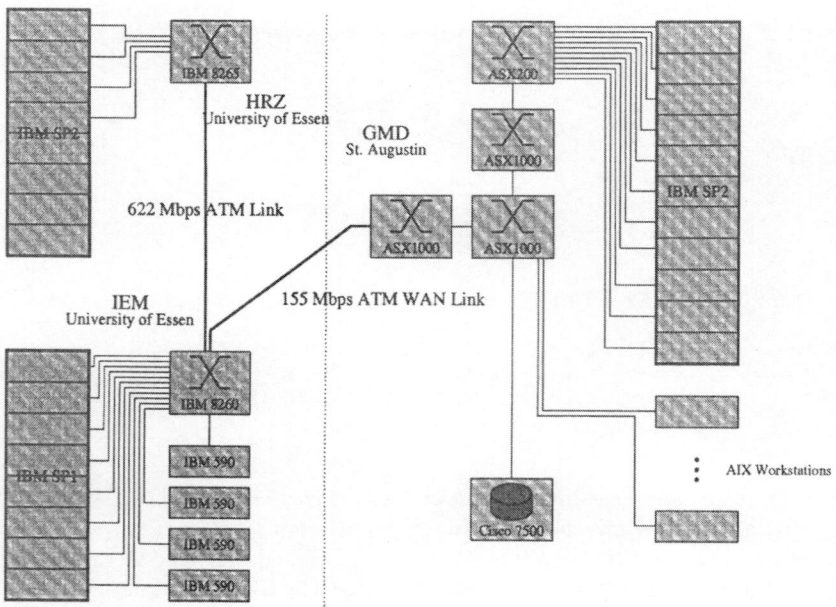

Fig. 3. The Essen — St. Augustin o.tel.o ATM-Testbed

The Institute of Experimental Mathematics (IEM) is connected to an ATM Testbed as shown in Fig. 3. The carrier o.tel.o provides a 100km

155Mbps connection to the GMD in St. Augustin and a short 622Mbps connection through the city of Essen into the building of the main computer centre (HRZ) of the university of Essen. Classical IP over ATM is used to run TCP/IP over this network. It turned out that any routers in this network have to be avoided and LAN Emulation is not to be used to achieve sensible performance. This way we can reach up to 80Mbps file transfers (each node has a 155Mbps adapter).

It is now possible to use the standard libraries for distributed computing over TCP/IP for parallel computing in such an environment. However, we only obtained bad performance this way. Only a Peak-Bandwidth of 7.5Mbps could be measured when using PVM. MPI-CH using P4 for communication only achieved 0.6Mbps of maximum transmission bandwidth. There also exists a package PLUS [1] which interconnects MPI implementations of different workstation clusters or parallel computers over TCP/IP. It could not be tested in the experiment of Fig. 3 as it is not yet available for AIX, but its creators already tested it in a 34Mbps ATM environment. Scaling the performance there to 155Mbps we would achieve only 11.45Mbps which is still less than a plain TCP/IP file transfer can achieve.

After the period of low-level ATM tests in the joint project with the lab of our carrier o.tel.o and support of Siemens and GN Nettest described below the dark fibre to the main computer centre of Essen University was built and we did our experiments again. This time running them on several nodes of all three institutions of Fig. 3 at the same time. Doing so, we found no significant, technical difference in running our application on nodes of one, two or three institutions. However, as more people and institutions were involved, it became much more difficult to ensure a correct setup of routing, host names and network for the computations.

Also, as some time had passed, firmware upgrades took place on the switch of the institute (V.3.1.0 to V.3.2.1), and the GMD machines were upgraded to AIX 4.3 from 4.2. Other ATM-related fixes of IBM were applied to the AIX 4.2 machines in Essen.

This way, we were able to measure a peak bandwidth under MPI-CH using P4 of 93Mbps during the initialization of a parallel application actually performing an MPI_Bcast. We used the same program, even the same binary, used for the former tests. Only switch firmware and the operating system on some of the nodes had changed. This program used to need 47 minutes (!) to broadcast 75MB of data to six–eight nodes. After the upgrades this time was reduced to 30 seconds.

Scattering an equation system of 800MB on 6 nodes still required more than 8000 seconds. But this data is scattered manually by the program in mostly small chunks. Hence it might not profit from the advantages of the newer firmware.

On the other hand, using native ATM communication, we were able to transfer up to 123Mbps (AAL and other overhead is already subtracted) from point to point in this network. The highest performance was achieved using large packets of 40KB which was the largest size the AIX operating system was able to handle.

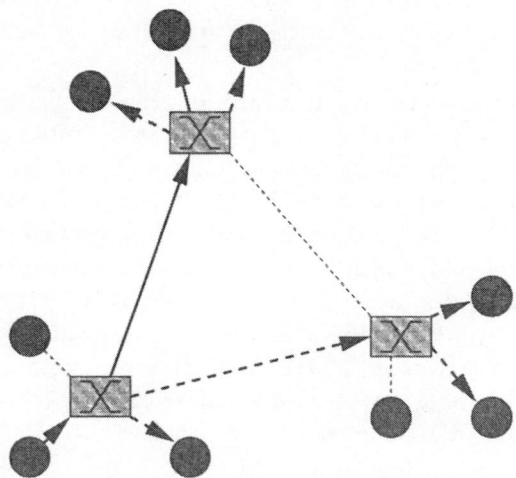

Fig. 4. Single to Multipoint ATM communication: Based on a point-to-point connection (solid line) from the lower left to the upper left ATM allows to add further recipients (dashed lines). Still data is only transmitted once over the WAN lines between the three switches. The receiving switches to the right and upper left distribute the data locally to all receiving parties

In addition, there is another interesting feature of ATM. As sketched in Fig. 4 ATM is able to broadcast. Although ATM per se is a strictly point to point connection oriented protocol, it is possible to specify multiple receivers of the same data stream which splits at the ATM switches as close to the receivers as possible. When connecting two clusters of workstations over an ATM WAN link, this means that data will only be transmitted once through the WAN link and then distributed automatically by the switch in the remote ATM cluster.

In our experiments, we were able to achieve the same peak transmission rate of 123Mbps (not counting any ATM or SDH overhead) even when distributing data to the GMD and the IEM. Unfortunately we experienced a very high rate of 1–2% data loss. Latency or delay variation could not be measured reliably with the workstations as they had no exact synchronous time source.

In a joint project (Fig. 5) with the lab of our carrier o.tel.o and support of Siemens and GN Nettest we found that there is actually no data loss in the network or the switches. It appears that these are either due to failure of the workstations to be unable to accept the data fast enough or not being

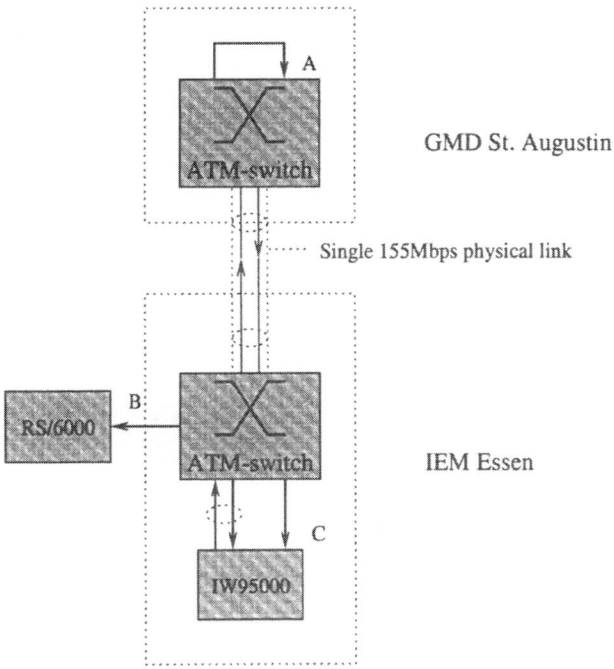

Fig. 5. Testing a single to multipoint connection: The data was sent through a unidirectional PVC to a loop on the remote switch and from there back to points B & C at once (using a single to multipoint connection). As the bandwidth on the IEM–GMD link was limited, it is guaranteed that the data was duplicated at the switch at the IEM, not at the other side

able not to overcommit a specified link bandwidth. As the data loss was even higher for very slow links, it appears that the latter might be the reason.

With GN Nettest equipment provided by the carrier and Siemens we could measure network latencies and delay variations which we found to be very small and below a millisecond even in the WAN segment (see also Fig. 6). Definitely these should be of no relevance for distributed computing. However, ATM is connection oriented and the time required to setup connections cannot be ignored. We found that a typical ATM LAN switch can stand at most bursts of about 100 connection setups per second (assuming there is no other traffic). Therefore, for a typical distributed application, it will be too slow to initiate the necessary connections when a data packet is to be sent. One can consider initiating all required connections at the beginning of the program, but there will be many such connections and the resources of the ATM switches are usually limited to a few thousand connections per interface. Thus, an interesting approach could be to initiate each connection in advance, before the data to be sent is actually available. Of course,

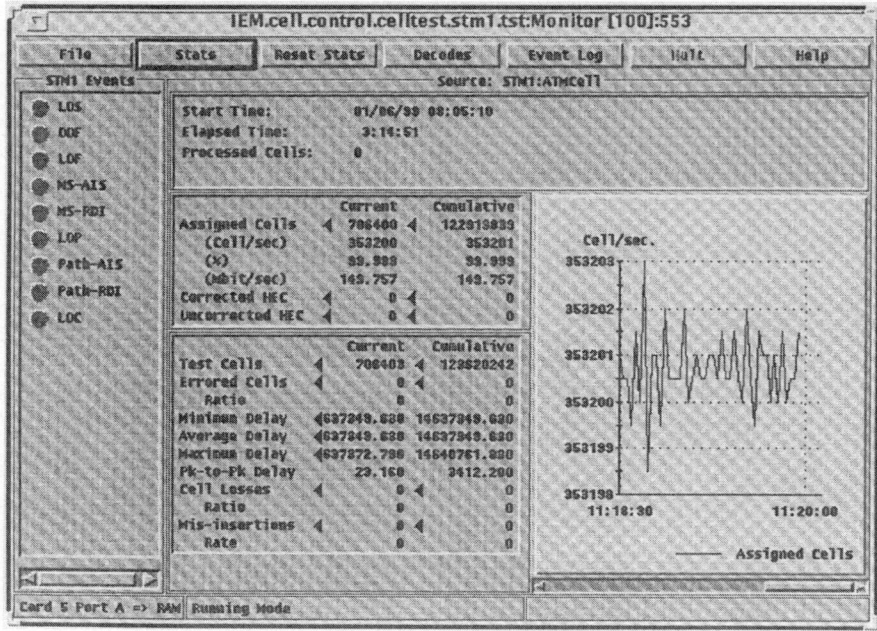

Fig. 6. Delay variation and throughput measured on one leave of a single to multipoint connection. The GN Nettest Interwatch does not allow to measure absolute delays in this configuration as the data was not received on the card which generated the traffic. Hence there are no sensible cell delay values reported. The values for delay variation and throughput are reported correctly. We had no access to a synchronous clock driving the switches, SDH equipment and traffic generators. This might result in increased jitter effects. The values measured on the other leaves did not differ significantly.

this requires that the application can foresee the recipient at such an early point and that the application programming interface of the message passing library allows to prepare sending messages in such a way.

The tests also included the use of network management systems. These are an indispensable tool to administer any large ATM network. Without them, a connection has to be configured separately on each switch it crosses. Using such a management system it is possible just to specify endpoints and have the system find a route through the network. However, distributed computing would require switched virtual connections which are normally not handled by the management system and were not a major part of this project. Still, the management system in question allows a network carrier to assign some backbone bandwidth to a customer which can then use this bandwidth at his own disposal. He can also book ATM-links in advance which are then made available on automatic at a later point (for example at night, when the traffic is low and the ATM connection is cheaper), maybe for a batch computation.

In conclusion, it appears to be worthwhile to perform distributed high performance computing over an ATM WAN network. However, currently a message passing library utilizing native ATM connections is not available but appears to be necessary to achieve the required data rates. In addition, such a library must be able to deal with the data loss which cannot be avoided when using ATM, although the network itself appears to be very reliable in this respect, as long as the ATM interfaces of the computing nodes are fast enough and don't exceed the traffic contracts.

4 Acknowledgements

The author kindly acknowledges financial support by the Ministry of Science and Education North Rhine Westfalia, Essen University and o.tel.o Düsseldorf. His work was also supported by the DFG-NSF exchange program, DFG grant# Mi-89/24-1. He is also grateful to Siemens Essen for providing access to a wide area switch and a management system, and to GN Nettest Munich for technical support.

Finally he would like to thank the GMD National Research Center for Information Technology, St. Augustin, and the Computer Center of the University of Essen for the permission to use their resources and the help of their maintenance staff in setting up the systems for the wide-area distributed computations.

References

1. Matthias Brune, Jörn Gehring, and Alexander Reinefeld. Communicating across parallel message-passing environments. Preprint submitted to Elsevier, February 1998.
2. P. Fleischmann, G. O. Michler, P. Roelse, J. Rosenboom, R. Staszewski, C. Wagner, and M. Weller. *Linear Algebra over Small Finite Fields on Parallel Machines*, volume 23 of *Vorlesungen aus dem Fachbereich Mathematik*. University of Essen, 1995.
3. Peter Roelse. Factoring high-degree polynomials over F_2 with Niederreiter's algorithm on the IBM-SP/2. *Math. Comp.* to appear.
4. M. Weller. Parallel gaussian elimination over small finite fields. In *Procedings of the 9th International Conference on Parallel and Distributed Computing Systems*. ISCA, September 1996.
5. Michael Weller. Construction of large permutation representations for matrix groups. In E. Krause and W. Jäger, editors, *High Performance Computing in Science and Engineering '98*, pages 430–452. HLRS Stuttgart, Springer-Verlag Berlin, Heidelberg, New York, 1998.
6. Michael Weller. Construction of large permutation representations for matrix groups II. Submitted, 1999.

Highly Available Distributed Storage Systems

Lihao Xu[1] and Jehoshua Bruck[2]

[1] Department of Computer Science, Washington University, Campus Box 1045, Saint Louis, MO 63130, USA. Email: lihao@cs.wustl.edu. This work was done while this author was at the California Institute of Technology.
[2] Department of Electrical Engineering, California Institute of Technology, Mail Stop 136-93, Pasadena, CA 91125, USA. Email: bruck@paradise.caltech.edu.

1. Introduction

Information is generated, processed, transmitted and stored in various forms: text, voice, image, video and multimedia types. Here all these forms will be treated as general data. As the need for data increases exponentially with the passage of time and the increase of computing power, data storage becomes more and more important. From scientific computing to business transactions, data is the most precious part. How to store the data reliably and efficiently is the essential issue, that is the focus of this chapter.

As with distributed computing, distributed storage is coming of age as a good solution to achieve scalability, fault tolerance and efficiency. Historically, since the speed of storage devices, such as tapes and disks, is much slower than the speed of computing devices, e.g., CPUs, I/O is a bottleneck in computing systems. To improve the data throughput of storage devices, RAID (*Redundant Array of Independent Disks*) was proposed[14][16] to store data over multiple storage devices in a distributed way, so that the total I/O bandwidth is sum of the bandwidths of the individual storage devices. That was the start of distributed (networked) storage. Since then, storage technologies have been advancing rapidly; the capacity of magnetic devices continuously increases and access speed constantly improves. But as with CPUs, there are physical limits to the density of disks, seek time and rotational speed of the disk drives. These limits mean that the capacity and access speed of a single storage device can *not* be improved infinitely. The need for storage capacity and access speed can be met by improving storage systems at the architectural level, i.e., using multiple distributed storage devices connected via a fast network, such as the *Fiber Channel*, which reduces data latency incurred over the network to much less than the latency time of a single storage device. A distributed structure not only can increase the capacity and speed of storage systems, but also can bring fault tolerance and scalability.

As with computing, fault tolerance (or *reliability*) is increasingly important in storage systems. Some critical data should be available and some services should be provided even when faults occur in storage units. Besides, a storage system that allows some faulty units can be replaced on-the-fly would have great value for business transactions, such as airport management, banking systems, and internet service provider systems. Naturally, reliability of storage systems can be achieved more easily using distributed structure. Scalability is another natural feature of distributed systems: addition or replacement of components is much more flexible in a distributed system than in a central system. Thus distributed storage systems can adapt better to dynamic and growing data demands.

In this chapter, the reliability, efficiency and scalability of distributed storage systems are all considered aspects of availability. A highly available storage system has high reliability (or can tolerate more faults), high efficiency (or performance) and scalability. Achieving high availability in distributed storage systems is the main topic of this chapter.

This chapter mainly consists of two parts. The first part discusses the reliability issue. The reliability is usually achieved by introducing data redundancy into a storage system. The second part shows that the efficiency of a data storage system can be improved by properly using the data redundancy in the system. So the approaches of introducing data redundancy are very important to a storage system, for both reliability and efficiency. This chapter will describe a few *MDS array codes*, a class of *error-control* codes that are very suitable to be used to introduce data redundancy in storage systems.

2. MDS Array Codes for Reliability

2.1 Array Codes

Reliability of storage systems is often achieved by storing redundant data in the systems using error-control codes. Usually in storage systems, the failure of a single storage unit can be detected by the storage controllers and then can be masked. Thus *erasure-correcting* codes are often used, since the device failures can be marked as erasures. Erasure-correcting codes are a mathematical means of representing data so that lost information can be recovered. With an (n, k) erasure-correcting code, we represent k symbols of the *original* data with n symbols of *encoded* data ($n - k$ is called the amount of *redundancy* or *parity*). With an m-erasure-correcting code, the original data can be recovered even if m symbols of the encoded data are lost[13], and the *distance d* of this code is defined to be $d = m + 1$. A code is said to be *Maximum Distance Separable* (MDS) if $m = n - k$. An MDS code is optimal in terms of the amount of redundancy versus the erasure recovery capability. The *Reed-Solomon* code [13] is a well-known example of an MDS code.

The complexity of the computations needed to construct the encoded data (a process called *encoding*) and to recover the original data (a process called *decoding*) is an important consideration for practical systems. Array codes are ideal in this respect. Array codes have been studied extensively [2][3][4][8].

A common property of these codes is that the encoding and decoding procedures use only simple binary exclusive-or (XOR), which can be implemented easily in hardware and/or software; thus these codes are much more efficient than Reed-Solomon codes in terms of computation complexity and are very suitable to be used in storage systems, for both reliability and efficiency.

In an array code, the information (original) and parity (redundant) bits are placed in a 2-dimensional array of size $l \times n$. In a distributed storage system, the bits in the same column are stored in the same disk. If any bit in a disk is corrupted, then the disk is considered to be a failure disk and needs repair, i.e., the corresponding column of the code is considered to be an erasure.

Current RAID (Redundant Array of Independent Disks) systems can tolerate at most *one* disk failure at a time, i.e., the code used is only a 1-erasure-correcting code. In more and more applications, fault-tolerance of only one single disk is not enough. A system that can tolerate more than one failure at the same time would be more robust and flexible. For example, when one disk fails, the system can still have some non-stop fault-tolerance capability while the bad disk is being replaced by a good one. This level of fault tolerance requires codes with higher erasure-correcting capability. A 2-erasure-correcting code can provide a much longer nonstop functioning time to a distributed storage system than a 1-erasure-correcting code.

Considering all the above factors, i.e., computation complexity, optimal redundancy and high level fault tolerance, we will focus on three classes of 2-erasure-correcting MDS array codes: the EVENODD code[2][3], the X-Code[21] and the B-Code[22]. These codes can be used effectively to achieve the reliability and efficiency of storage systems.

2.2 EVENODD Code

The EVENODD code has a very simple structure: all bits are placed in an array of size $(p - 1) \times (p + 2)$, where p is a prime number, i.e., it has $p + 2$ columns. All the information bits are placed in the first p columns, and the last 2 columns contain all parity bits. The 2 parity columns are constructed using the diagonals of slope 0 and slope -1, respectively. Details about the construction of the EVENODD code can be found in [2]. Following example shows a construction for a (7,5) EVENODD code:

Example 2.1. A (7,5) EVENODD code
Table 2.1 shows an encoding rule of a (7,5) EVENODD code, and Table 2.2 is a numerical example of Table 2.1.

□

It was proven that [2] that the EVENODD code is a 2-erasure-correcting array code, i.e., it is MDS. An algorithm for recovering 2 erased columns for the EVENODD code and other details can be found in [2]. A generalization

Table 2.1. Encoding of a (7,5) EVENODD code, where $s = a_5 + b_4 + c_3 + d_2$

a_1	a_2	a_3	a_4	a_5	$a_1 + a_2 + a_3 + a_4 + a_5$	$s + a_1 + b_5 + c_4 + d_3$
b_1	b_2	b_3	b_4	b_5	$b_1 + b_2 + b_3 + b_4 + b_5$	$s + a_2 + b_1 + c_5 + d_4$
c_1	c_2	c_3	c_4	c_5	$c_1 + c_2 + c_3 + c_4 + c_5$	$s + a_3 + b_2 + c_1 + d_5$
d_1	d_2	d_3	d_4	d_5	$d_1 + d_2 + d_3 + d_4 + d_5$	$s + a_4 + b_3 + c_2 + d_1$

Table 2.2. Numerical example of a (7,5) EVENODD code

1	0	1	1	0	1	0
0	1	1	0	0	0	0
1	1	0	0	0	0	1
0	1	0	1	1	1	0

of the EVENODD code to recover more erasures while still to maintain the MDS property is described in [3].

2.3 Update Complexity

One important parameter of array codes is the average number of parity bits affected by a change of a single information bit; this parameter is called the *update complexity* here. The update complexity is particularly crucial when the codes are used in storage applications that update information frequently. It also measures of the encoding complexity of the code. The lower this parameter is, the simpler the encoding operations are. If a code is described by a *parity check matrix*[13], then this parameter is the average *row density* — the number of nonzero entries in a row — of the parity check matrix. Research has been done to reduce this parameter or to make the density of parity check matrix of codes as low as possible [9][17]. The obvious lower bound of the update complexity of any 2-erasure-correcting code is 2. The update complexity of *EVENODD* codes approaches 2 as the length (number of the columns) of the codes increases. But it was proven in [3] that for any linear array code with separate information and parity columns, the update complexity is always *strictly* larger than 2. Then a natural question is whether the update complexity of 2 is achievable for general array codes. The answer is, fortunately, yes. The next two subsections will describe two classes of codes, called X-Code and B-Code respectively, whose update complexity is exactly 2.

2.4 X-Code

The X-Code is a class of 2-erasure-correcting MDS array code. Its update complexity is exactly 2, i.e., it has the optimal encoding (update) property. It has a very simple geometrical construction structure.

2.4.1 Structure of the X-Code. In X-Code, information bits are placed in an array of size $(n-2) \times n$. Like other array codes [2][3][5][11], parity bits are constructed by adding the information bits along several *parity check lines* or *diagonals* of given *slopes*. The addition operation is just the binary XOR. But instead of being put in separate columns, the parity bits of the X-Code are placed in *two* additional *rows*. So the coded array is of size $n \times n$, with the first $n-2$ rows containing information bits, and the last two rows containing parity bits. Notice that each column has information bits as well as parity bits, i.e., information bits and parity bits are mixed in each column. By the structure of the code, if two columns are erased, the number of remaining bits is $n(n-2)$, which is equal to the number of original information bits, making it possible to recover the two erased columns.

2.4.2 Encoding Rules. The encoding rule of the X-Code is simple: let $C_{i,j}$ be the bit at the ith row and jth bit, then parity bits are constructed according to the following rules:

$$C_{n-2,i} = \sum_{k=0}^{n-3} C_{k,\langle i+k+2 \rangle_n}$$

$$C_{n-1,i} = \sum_{k=0}^{n-3} C_{k,\langle i-k-2 \rangle_n} \tag{2.1}$$

where $i = 0, 1, \cdots, n-1$, and $\langle x \rangle_n = x \bmod n$. Geometrically speaking, the two parity rows are just the checksums along diagonals of slopes 1 and -1 respectively.

The following example shows the encoding of a $(7,5)$ X-Code:

Example 2.2. A $(7,5)$ X-Code

The first parity row is calculated along the diagonals of slope 1, with the last row being an imaginary 0-row, as follows:

□	◊	△	♣	◊	♡	♠	
♠	□	◊	△	♣	◊	♡	
♡	♠	□	◊	△	♣	◊	
◊	♡	♠	□	◊	△	♣	
♣	◊	♡	♠	□	◊	△	
△	♣	◊	♡	♠	□	◊	← *1st parity row*
◊	△	♣	◊	♡	♠	□	← *imaginary 0-row*

The second parity row is calculated along the diagonals of slope -1, with the last row being an imaginary 0-row, as follows:

□	◇	△	♣	◇	♡	♠	
◇	△	♣	◇	♡	♠	□	
△	♣	◇	♡	♠	□	◇	
♣	◇	♡	♠	□	◇	△	
◇	♡	♠	□	◇	△	♣	
♡	♠	□	◇	△	♣	◇	← *2nd parity row*
♠	□	◇	△	♣	◇	♡	← *imaginary 0-row*

Table 2.3 shows a numerical example of a (7,5) X-Code.   □

Table 2.3. Numerical example of a (7,5) X-Code

1	0	1	1	0	1	0
0	1	1	0	0	0	0
1	1	0	0	0	0	1
0	1	0	1	1	1	0
1	0	0	1	0	1	0
0	0	1	1	0	1	1
1	1	1	0	0	1	0

From the construction of the X-Code, it is easy to see that the two parity rows are obtained independently, more specifically, each information bit affects exactly *one* parity bit in each parity row. All parity bits depend only on information bits, but *not* on each other. So, updating a single information bit results in updating only *two* parity bits. Thus the X-Code has the optimal encoding (or update) property, i.e., its update complexity of 2 matches the lower bound for any 2-erasure-correcting code.

In addition, notice that each column has two parity bits, each of which is the checksum of $n-2$ information bits. Thus computing parity bits at each column needs $2(n-3)$ XORs. This balanced computation property of X-Code is very useful in applications that require evenly distributed computations.

It was proven in [21] that

Theorem 2.1. (MDS Property of the X-Code)
X-Code can recover up to two erased column, i.e., it is MDS, if and only if n is a prime number.

The procedure to recover up to two erased columns is called *erasure-correcting* or *erasure-decoding*. A formal description and correctness proof of the erasure-decoding algorithm for the X-Code can be found in [21]. A pseudo-code description of the algorithm can also be found in [23]. More details about the X-Code, including examples of its erasure-decoding algorithm are discussed in [21] and [23].

2.5 B-Code

Now we describe another class of 2-erasure-correcting MDS array code, called B-Code. Its update complexity is also exactly 2, i.e., it has the optimal encoding (update) property. Construction for the B-Code is not as direct as for the X-Code. It is related to a classical graph theory problem.

2.5.1 Structure of the B-Code. B-Code is of size $n \times l$, where $l = 2n$ or $2n + 1$. Denote such a B-Code B_l. As the X-Code, parity bits are placed in a row rather than columns. For B_{2n}, the first $n - 1$ rows are information rows, and the last row is a parity row, i.e., all the bits in the first $n - 1$ rows are information bits, while the $2n$ bits in the last row are parity bits. The structure of B_{2n+1} can be derived from that of the B_{2n} simply by adding one more information column as the last column. Their structures are shown in Figure 2.1.

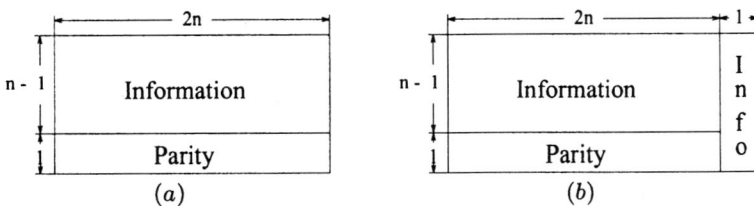

Fig. 2.1. Structures of (a) B_{2n} and (b) B_{2n+1}.

Intuitively, if the roles of the information and parity bits of the B-Code are exchanged, i.e., the parity bits are placed in the entries which originally were for the information bits and vice versa, then we get the dual code of the B-Code. Denote the dual B-Code of length l $\hat{B}_l$. A more rigorous definition of the dual code for general array codes can be found in [22]. It was also proven that [22]

Theorem 2.2. *The dual code of an MDS array code is also MDS.*

So the dual B-Code is also an MDS array code; it has distance $l - 1$, i.e., the dual B-Code can be recovered from any two of its columns. Figure 2.2 shows structures of $\hat{B}_{2n}$ and $\hat{B}_{2n+1}$.

2.5.2 A New Graph Description of the B-Code. Typically, an array code is described by its *geometrical constructions lines* or *diagonals*[2][3][5][11], such as the X-Code. Constructions of array codes are difficult to get using this description. Here we describe the B-Code and its dual using a new graph approach, which allows us to get the construction of the B-Code easily.

For any array code, each parity bit is the sum of some information bits. For binary codes, the addition is just the simple XOR (binary *exclusive OR*)

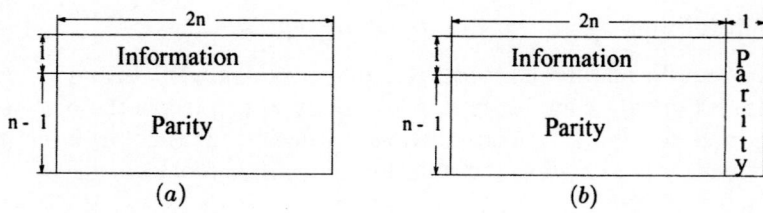

Fig. 2.2. Structures of (a) $\hat{B}_{2n}$ and (b) $\hat{B}_{2n+1}$

operation. Now consider the dual B-Code $\hat{B}_l$. Simple counting [22] shows that each parity bit must be the sum of *exactly* 2 information bits. Thus if we represent an information bit as a vertex, then a parity bit can be represented by an edge, where the parity bit is the sum of the two information bits whose vertices form the edge. This is the key idea of describing the B-Code and its dual with graphs.

Since the construction of $\hat{B}_{2n}$, B_{2n} and B_{2n+1} can be easily obtained from the $\hat{B}_{2n+1}$, here we give a graph description of the $\hat{B}_{2n+1}$. Detailed justifications of this description can be found in [22].

Description 2.1. *Graph Description of* $\hat{B}_{2n+1}$

Given a complete graph K_{2n} with 2n vertices, which are labeled with integers from 1 to 2n, find an edge labeling scheme such that

1) each edge is labeled exactly once by an integer from 1 to 2n+1

2) For any pair of vertices (i, j) and any other vertex k, where $i, j, k \in [1, 2n]$, there is always a path to k from either i or j, using only the edges labeled with i or j.

3) For any vertex i and any other vertex k, where $i, k \in [1, 2n]$, there is always a path from i to k, using only the edges labeled with i or $2n + 1$.

With the above description, it is easy to see that the vertex and edges with the label i in the K_{2n} represent the information bit and parity bits in the ith column of the $\hat{B}_{2n+1}$. The properties 2) and 3) ensure that any two columns of the code can recover the information bits in all other columns, thus the code is of column distance 2n. Figure 2.3 shows such a labeling of K_4 and the corresponding $\hat{B}_5$, where a_1 through a_4 are the information bits.

2.5.3 Construction of the B-Code. As already described above, constructing the B-Code amounts to the same problem as designing an edge labeling scheme such as in Description 2.1 for a complete graph K_{2n}. Fortunately this can be related to another graph theory problem, namely the *perfect one-factorization* (P1F) problem.

Definition 2.1. [19] Let $G=(V,E)$ be a graph. A *factor* or *spanning subgraph* of G is a subgraph with vertex set V. In particular a *one-factor* is a factor

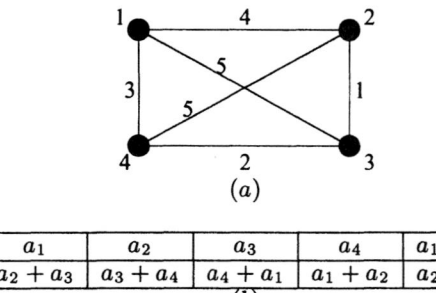

a_1	a_2	a_3	a_4	$a_1 + a_3$
$a_2 + a_3$	$a_3 + a_4$	$a_4 + a_1$	$a_1 + a_2$	$a_2 + a_4$

(b)

Fig. 2.3. (a) graph and (b) array representations of $\hat{B}_5$

which is a regular graph of degree 1. A *factorization* of G is a set of factors of G which are pairwise *edge disjoint*, and whose union is all of G. A *one-factorization* of G is a factorization of G whose factors are all one-factors. In particular, a one-factorization is *perfect* if the union of any pair of its one-factors is a *Hamilton cycle*, a cycle that passes through every vertex of G.

Figure 2.4 shows a perfect one-factorization of K_4.

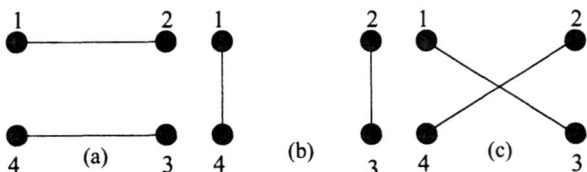

Fig. 2.4. (a)(b)(c) are 3 one-factors, that together form a perfect one-factorization of K_4

The perfect one-factorization of complete graphs has been studied for many years since its introduction in 1960's in [12]. It is now known that[19]:

Theorem 2.3. *If p is an odd prime, then K_{p+1} and K_{2p} have perfect one-factorizations.*

Constructions of P1F for K_{p+1} and K_{2p} can be found in [1] and [18]. Additionally, constructions of P1F for K_{2n}'s whose n's are some other sporadic integers have also been found[18][19]. However it still remains a conjecture [18][19] that:

Conjecture 2.1. For any positive integer n, K_{2n} has perfect one-factorization(s).

It was proven in [22] that:

Theorem 2.4. *Let P_m be a P1F for K_m. Constructing $\hat{B}_{2n+1}$ (or equivalently B_{2n+1}) is equivalent to constructing P_{2n+2}.*

Theorem 2.4 was proven *constructively* in [22]. Combining Theorem 2.3 and Theorem 2.4, we get that:

Theorem 2.5. *For any odd prime p, a B-Code and its dual code of size $n \times l$ can be constructed, where n is either $\frac{p-1}{2}$ or $p-1$, and $l = 2n$ or $2n+1$.*

2.5.4 Erasure Recovery. Recall that the dual B-Code can recover all information bits from any two columns. Erasure decoding for the dual B-Code is almost obvious from its graph description (*Description 2.1*). The two paths, starting from i and j and leading to all the other vertices in the graph, give the decoding chain used in recovering a $\hat{B}$-Code from its ith and jth columns. Figure 2.5 shows the decoding chain used in recovering $\hat{B}_5$ from its 1st column and its 2nd, 3rd and 5th columns, respectively.

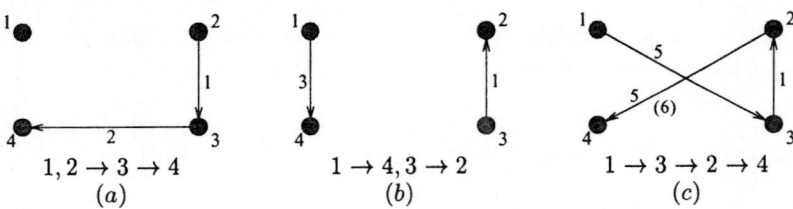

Fig. 2.5. Erasure decoding of $\hat{B}_5$: recovering from its 1st and (a) 2nd (b) 3rd and (c) 5th columns. The decoding chains for each case are also listed. 1 through 4 are the information bits in the corresponding columns.

Formal erasure recovery algorithms for the B-Code and its dual code, and more details about the B-Code can be found in [22] and [23].

2.6 Comparisons of Array Codes

As already seen above, the X-Code and B-Code have the optimal update property, i.e., their update complexity is exactly 2. The B-Code also achieves the *maximum* length possible for MDS codes with the optimal update, thus the B-Code has *optimal length*, twice that of the X-Code with the same column size. In addition, the parity bits are evenly distributed over all columns, and each parity bit requires the same number of *XOR* operations. Consequently, the computational complexity for computing parity bits is *balanced*, i.e., the X-Code and B-Code feature *balanced computation* as well. This property is quite useful in distributed storage systems, since the computational loads are naturally distributed to all disks evenly, eliminating another bottleneck. The properties of the X-Code and the B-Code are summarized in Table 2.4, together with a comparison with *Reed-Solomon* and *EVENODD* codes.

Table 2.4. X-Code, B-Code vs. Reed-Solomon and EVENODD.

Codes \ Properties	MDS	XOR	Optimal Update	Optimal Length	Balanced Computation
Reed Solomon	Yes	No	No	Yes	No
EVENODD	Yes	Yes	No	No	No
X-Code	Yes	Yes	Yes	No	Yes
B-Code	Yes	Yes	Yes	Yes	Yes

3. Efficiency through Redundancy

While it is conventional wisdom that redundancy is necessary for fault tolerance, redundancy is in general regarded as a passive cost (overhead) to achieve reliability. However, in this section, it will be shown that in a distributed storage system, redundancy is an active part of the system in the sense that proper data redundancy can help to improve the performance (data throughput) of storage systems. Thus data redundancy will improve not only the reliability of a system, but also the efficiency of a system. A similar idea was first shown in [7], namely that redundant data can make packet routing more efficient by reducing the mean and variance of the routing delay. Recently, more scalable and efficient reliable multicast schemes have been proposed, based on data redundancy in the messages to be multicast[10]. We will show here a more systematic way of using proper redundancy, based on error-correcting codes (particularly the MDS array codes described above), to improve the performance of data server systems, which are a superset of storage systems.

Our data server system setup is shown in Figure 3.1: a cluster of servers is connected via some reliable communication network. In addition, broadcast is supported over the network, so that a client can broadcast its request for certain data to some or all of the n servers in the system. The data is distributed over the servers in such a way that a client can recover the complete requested data after it gets data from at least k of the n servers and this is true for *any* k servers. Such a distributed data server system is called an (n, k) server system. Again, such (n, k) systems can be implemented by using error-correcting codes, particularly MDS array codes.

For the above data server system, there are a couple problems to solve: (1) What is the *proper* redundancy when the total number of the servers is given? Or how should k be determined when n is given, in order to achieve the best system performance? This is the so-called *data distribution* problem at the server side. (2) Once data redundancy is properly distributed among the servers, how should matching read approaches be chosen to optimize mean service time? This is the problem called the *data acquisition* at the client side. Both problems will be explored in this section, mostly theoretically.

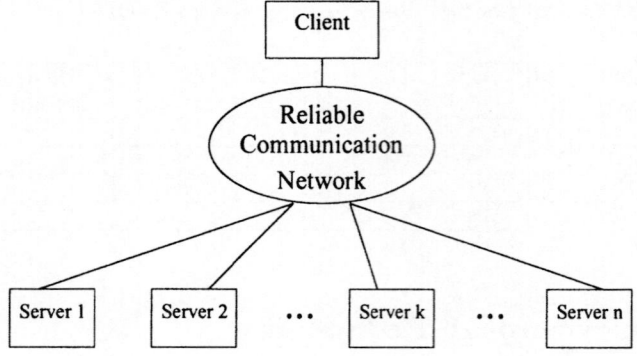

Fig. 3.1. An (n, k) server system

3.1 Preliminary Analysis

Before we seek the solutions to the above problems, we first define a server system model we will be using, based on probability analysis. Then we give some basic analytical results that can be used further to solve the data distribution and the data acquisition problems.

3.1.1 System Model. Define the service time T_i of the server i $(1 \leq i \leq n)$ to be the elapsed time from when the client sends its request to the server i to when it receives data from the server i. Notice that T_i does *not* include the time needed at the client side to do any necessary *computations* to recover the final data, since here we assume that the computations are rather simple and thus take much less time than does the data delivery through communication media. We model T_i as a continuous random variable with *probability density function* (pdf) $f_i(t)$[15]. For simplicity of analysis, we assume that all T_is are i.i.d (*independent, identically distributed*) random variables, i.e., $f_i(t) = f(t)$, $1 \leq i \leq n$.

3.1.2 Analysis Results. Let $F_i(t)$ be the *cumulative distribution function* (cdf) of T_i, i.e.[15],

$$F_i(t) = Probability(T_i \leq t) = \int_0^t f_i(x)dx$$

Now let $T(n, k)$ be the elapsed time from when the client broadcasts its data request to the servers to when it receives data from at least k out of the n servers. Then $T(n, k)$ is another random variable and is a simple function of all the T_is:

$$T(n, k) \geq T_i, \quad where \ \|\{i\}\| \geq k$$

In the above equation, $\|S\|$ is the number of the elements in the set S.

Let $f_{(n,k)}(t)$ and $F_{(n,k)}(t)$ be the pdf and cdf of $T(n,k)$ respectively, then it is easy to relate $F_{(n,k)}(t)$ and $f_{(n,k)}(t)$ to $F(t)$ and $f(t)$ [7]:

$$F_{(n,k)}(t) = \sum_{i=k}^{n} \binom{n}{i} F(t)^i [1 - F(t)]^{n-i} \tag{3.1}$$

or [7][20]:

$$f_{(n,k)}(t) = \frac{dF_{(n,k)}(t)}{dt} = k \binom{n}{k} F(t)^{k-1} [1 - F(t)]^{n-k} f(t) \tag{3.2}$$

The *mean* of $T(n,k)$ $E[T(n,k)]$ is a good measurement of the server system's performance. It can be calculated once the $f_{(n,k)}(t)$ is known:

$$E[T(n,k)] = \int_0^\infty t f_{(n,k)}(t) dt \tag{3.3}$$

3.1.3 Properties of Mean Service Time. Though it is usually hard to get a clean closed form of $E[T(n,k)]$ for a general pdf $f(t)$, it is still possible to get some of its properties with respect to n and k. Intuitively, for a fixed pdf $f(t)$, a bigger n and/or a smaller k leads to a smaller $E[T(n,k)]$ and this can be proven mathematically[23]:

Theorem 3.1. *For a random variable T with a fixed pdf $f(t)$, the following inequalities hold for $1 \le k \le n$:*

1. $E[T(n,k)] > E[T(n+m,k)]$, *for* $m \ge 1$;
2. $E[T(n,k)] < E[T(n,k+m)]$, *for* $m \ge 1$;
3. $E[T(n,k)] < E[T(n+m,k+m)]$, *for* $m \ge 1$;
4. $E[T(i,j)] \ge E[T(n,k)]$, *if* $n \ge i$ *and* $k \le j$, *equality holds only when* $n = i$ *and* $k = j$;
5. $E[T(i,j)] < E[T(n,k)]$, *if* $n \ge i$, $k > j$ *and* $n - k \le i - j$.

□

We will use these properties above as guidelines for the data distribution and the data acquisition problems. One would hope that the *variances* of random variables also had the similar properties. Unfortunately, however, the above properties do *not* hold for the variances. One such an example is shown in [23].

3.2 Server Performance Model

From Eq.(3.2) and Eq.(3.3), $E[T(n,k)]$ is a function of the pdf $f(t)$ of an individual server's data service time. The goal of the data distribution and the data acquisition problems is to reduce $E[T(n,k)]$ under various conditions. Before we analyze the data distribution and the data acquisition problems, it is necessary to establish some model of $f(t)$.

3.2.1 Abstraction from Experiments. The data service time T depends on many factors in a practical server system, such as computing power (i.e., CPU speed) of the servers and the client, local disk I/O speed of the servers and bandwidth and latency of the communication medium (usually including a reliable communication software layer) connecting the servers and the client. A model considering all the factors will be fairly complex. In this section, we will try to model the data service time as a simple probability distribution, that can be analyzed rather easily, and yet can approximate the real data service time closely. Such a model will be abstracted from experimental results of a real data server system.

Our experimental server system consists of several servers, which are PCs running Linux. Each server has data stored on its local hard disk. Data is accessed via the Linux file system. The client is also a PC running the same Linux. The nodes are connected via Myrinet switches. A *sliding window* protocol is used to ensure reliable communication. Experiments are conducted in such a real system to measure the service time for data of different sizes. The procedure of the experiment is as follows: (1) the client sends a request for a certain amount of data to a server; (2) the server reads the data from its local disk and sends it to the client through the reliable communication layer; (3) the data is delivered to the client through the reliable communication layer. The data service time is measured from the instant that the client finishes sending its request to the instant that the client gets the data. We run the above procedure a few thousand times for data of a given size, and get the service time pdf according to the observed frequencies of different ranges of service time. Figure 3.2 shows empirical service time pdfs for data sizes (a) 32 Kbytes, (b) 320 Kbytes and (c) 3200 Kbytes.

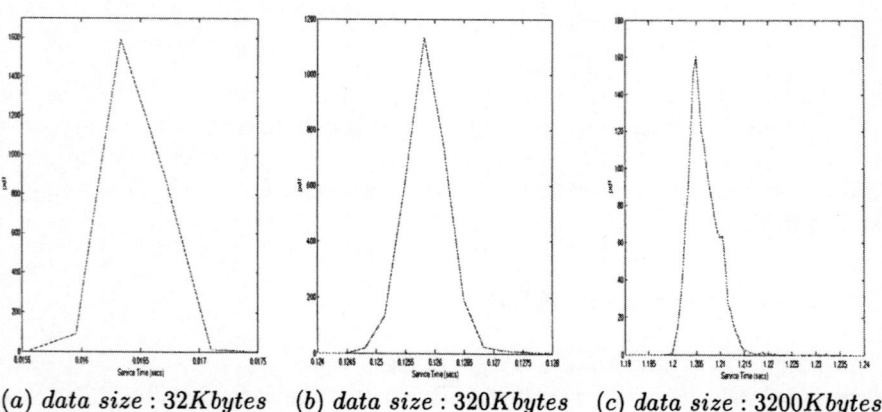

(a) data size : 32Kbytes   (b) data size : 320Kbytes   (c) data size : 3200Kbytes

Fig. 3.2. Empirical pdfs of service time for data of different sizes

The effective data bandwidths in this experiments are quite low, since they are the concatenation of the local disk bandwidth and the reliable communication layer bandwidth. But the shape of the bandwidth pdfs is more interesting. The experiment results show that the shape of empirical pdfs of different data size can be approximated by the same distribution. A closer look shows that the width of the distribution base is approximately *proportional to* the data size. More complex distributions, such as the Gamma distribution or the Beta distribution, might be more accuracy. But to simplify the analysis, that follows, we will regard the data service time T as a random variable defined on $[a, b]$ (a and b are two parameters of a real system), which follows a triangular distribution, denoted $Tr[a, b]$:

$$f(t) = \begin{cases} \frac{4(t-a)}{(b-a)^2} & a \leq t \leq \frac{a+b}{2} \\ \frac{4(b-t)}{(b-a)^2} & \frac{a+b}{2} < t \leq b \end{cases} \qquad (3.4)$$

Its cdf (*cumulative distribution function*) is

$$F(t) = \begin{cases} \frac{2(t-a)^2}{(b-a)^2} & a \leq t \leq \frac{a+b}{2} \\ 1 - \frac{2(b-t)^2}{(b-a)^2} & \frac{a+b}{2} < t \leq b \end{cases} \qquad (3.5)$$

One explanation for this model is as follows: in a real system, data is delivered in packets of some small size. The delivery time of the ith packet is a random variable t_i, whose probability distribution can be characterized by a uniform distribution over some time span; the t_i's are assumed to be i.i.d. random variables. Then the service time T of the whole data is: $T = s + \sum_i t_i$, where s is another uniform random variable describing the setup (or overhead) time for sending a certain amount of data. Thus the pdf of T is a *Gaussian-like* function, whose base width is approximately proportional to the number of the packets in the data, which in turn is proportional to the data size. For simplicity, we approximate the Gaussian-like function by a suitable triangular function. The distributions are shown in Figure 3.3.

3.2.2 Verification with T(n,1). Intuitively, having more servers should provide better performance when the amount of data stored on each server is fixed, i.e., $E[T(n, k)]$ decreases as n increases and/or k decreases. We can get pdfs of the $T(n, k)$ for a data server system by evaluating Eq.(3.2) for the service time distribution in Eq.(3.4) and Eq.(3.5). Figure 3.4(a) shows the pdfs of $T(n, 1)$, where $1 \leq n \leq 3$ and T is of the triangular distribution $Tr[1, 2]$. Here we can see the pdf of $T(n, k)$ shifts left as n increases, which indicates that the average of the random variable $T(n, k)$ decreases as n increases.

To further verify the properties of $E[T(n, k)]$, simple experiments to measure $T(n, 1)$ were done on the experimental server system described in previous subsection. The system consists of three servers. In order to remove other factors that also affect data service time, such as contention in the

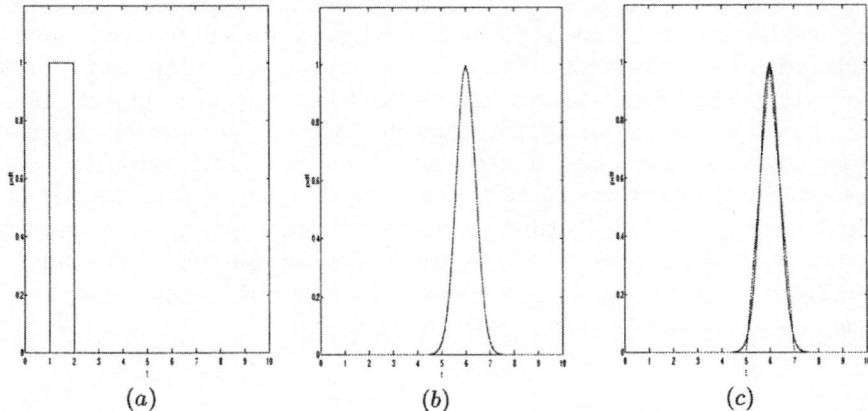

Fig. 3.3. Probability distributions of data service time of (a) single packet, (b) the whole data, (c) the approximation with $Tr[a, b]$

communication medium (including the reliable communication layer, which is a bottleneck if we use a single client which communicates with the three servers), we use three clients, each of which is served by a separate server. Conceptually the three clients are regarded as a single client, thus the whole data service time is the minimum of the three individual service time of the server-client pairs. Figure 3.4(b) shows the service times $(T_1, T_2,$ and $T_3)$ of the three individual server-client pairs for 3200 Kbytes data each. Since the variance among the three pairs is bigger than the variance within each pair, the whole service time (T_{min}), which is the minimum of the three, is determined by the service time of the best client-server pair as can be seen in the experimental results. In this case, the pdf of T_{min} is very close to that of T_1. To make the experimental results more interesting, some random loads are added to each server, so that the variance among the three client-server pairs is less than the variance within each pair, i.e., each pair behaves more similarly. The service times of three individual pairs $(T_1, T_2,$ and $T_3)$ and the whole service time (T_{min}) are shown in Figure 3.4(c). Of those four pdfs $(T_{min}, T_1, T_2$ and $T_3)$, that of T_{min} is the leftmost, which supports the analytical properties of $T(n, k)$ and the pdf model of T.

3.3 Data Distribution Scheme

Now let's turn to the data distribution problem: in a server system, with a given total number of servers, n, we need to determine the number k of the servers which store the *raw* data in order to maximize the performance of the whole system (i.e., to minimize the mean service time of client's data request); given k, the rest of the servers can store the *redundant* data. When n and the pdf $f(t)$ are fixed, $E[T(n, k)]$ decreases monotonically as k decreases.

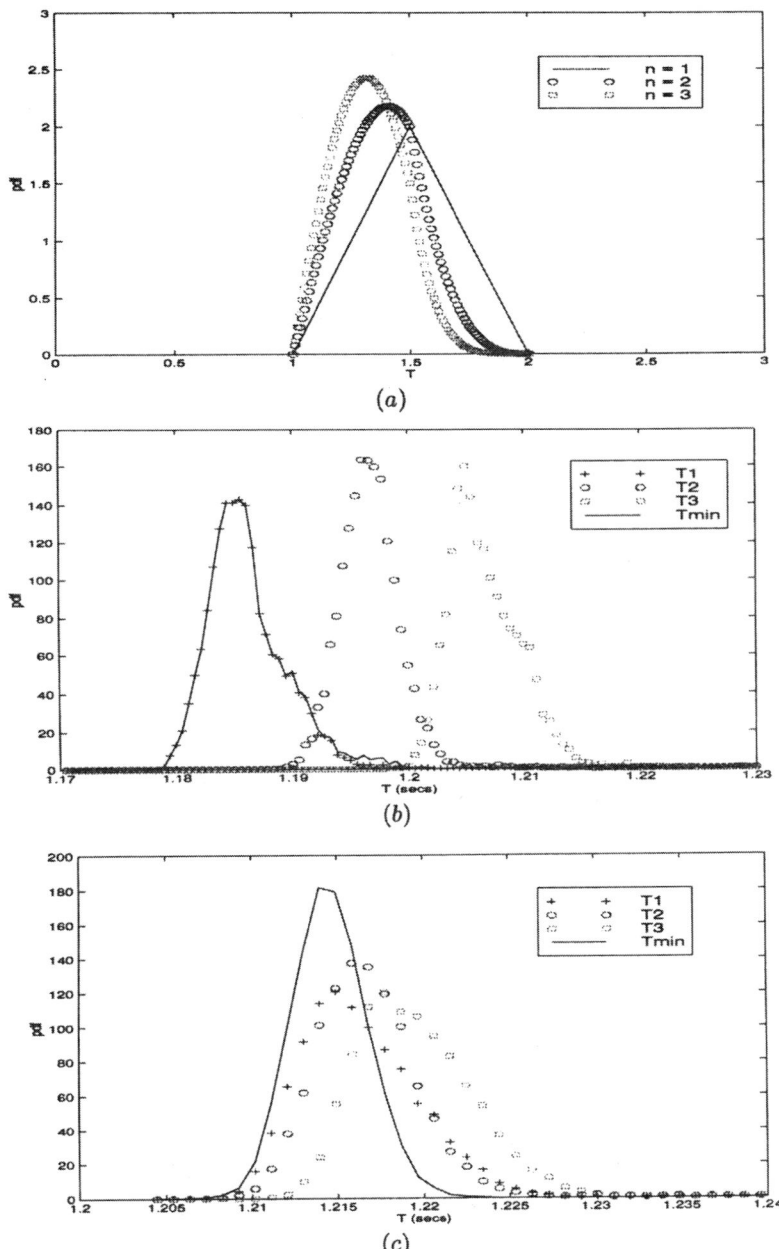

Fig. 3.4. pdfs of $T(n, 1)$: (a) analytical result, where the pdf of T is $Tr[1, 2]$, and experimental service time for data of size 3200 Kbytes, where (b) no other loads on the servers, and (c) other random loads on the servers

This means that in order to make $E[T(n, k)]$ small, k should be as small as possible. On the other hand, however, the smaller k is, the more data needs to be stored on each server, since the total amount of the data a client needs is always fixed; this means higher service time from each server. Our goal is to find such a k that when both sides of the problem are considered, $E[T(n, k)]$ is minimized.

After the parameter k is determined, in order to achieve optimal performance in terms of $E[T(n, k)]$, we can use MDS array codes to distribute the redundant data so that data from *any* k servers can be assembled to form the whole of the requested data, as was shown in the previous section. The only remaining problem is to determine k to minimize $E[T(n, k)]$. Applying the pdf model of each server's service time, T, and using MDS codes for distributing the redundant data, we get that if the pdf of T is $Tr[a, b]$ when $k=1$, then for general k, the corresponding pdf is $Tr[\frac{a}{k}, \frac{b}{k}]$, since the base width of the pdf is proportional to the data size. Theoretically, the optimal k can be calculated as follows:

$$k_{min} = argmin_k \int_0^\infty k \binom{n}{k} F(t)^{k-1} [1 - F(t)]^{n-k} t f(t) dt \qquad (3.6)$$

where $f(t)$ and $F(t)$ are as in Eq.(3.4) and Eq.(3.5), except that a and b should be replaced by $\frac{a}{k}$ and $\frac{b}{k}$ respectively. Notice that k_{min} is a function of the entire pdf $f(t)$, not only the mean $E(T)$ and the variance $Var[T]$.

Even for a simple pdf such as $Tr[a, b]$, the above equation can not be solved in closed form. But in practice, the system parameters a and b can be determined by experiments, then the above equation can be solved numerically. Figure 3.5 gives several examples of solving the above equation. In the examples, $a = 1$ and $b = 5$. For $n = 10$, 20, and 40, $E[T(n, k)]$ is calculated for $1 \leq k \leq n$. The results are shown in Figure 3.5(a)(b)(c), where (b) and (c) only show the last few values for k, since for small k $E[T(n, k)]$ decreases monotonically as k. From the results, we can see (a) $k_{min} = 10$, when $n = 10$, (b) $k_{min} = 19$, when $n = 20$, and (c) $k_{min} = 37$, when n = 40.

Even though the above examples use specific pdfs, the same method also apply with other pdfs by plugging suitable $f(t)$ into Eq.(3.6). Thus, for a given server system, such a k_{min} can always be found. Proper MDS array codes can then be used based on the (n, k) pair. Thus we get an optimal data distribution scheme for a given server system.

3.4 Data Acquisition Scheme

Once the data distribution scheme is set, i.e., k is determined and the proper MDS array code is chosen, the client needs to decide how to request (or read) data. In general, a client should send its request to as many servers as possible and also make the amount of data it needs from *each* server as small as possible, since the properties of $E[T(n, k)]$ show that more redundancy

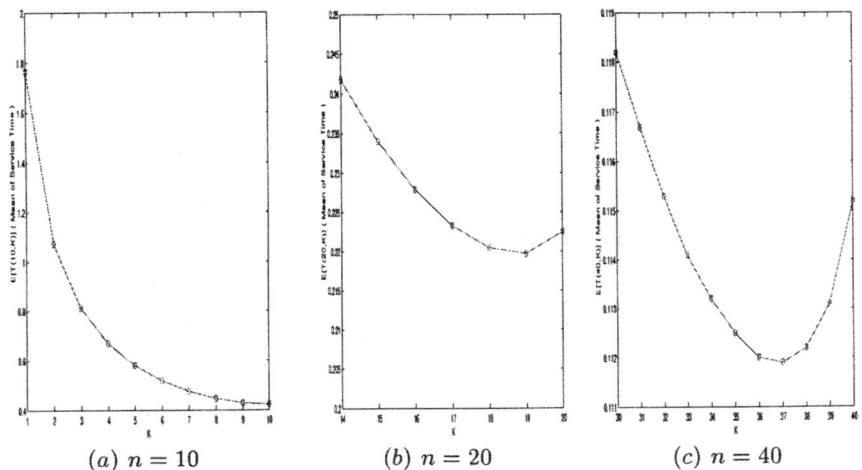

(a) $n = 10$ (b) $n = 20$ (c) $n = 40$

Fig. 3.5. $E[T(n,k)]$ vs. k for different n, where $a = 1$ and $b = 5$

brings better performance. For a specific distribution scheme, the client needs to calculate the pdfs of all possible data read schemes, and then choose an optimal read scheme. Since the read schemes are closely related to the MDS array code being used, here we will give an example using a specific code to show the guidelines for choosing an optimal read scheme.

In this example, the server system has $2n$ servers, and the data that the client requests can be assembled from any $2n - 2$ servers, i.e., this is a $(2n, 2n - 2)$ system. The B-Code can be used to implement this system. The data distribution using the B-Code is as follows: (1) the whole raw (*information*) data is partitioned into $2n(n - 1)$ blocks of equal size (some paddings are added if necessary); (2) each of the $2n$ servers stores $n - 1$ blocks of the data; (3) $2n$ blocks of redundant (or *parity*) data are calculated according to the encoding rules of the B-Code, i.e., each parity block is an XOR of suitable $2n - 2$ raw data blocks, and then each server stores 1 parity block. The structure of the B-Code is shown in Figure 2.1.

The MDS property of the B-Code gives 3 schemes for reconstructing the whole raw data from the data stored on $2n$ servers, each of which has $n - 1$ blocks of raw data and 1 block of parity data: (1) read from all of the $2n$ servers, each of which sends its $n - 1$ blocks of raw data; (2) read from any $2n - 2$ servers, each of which sends all of its n blocks of data (including raw and parity data); (3) read from all of the $2n$ servers, each of which sends all of its n blocks of data. The 3 schemes are shown in Figure 3.6, where the shaded parts are the data to be read.

Notice that there is no redundant data in scheme (1) or scheme (2), so the client must wait until it receives all the data from all the servers. But in

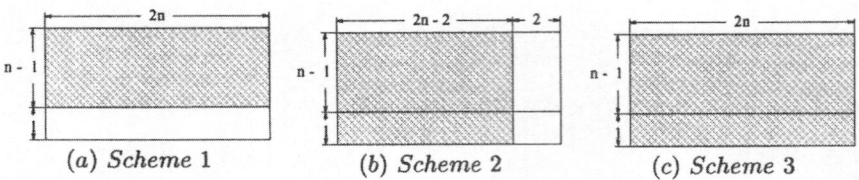

(a) *Scheme 1*   (b) *Scheme 2*   (c) *Scheme 3*

Fig. 3.6. Three read schemes using the B-Code

scheme (3), there is redundant data, then the client only needs to receive data from any $2n - 2$ of the $2n$ requested servers. Let $E[T(2n, 2n)]_{n-1}$, $E[T(2n - 2, 2n - 2)]_n$ and $E[T(2n, 2n - 2)]_n$ denote the mean data service time of the three schemes respectively. From Property 1 of Theorem 3.1, $E[T(2n - 2, 2n - 2)]_n > E[T(2n, 2n - 2)]_n$. But the relation between $E[T(2n, 2n)]_{n-1}$ and either $E[T(2n - 2, 2n - 2)]_n$ or $E[T(2n, 2n - 2)]_n$ is not so obvious, since in scheme (1) the client needs to wait for more servers, but needs less data (thus less service time) from each server. So to determine which scheme is best scheme for a given system, we need to calculate the pdf of the whole service time for all possible the schemes, which are scheme (1) and scheme (3) in this case.

Assume that the pdf of the time T for each server to send n blocks of data to the client is $Tr[a, b]$; then the pdf of T in scheme (1) is $Tr[\frac{n-1}{n}a, \frac{n-1}{n}b]$, since each server only needs to send $n - 1$ blocks of data, and the pdf of T in scheme (2) or (3) is $Tr[a, b]$. Now the pdfs of the whole service time in the different schemes can be calculated according to Eq.(3.2), Eq.(3.4) and Eq.(3.5). Figure 3.7 shows the pdfs for different values of n, where $a = 1$ and $b = 10$.

Using Eq.(3.3), the mean of the whole service time of different schemes can be calculated. These means are listed in Table 3.1, for $a = 1$ and $b = 10$.

Table 3.1. Mean service time of different data read schemes, where $a = 1$, and $b = 10$

n	3	7	10
$E[T(2n, 2n)]_{n-1}$	**5.2195**	7.3128	7.8857
$E[T(2n - 2, 2n - 2)]_n$	7.4089	8.4207	8.6976
$E[T(2n, 2n - 2)]_n$	5.8910	**7.2466**	**7.6786**

The above calculations show that the performance of the three schemes depends on the system parameter n (when a and b are fixed). In a small server system, scheme (1) is the best. As n increases, scheme (3) becomes better. For a system of 6 servers $(n = 3)$, scheme (1) is the best, but for systems of 14 servers $(n = 7)$ and 20 servers $(n = 10)$, scheme (3) is the best.

Though quite simple, the above example shows that after the data distribution is set at the server side, the client has different ways of reading data from the servers. For a given system (i.e., a certain pdf of T, a fixed (n, k) pair and a particular code), there always exists an optimal read scheme for the client. Finding this scheme requires careful calculation. Since the read schemes are highly related to the codes used, exploring codes that offer more read choices is an interesting research problem. It is conjectured in [23] that all MDS codes have a so-called *strong MDS* property, which provides the flexibility of reading schemes.

4. Summary

This chapter deals with two issues in highly available distributed storage systems: reliability and efficiency. To achieve reliability, three classes of MDS array codes are described. They are suitable for storage applications because of their simple computations for encoding and decoding, their MDS property and their low (or optimal) update complexity. Two problems, namely the data distribution problem and the data acquisition problem, and their solutions are proposed to use the redundancy in storage systems properly to improve their performance.

A practical distributed storage system is implemented as part of the RAIN (Reliable Array of Independent Nodes) system, a reliable and efficient computing environment at the Parallel and Distributed Computing Lab of Caltech, using the approaches discussed in this chapter. A detailed description about the RAIN system can be found in [6].

328

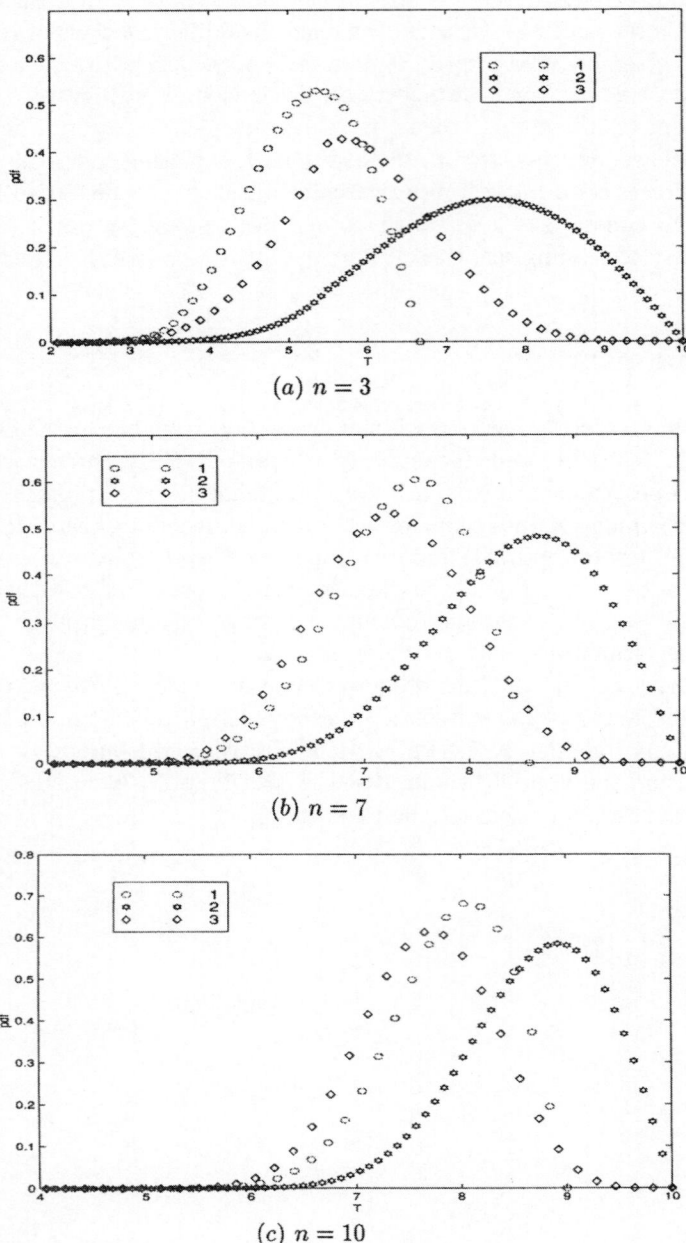

Fig. 3.7. PDFs of different data read scheme, where $a = 1$, $b = 10$; 1, 2 and 3 represent scheme (1), (2) and (3) respectively.

References

1. B. A. Anderson, "Symmetry Groups of Some Perfect 1-Factorizations of Complete Graphs," *Discrete Mathematics*, 18, 227-234, 1977.
2. M. Blaum, J. Brady, J. Bruck and J. Menon, "EVENODD: An Efficient Scheme for Tolerating Double Disk Failures in RAID Architectures," *IEEE Trans. on Computers*, 44(2), 192-202, Feb. 1995.
3. M. Blaum, J. Bruck, A. Vardy, "MDS Array Codes with Independent Parity Symbols," *IEEE Trans. on Information Theory*, 42(2), 529-542, March 1996.
4. M. Blaum, P. G. Farrell and H. C. A. van Tilborg, "Chapter on Array Codes", Handbook of Coding Theory, edited by V. S. Pless and W. C. Huffman, to appear.
5. M. Blaum, R. M. Roth, "New Array Codes for Multiple Phased Burst Correction," *IEEE Trans. on Information Theory*, 39(1), 66-77, Jan. 1993.
6. V. Bohossian, C. Fan, P. LeMahieu, M. Riedel, L. Xu and J. Bruck, "Computing in the RAIN: A Reliable Array of Independent Nodes", Caltech Technical Report, 1998.
   Available at: `http://paradise.caltech.edu/papers/etr029.ps`.
7. M. N. Frank, "Dispersity Routing in Store-and-Forward Networks," Ph.D. thesis, University of Pennsylvania, 1975.
8. P. G. Farrell, "A Survey of Array Error Control Codes," *ETT* , 3(5), 441-454, 1992.
9. R. G. Gallager, "Low-Density Parity-Check Codes," MIT Press, Cambridge, Massachusetts, 1963.
10. J. Gemmell, "Scalable Reliable Multicast Using Erasuring-Correcting Re-Sends," Technical Report MSR-TR-97-20, Microsoft Research, June, 1997.
11. R. M. Goodman, R. J. McEliece and M. Sayano, "Phased Burst Error Correcting Arrays Codes," *IEEE Trans. on Information Theory*, 39, 684-693,1993.
12. A. Kotzig, "Hamilton Graphs and Hamilton Circuits," *Theory of Graphs and Its Applications* (Proc. Sympos. Smolenice), 63-82, 1963.
13. F. J. MacWilliams and N. J. A. Sloane, *The Theory of Error Correcting Codes*, Amsterdam: North-Holland, 1977.
14. Norman K. Ouchi, "System for Recovering Data Stored in Failed Memory Unit," US Patent 4092732, May 30, 1978.
15. A. Papoulis, *Probability, Random Variables, and Stochastic Processes*, 2nd Edition, McGraw-Hill, Inc., 1984.
16. D. A. Patterson, G. A. Gibson and R. H. Katz, "A Case for Redundant Arrays of Inexpensive Disks," *Proc. SIGMOD Int. Conf. Data Management*, 109-116, Chicago, IL, 1988.
17. R. M. Tanner, "A Recursive Approach to Low Complexity Codes," *IEEE Trans. on Information Theory*, 27(5), 533-547, Sep. 1981.
18. D. G. Wagner, "On the Perfect One-Factorization Conjecture," *Discrete Mathematics*, 104, 211-215, 1992.
19. W. D. Wallis, *One-Factorizations*, Kluwer Academic Publisher, 1997.
20. Samuel S. Wilks, *Mathematical Statistics*, John Wiley & Sons, Inc., 1963.
21. L. Xu and J. Bruck, "X-Code: MDS Array Codes with Optimal Encoding," *IEEE Trans. on Information Theory*, 45(1), 272-276, Jan., 1999.
    Also available at: `http://paradise.caltech.edu/papers/etr020.ps`.
22. L. Xu, V. Bohossian, J. Bruck and D. Wagner, "Low Density MDS Codes and Factors of Complete Graphs," Proceedings of 1998 IEEE Symposium on Information Theory, Aug., 1998; Revised version to appear in *IEEE Trans. on*

Information Theory, Sep. 1999.
Also available at: `http://paradise.caltech.edu/papers/etr025.ps`.

23. L. Xu, "Highly Available Distributed Storage Systems," Ph.D. thesis, California Institute of Technology, 1998.
Also available at: `http://paradise.caltech.edu/~lihao/thesis.html`.

List of Lectures

K. Abdali	Advanced Computing and Communi-cation Research under NSF Support
W. Almesberger	SRP – a Scalable Resource Reservation Protocol for the Internet
J. Blum	Para-Station: High Performance Environment for Clusters
T. Braun	Differentiated Internet Services
J. Bruck	Reliable Distributed High Performance Computing
H. Busch	BRAIN – Berlin Research Area Information Network
G. Cooperman	Parallel TOP-C and Scaling Up with DSM
T. Eickermann	Metacomputing in the Gigabit Testbed West
L. Finkelstein	Experiences at Northeastern University Connecting to a High Performance National Network (presented by G. Cooperman)
E. Gabriel	High Performance Metacomputing in a Transatlantic Wide Area Application Testbed
G. Havas	Some Performance Studies in Exact Linear Algebra
A. Hoisie	Performance and Scalability Analysis of Applications on Teraflop-class Distributed Architectures
P. Holleczek	Controlling the Quality of Service in Wide Area ATM Networks
E. Jessen	The Gigabitwissenschaftsnetz of DFN
M. Köster	High Performance Computing across the ATM-WAN Essen-Bonn
H. Lederer	Visual Supercomputing and Metacomputing – Gigabit Testbed Projects with Contributions of the Max Planck Society

List of Registered Participants

Dr. Kamal Abdali
National Science Foundation
1800 G Street
N. W. Washington DC 20550
USA
Kabdali@ndf.gov

Dr. W. Almesberger
Département d'informatique
Laboratoire de réseaux de
communication- LRC
IN- Ecublens
CH-1015 Lausanne
Werner.almesberger@di.epfl.ch

J. Blum
Fakultät für Informatik
Universität Karlsruhe
Am Fasanengarten5
76131 Karlsruhe
blum@ira.uka.demailto:blum@ira.uka.
de

Prof. Dr. T. Braun
Universität Bern
Institut für Informatik
und angewandte Mathematik
Neubrückstr. 10
3012 Bern
braun@iam.unibe.ch

Prof. J. Bruck
Department of Electrical Engineering
California Institute of Technology
Pasadena, CA 91125
USA
bruck@vangogh.paradise.caltech.edu

H. Busch
Konrad-Zuse-Zentrum für
Informationstechnik Berlin
Bereich Rechenzentren
Abteilung Höchstleistungsrechner –
Leiter, Takustr. 7
14195 Berlin-Dahlem
busch@zib.de

E. Gabriel
Rechenzentrum Universität Stuttgart
Abteilung Paralleles Rechnen
Allmandring 30
70550 Stuttgart
gabriel@hlrs.de

H. Gollan
Institut für Experimentelle Mathematik
Universität GH Essen
Ellernstr. 29
45326 Essen
holger@exp-math.uni-essen.de

Prof. George Havas
Dept. of Computer Science
University of Queensland
Queensland 4072
Australien
havas@cs.uq.edu.au

Dr. Adolfy Hoisie
Scientific Computing, CIC-19
MS B256
Los Alamos National Laboratory
Los Alamos, NM 87545
hoisie@lanl.gov

Dr. P. Holleczek
Abt. Kommunikationssysteme
Regionales Rechenzentrum Erlangen
Martensstr. 1
91058 Erlangen
peter.holleczek@rrze.uni-erlangen.de

Prof. Dr.-Ing. E. Jessen
Institut für Informatik
TU München
Augustenstr. 77
80290 München
jessen@informatik.tu-muenchen.de

334

Prof. Gene Cooperman
College of Computer Science
Northeastern University
M/S 215 CN
Boston, MA 02115
gene@ccs.neu.edu

Dr. T. Eickermann
Forschungszentrum Jülich GmbH
ZAM
Leo-Brandt-Straße
52428 Jülich
th.eickermann@fz-juelich.de

Dr. B. Lix
Hochschulrechenzentrum
Universität GH Essen
Schützenbahn 70
45141 Essen
lix@hrz.uni-essen.de

Dr. M. Mähler
IBM Deutschland GmbH
European Network Center, Heidelberg
Vangerowstr. 18
69115 Heidelberg
maehler@heidelbg.ibm.com

Prof. Dr. P. Martini
Rheinische Friedrich-Wilhelms-
Universität Bonn
Institut für Informatik IV
Römerstrasse 164
D-53117 Bonn
Peter.Martini@cs.uni-bonn.de

Prof. Dr. G. Michler
Institut für Experimentelle
Mathematik
Universität GH Essen
Ellernstr. 29
45326 Essen
archiv@exp-math.uni-essen.de

Dipl.-Inf. M. Köster
Rheinische Friedrich-Wilhelms-Universität
Bonn
Institut für Informatik IV
Römerstrasse 164
D-53117 Bonn
koester@cs.uni-bonn.de

Dr. H. Lederer
Rechenzentrum Garching der Max-Planck-
Gesell-schaft
Max-Planck-Institut für Plasmaphysik
Boltzmannstr.2
85748 Garching
lederer@rzg.mpg.de

Prof. E. Quintana-Orti
Departamento de Informatica
Universidad Jaime I
Campus Penyeta Roja
E- 12071 Castellon, Spanien
Quintana@nuvol.uji.es

Prof. Dr.-Ing. E. P. Rathgeb
Institut für Experimentelle Mathematik
Universität GH Essen
Ellernstr. 29
45326 Essen
erwin.rathgeb@exp-math.uni-essen.de

Andreas Rieke
Lehrstuhl für Kommunikationssysteme
Fachbereich Elektrotechnik
Fernuniversität Hagen
Feithstr. 142
58084 Hagen
andreas.rieke@fernuni-hagen.de

Prof. Dr. G. Schneider
Gesellschaft für wissenschaftliche
Datenverarbeitung
mbH Göttingen (GWDG)
Am Faßberg
37077 Göttingen
gschnei2@gwdg.de

D. Nastoll
Hochschulrechenzentrum
Universität GH Essen
Schützenbahn 70
45141 Essen
nastoll@hrz.uni-essen.de

Prof. Dr. H. Obrecht
Lehrstuhl für Baumechanik-Statik
Universität Dortmund
Fakultät Bauwesen
August-Schmidt-Straße 8
44221 Dortmund
msobr@busch.bauwesen.uni-
dortmund.de

Dr. D. V. Pasechnik
Faculty of Technical Mathematics and
Informatics
Department of Statistics, Probability
and Operations
Mekelweg 4
NL-2628 CD Delft
D.Pasechnik@twi.tudelft.nl

Dr. T. Plagemann
University of Oslo, UNIK
Granveien 33, P.O Box 70
N-2007 Kjeller
plagemann@unik.no

Dr. T. Warschko
Fakultät für Informatik
Universität Karlsruhe
Am Fasanengarten5
76131 Karlsruhe
warschko@ira.uka.de

Dr. M. Weller
Institut für Experimentelle
Mathematik
Universität GH Essen
Ellernstr. 29
45326 Essen
eowmob@exp-math.uni-essen.de

Prof. Dr. U. Schwiegelshohn
Universität Dortmund
Lehrstuhl Datenverarbeitungssysteme
Otto-Hahn-Str. 4
44221 Dortmund
uwe.@ds.e-technik.uni-dortmund.de

Prof. Dr. U. Stammbach
ETH Zürich
Forschungsinstitut für Mathematik
ETH Zentrum
CH-8092 Zürich
Stammb@math.ethz.ch

Dr. R. Staszewski
Institut für Experimentelle Mathematik
Universität GH Essen
Ellernstr. 29
45326 Essen
reiner@exp-math.uni-essen.de

Dr. R. Völpel
GMD
SCAI (Institut für Wissenschaftliches
Rechnen)
Schloss Birlinghoven
D-53754 Sankt Augustin
Roland.voelpel@gmd.de

P. Wunderling
GMD
IMK (Institut für Medienkommunikation)
Schloss Birlinghoven
D-53754 Sankt Augustin
wunderling@gmd.de

R. Yahyapour
Fakultät für Elektrotechnik
Lehrstuhl für Datenverarbeitungssysteme
Otto-Hahn-Str. 4
44221 Dortmund
yahya@peggy.E-technik.Uni-Dortmund.de

Lecture Notes in Control and Information Sciences

Edited by M. Thoma

1993–1999 Published Titles:

Vol. 186: Sreenath, N.
Systems Representation of Global Climate
Change Models. Foundation for a Systems
Science Approach.
288 pp. 1993 [3-540-19824-5]

Vol. 187: Morecki, A.; Bianchi, G.;
Jaworeck, K. (Eds)
RoManSy 9: Proceedings of the Ninth
CISM-IFToMM Symposium on Theory and
Practice of Robots and Manipulators.
476 pp. 1993 [3-540-19834-2]

Vol. 188: Naidu, D. Subbaram
Aeroassisted Orbital Transfer: Guidance
and Control Strategies
192 pp. 1993 [3-540-19819-9]

Vol. 189: Ilchmann, A.
Non-Identifier-Based High-Gain Adaptive
Control
220 pp. 1993 [3-540-19845-8]

Vol. 190: Chatila, R.; Hirzinger, G. (Eds)
Experimental Robotics II: The 2nd
International Symposium, Toulouse,
France, June 25-27 1991
580 pp. 1993 [3-540-19851-2]

Vol. 191: Blondel, V.
Simultaneous Stabilization of Linear
Systems
212 pp. 1993 [3-540-19862-8]

Vol. 192: Smith, R.S.; Dahleh, M. (Eds)
The Modeling of Uncertainty in Control
Systems
412 pp. 1993 [3-540-19870-9]

Vol. 193: Zinober, A.S.I. (Ed.)
Variable Structure and Lyapunov Control
428 pp. 1993 [3-540-19869-5]

Vol. 194: Cao, Xi-Ren
Realization Probabilities: The Dynamics of
Queuing Systems
336 pp. 1993 [3-540-19872-5]

Vol. 195: Liu, D.; Michel, A.N.
Dynamical Systems with Saturation
Nonlinearities: Analysis and Design
212 pp. 1994 [3-540-19888-1]

Vol. 196: Battilotti, S.
Noninteracting Control with Stability for
Nonlinear Systems
196 pp. 1994 [3-540-19891-1]

Vol. 197: Henry, J.; Yvon, J.P. (Eds)
System Modelling and Optimization
975 pp approx. 1994 [3-540-19893-8]

Vol. 198: Winter, H.; Nüßer, H.-G. (Eds)
Advanced Technologies for Air Traffic Flow
Management
225 pp approx. 1994 [3-540-19895-4]

Vol. 199: Cohen, G.; Quadrat, J.-P. (Eds)
11th International Conference on
Analysis and Optimization of Systems –
Discrete Event Systems: Sophia-Antipolis,
June 15–16–17, 1994
648 pp. 1994 [3-540-19896-2]

Vol. 200: Yoshikawa, T.; Miyazaki, F. (Eds)
Experimental Robotics III: The 3rd
International Symposium, Kyoto, Japan,
October 28-30, 1993
624 pp. 1994 [3-540-19905-5]

Vol. 201: Kogan, J.
Robust Stability and Convexity
192 pp. 1994 [3-540-19919-5]

Vol. 202: Francis, B.A.; Tannenbaum, A.R.
(Eds)
Feedback Control, Nonlinear Systems,
and Complexity
288 pp. 1995 [3-540-19943-8]

Vol. 245: Garulli, A.; Tesi, A.; Vicino, A. (Eds)
Robustness in Identification and Control
448pp: 1999 [1-85233-179-8]

Vol. 246: Aeyels, D.;
Lamnabhi-Lagarrigue, F.; van der Schaft, A. (Eds)
Stability and Stabilization of Nonlinear Systems
408pp: 1999 [1-85233-638-2]

Vol. 247: Young, K.D.; Özgüner, Ü. (Eds)
Variable Structure Systems, Sliding Mode
and Nonlinear Control
400pp: 1999 [1-85233-197-6]

Vol. 248: Chen, Y.; Wen C.
Iterative Learning Control
216pp: 1999 [1-85233-190-9]